T0311841

Finding Your Career in the Modern Audio Industry

Finding Your Career in the Modern Audio Industry equips the reader with the skills they need to turn an interest in audio, sound, or music technology into a career.

This book provides insight for aspiring professionals seeking audio-related opportunities in entertainment, technology, education, and more. In the audio industry, there is typically a gap between those in-training and those with professional (and financially sustainable) careers. This book bridges the information gap, offering practical and real-world advice to those in this volatile stage of their career. Including 70+ interviews with professionals from over 20 countries, *Finding Your Career in the Modern Audio Industry* offers insight into how others (across the industry and the world) have applied entrepreneurial thinking, problem-solving, and creative solutions to build their careers.

Including international case studies and interviews with diverse professionals, *Finding Your Career in the Modern Audio Industry* is essential reading for anyone taking their first steps into an audio-related field.

April Tucker is an audio engineer with nearly two decades of experience in post-production sound. April's mixes have been heard by television audiences with over 10 million viewers and at major film festivals including Sundance, Cannes, Toronto, and Tribeca. Based in Los Angeles, April's clients have ranged from major studios and corporations to composers and concert halls. April has written for audio industry publications and blogs since 2015.

"When I received word about this book, I first thought "How could anyone write about finding a career in the audio biz?" The possibilities and personal experiences of this lifelong process are simply too big and varied to write about. Then, I started reading and, WOW, through the use of straightforward talk and first-hand excerpts from industry pro experiences, I think she nailed it! I'm a huge fan of anyone who'll take the time to help others find their way in the audio and music production industry. We all need guidance to make it through our own personal maze in life. I'd truly recommend it to anyone who wants to devote their life towards getting and maintaining a career in the music industry."

David Miles Huber, *4X Grammy-nominated musician and producer*

"April's book contains immensely valuable information and detailed advice for anyone trying to carve out a career in the audio industry where freelancing is now the norm. The interviews with so many varied professionals are especially informative – they form a tapestry of different experiences from which April draws out the key messages and common themes. It is refreshing to hear from so many different voices in the audio world, and I really wish I had been able to read this book 20 years ago! I can recommend this to anyone currently working in audio, whatever their career stage, but especially to those trying to start out."

Caroline Haigh, *Institute of Sound Recording, University of Surrey*

"You need non-audio skills to succeed in the audio industry. April Tucker's *Finding Your Career in the Modern Audio Industry* spells it out, showing a structure and a process for introspective assessments of your goals, your limits, your dreams, as well as essential outward-facing action: networking, job searching, marketing yourself, and more. Business acumen can't be ignored, and can be mastered: negotiating rates and salaries, contracts, accounting, etc. You might be focused on one career path, but this book widens your view to full range of possibilities in the wide world of audio. With quotes and advice from folks across all facets of the industry throughout each chapter, Tucker's book is a pleasure to read, and you'll see there is no mystery to beginning your career in the modern audio industry."

Alex U. Case, *Sound Recording Technology,*
University of Massachusetts Lowell

Finding Your Career in the Modern Audio Industry

April Tucker

Routledge
Taylor & Francis Group

LONDON AND NEW YORK

Cover image: Ryan Tucker

First published 2023
by Routledge
4 Park Square, Milton Park, Abingdon, Oxon OX14 4RN

and by Routledge
605 Third Avenue, New York, NY 10158

Routledge is an imprint of the Taylor & Francis Group, an informa business

British Library Cataloguing-in-Publication Data
A catalogue record for this book is available from the British Library

Library of Congress Cataloging-in-Publication Data
A catalog record has been requested for this book

ISBN: 978-0-367-50557-8 (hbk)
ISBN: 978-0-367-50555-4 (pbk)
ISBN: 978-1-003-05034-6 (ebk)

DOI: 10.4324/9781003050346

Typeset in Goudy
by Newgen Publishing UK

Access the companion website: apriltucker.com/books/career-paths/

For Jackson – and all the other kids creating sounds and music today – who someday may ask, "Could I actually do this for a living?"

Contents

Part 1   Planning   1

Part 1 helps the reader define their personal preferences, challenges, barriers, and how these can affect the direction of a career path. These chapters teach the fundamentals of networking, business relationships, marketing, sales, and finding opportunities, all of which are critical foundational skills for any career in the entertainment or audio industry.

Part 2 Operations   207

Part 2 addresses the logistics of operating a business, whether doing an occasional freelance gig or making a living off of freelance work. These chapters help readers with common business concerns: learning how to set rates, create an invoice, offers and contracts, and managing financial information.

Part 3 Career Paths   275

Part 3 takes the principles and ideas of the earlier chapters and puts them into practice. Each chapter addresses the needs specific or unique to these industries. These chapters are largely based on quotes, interviews, stories, and case studies. What business advice would someone who's been through it give to someone who's just getting started?

Part 4 Industry Perspectives   441

Preface

When I moved to Los Angeles in 2004, I only knew three people, but it was LA so I thought, "How hard could it be?" My goal was to find a full-time job at a major music studio or film scoring stage where I could apprentice and learn the craft from top industry professionals. George Massenburg, one of my professors during my master's at McGill University (in Canada), told me, "Just go to the major studios and tell them I sent you."

I quickly discovered it wasn't that simple. Years later I wrote in a blog post (for soundgirls.org) about the experience:

> Since I didn't move to LA with a job, my first couple of months were spent taking "meetings" (a.k.a. lunch, coffee, or a drink) with "connections" (a.k.a. friends of friends or people I met [online]). I tried the regular job channels like applying to job listings or sending my resume to studios but got no response. It was so bizarre that all of this was the reality of starting a career. It felt more like a game making friends using code words versus job seeking.

I spent a year meandering through audio-related jobs from recording classical music to working on ribbon microphones. When I was offered an entry-level job in post-production sound for television, I thought, "Why not try it?" It was a discipline I never saw myself working in but turned out to be a great fit. Alex Trebek made fun of my grammar during an ADR session. I knew who won the final rose on *The Bachelor* every week while millions of fans were waiting to watch. One client was being monitored by a foreign government, which they shared while we were doing their mix. Other times, I get notes from producers like, "Can we add more farts?"

During my first few years in post, I still needed side hustles (like recording classical music and testing audio software) to help pay the bills. I took business courses at a local community college when I recognized I would probably *not* spend most of my career in full-time jobs (which have become a rarity). Those courses were as valuable as my six years of university education (across two audio degrees), and I used those skills regularly during the decade I ran my own business. In 2015, I started writing for industry blogs, publications, and my website (proaudiogirl.com – now apriltucker.com) with a special focus on business skills. Many of those posts shaped the foundation of this book, but with an emphasis on the stories, experiences, and career paths of other professionals.

This book does not cover specific software, technical skills, or the art behind the job. The goal is to learn how others (across the industry and around the world) have applied entrepreneurial thinking, problem-solving, and creative solutions to build their careers. The stories and advice given in this book will not be exhaustive of all the pathways to a career (or careers that exist today or in the future), but the framework behind it will still apply. It is important to recognize that the "audio industry" is not just working in entertainment or content creation, and what you find to be a great career fit may not be what you expected.

The book was not titled "Creating Your Career" for a reason; the path to a career tends to be a balance between exploration and survival versus forging a path of your choosing.

I hope this book will provide some ideas on how to find a path of your own. Additional materials and related articles are available at: apriltucker.com/books/career-paths/

Acknowledgments

The interviews for this book (over 70 professionals) took place during the COVID-19 pandemic of 2020–2021. I am grateful to everyone who took the time to talk (or write) especially as they were coming to terms with layoffs, furloughs, canceled performances and tours, company changeovers, cross-country moves, adapting to working from home, or being at work uncertain of their safety. Recording engineer James Clemens-Seely put it perfectly in his 2020 interview: "Literally everyone who could in any way be remotely related to the book is in some kind of existential crisis." On top of it all, we had no idea what the audio industry would look like when all the work came back.

What kept me motivated through the pandemic (and connected to the outside world) was these interviews. If there is one word to sum up successful audio professionals, it is "perseverance." For a couple of hours here and there, talking to them (over Zoom) made the world feel totally normal.

Thank you to my husband, Ryan, for your ongoing support. We were sharing our home studio (both working full-time) and parenting our preschooler from home during the pandemic, so managing a book on top was not easy. Thank you for helping with interviews, graphics, the book cover, and for listening to my never-ending thoughts and inquiries about the industry.

Thank you to my parents, Tom and Grace Cech, who were reading the book and helped edit it all along the way. I am lucky to have a father who writes and edits textbooks and a mother who is a textbook editor and former librarian. If they understand all the concepts in this book, you (and your loved ones) can, also.

Special thanks to the professionals whose unique expertise and insight became the foundation of certain chapters: Susan Rogers (Chapter 2), Karol Urban (Chapter 4), Megan Frazier (Chapter 5), John McLucas (Chapter 8), and Andrea Espinoza (Chapter 12). Thanks to Leslie Gaston-Bird for your support and sharing what you learned writing *Women in Audio*, and to Karrie Keyes for interview suggestions, introductions, and letting me tap into the amazing content created by SoundGirls.

Thanks to Samantha Potter and Jenni Alpert for consulting and reviewing content. Thank you to Eric Cookson, Jeff Gross, Ghazi Hourani, Kyle Tilbury, and Justin Kurtz for offering your thoughts when I had random questions to ask. Thanks to Nathan Lively and Patrushkha Mierzwa for sharing your amazing books.

It was a huge undertaking to learn the ins and outs of disciplines beyond my expertise and also to learn about the audio industry around the world. Thank you to everyone who provided information supporting my informal research, including Iga Adela Zatorska, Ainjel Emme, George J. Hess Jr., David Liu, Caroline Losneck, Douglas McKinnie, Melissa and Nick Passanante, Mark Shevy, and Lee Yee. Hundreds of others offered their insight through social media posts, and many of the groups I utilized are listed in the Appendix.

I was often reflecting on my own experiences as an emerging professional while writing the book. Thank you to the professors I studied with at McGill University and the University of Hartford (especially Scott Metcalfe, Justin Kurtz, and Steve Bellamy); Joe Patrych, Wes and Sara Dooley, Matt Snyder, Ben Maas; and the team from Pacific Soundwaves (especially Steve Clark, Glenn Aulepp, and Doug Latislaw).

Lastly, thank you to Hannah Rowe at Routledge (who found me on an AES conference presenter list. Opportunities can come from anywhere!). For as crazy as the world has been, the process of writing this book has been rather smooth with your input and guidance. Thank you to Adam Woods, Emily Tagg, Jayashree Thirumaran, and the rest of the teams at Routledge and Newgen for helping to bring my writing into the world.

Disclaimer

The information presented in this book is for educational and informational purposes only. It does not constitute tax, legal, or accounting advice. If financial or legal advice is required (or other expert assistance), the services of a competent professional should be sought.

Part 1
Planning

1
Industry Introduction

When you choose to pursue a career in the audio industry, it means trying to do what you love every day. It's effortless to focus on the craft and learn skills that you are passionate about. Not only could you watch videos for months on end, there are courses, degree programs (up to the Ph.D. level), and people online around the clock (and worldwide) who share in what you enjoy doing. You may live, eat, breathe, and sleep audio – but none of that pays the bills. So, how do you turn that from an interest into a career?

A professional career cannot be developed and sustained without earning money, but learning how to do it can seem more of a chore than it is interesting or fun. "Everybody wants to play with the console, be in the studio and play with microphones," says live sound engineer Chris Bushong. "Nobody wants to learn the business and financial side (dealing with contracts and all that stuff) because it's not sexy. It's not fun. It's really just boring. But at times, it's most of the job."

Rob Jaczko, recording engineer and educator, shares:

> The first time you get burned and not get paid on a $10,000 invoice, you start to appreciate how important business acumen really is. I got burned spectacularly by a record company that shall remain unnamed. Thirty years later, I'm still pissed about it. If I had known about my rights, contracts, really understood the fine print and how these things worked earlier – and understood how the business really worked to get paid, I would be richer than I am today. These are essential skills.

This sentiment comes up from professionals across the industry who learned business skills in the field out of necessity. EveAnna Dauray Manley, president of Manley Laboratories, says:

> The hardest part about owning a small company is you can't afford to hire experts, so you *have* to learn how to do this s— yourself . . . You are always

DOI: 10.4324/9781003050346-2

in a catch-22: you're too small to hire so you have to learn to be a versatile person.

(qtd. in Gaston-Bird)

Music producer Jeff Gross adds:

> If you really want to know the craft and know the business, it's not just "I know how to program a beat, stick a mic in front of a guitar cabinet, and tune a vocal." You've also got the business side of it, you've got the personal side of it, and the personal and the business sides mix.

What Is a Professional (in the Audio Industry)?

An **amateur** (or **novice**) is someone who is self-trained, lacks professional experience, and has gained experience primarily on personal projects or doing projects with other amateurs. An amateur is generally earning no income (or low income) for these projects.

An **emerging professional** is someone who aspires to have a professional-level career, has a professional attitude, is using or learning professional tools, but is not financially supported by their work yet. This may be due to limited job experience, emerging skills, working entry-level positions, low-income work opportunities, and/or being a student. Emerging professionals can be any age (not only young adults).

A **professional** (or **established professional**) is someone whose primary source of income is from working in the audio industry, and whose work sustains their lifestyle financially. A professional has a highly developed skillset, extensive experience, and work history over many years or decades. In some disciplines of the industry, it is necessary to use professional tools and know industry standards and norms to be considered a professional.

A **hobbyist** is someone performing work for personal enjoyment or recreation (not as a primary source of income). A hobbyist is not necessarily an amateur and could have professional training and experience. For example, a full-time professor who does mixing outside their job may have been a full-time mixer prior. Audio professionals have pivoted into film producing, real estate, and even winemaking – some who continue to work on professional audio work and projects for their own enjoyment (or extra income).

Podcast editor and consultant Britany Felix shares, "If you don't have a desire to learn those things, you could be the best audio editor in the world. But, if no one knows you exist . . . it's going to serve you no good. It's going to be pointless."

The Early Phases of a Career

While everyone's career path will be unique, there are three distinct phases at the beginning of a career: **exploration**, **preparation**, and **establishment**. One of the most common misunderstandings about the audio industry is the time it can take to become an established professional – especially for those pursuing entertainment-related work. "It takes between one to three years to even start," says VR sound designer Megan Frazier. "I believe that it's closer to five years if you really want to have staying power in the industry, not just 'I can get a job.' When you're coming from nothing, receiving employment is everything."

A career begins with a period of exploration and learning. The purpose of the **exploration phase** is to discover what you like (and do not like) to do, if you enjoy the lifestyle of the work you are doing, where you live (or foresee moving), and if you connect with the people you would be working with in the field. While this phase is about personal exploration, income can be a challenge, especially in entertainment-related jobs.

The exploration phase can take place formally or informally. Formal exploration happens in a learning environment such as a school, training program, or internship. Informal exploration is learning on your own about different job roles, different disciplines, taking on your own work, or working for others. "You don't know until you get your feet wet," says Gabrielle Fisher, Audio Designer at Disneyland (California, USA). "You have to try it out. It's okay that at any point in time it's not for you. Or, you might learn that some portion of it might be where you want to go."

In the exploration phase, an emerging professional may be learning (or seeking work) in a variety of jobs and disciplines all at once, or move from one to another over time. Kevin McCoy (Head Audio Engineer, *Hamilton* "And Peggy" Company) says of exploring:

> You tend to specialize as you get further into your career. So, the time to do all the things you're even remotely interested in is when you're early in your career – especially if you go to school [where] there are far less stakes. You can do lighting if you're a sound person without collapsing your whole career because you failed at lighting. I think it's great to do all the things

you can in that part of your life because it will help you be a better sound person or whatever you wind up being. Maybe you think you're a sound person but then you wind up doing production management and think, this is actually really fun. Maybe I'd rather do this. It'll help you find those things, too.

The **preparation phase** begins when an emerging professional chooses to work toward a specific goal or specialization. The preparation phase could be focused on finding your way into a discipline or niche, such as live sound or broadcast mixing. A goal could be to build a career in a specific job role, such as a mastering engineer, podcast editor, or composer/artist. In this phase, an emerging professional may seek out strategic opportunities beyond audio, such as testing video games to start a path toward game sound design. In some countries, established professionals can't sustain financially from a single discipline or job role. In these cases, the goal of the preparation phase is to gain skills and experience that can help build a diverse (but financially sustainable) career. The preparation phase (for all) can last many years.

One goal of the preparation phase is to onboard into the industry (or a discipline). In business, **onboarding** is training and preparing a new employee to work for a company. In the audio industry, onboarding is learning all the "real-life stuff" firsthand (like how to handle an angry client or what you can't put down the toilet on a tour bus). Studio manager Tina Morris says of onboarding at the Village Studios (Los Angeles, California, USA):

> You learn the culture and you learn how to take care of the building. Then, when you're assisting, you're learning how to take care of the engineer and help them. Then, the producers. Once you're an engineer, you're learning how to take care of the clients. There are all these layers of gradual learning.

Some emerging professionals come into the field knowing what they want to pursue and can bypass the exploration phase. A degree or vocational program can also help an emerging professional come into the field in the preparation phase, as well. However, it does not replace the experience and time required to onboard into the industry. Italian composer/producer Giosuè Greco explains, "You get out of college and you think you're the shit, but you're not. For a long time, I thought that I was more prepared, but that actually wasn't true."

For many, the challenge of the preparation phase is pursuing your interests while earning enough money to survive. Shaun Farley (sound editor at Skywalker Sound) explains:

> I don't think there's any way to build a business quickly in this profession, and I regularly tell people just getting started to expect to be working multiple gigs (i.e., paying rent with non-sound work) until they can build

enough of a reputation to pull in the volume of work that will let them subsist on sound work alone. That's true if you want to build your own personal business or are happy working for someone else.

While there is some choice in what work you pursue, the preparation phase can be a period of taking whatever opportunities you can find. Fei Yu says of the films she worked on as an emerging music editor in China:

> At the very beginning, I wasn't choosing the director. It was more about the director choosing me. I didn't have that much choice. If they were a really good director, I wanted to work with them. But, nowadays, if I feel like our personalities are not really matching with each other, I have a choice . . . at the very beginning, you don't have a choice, and you do it because you want to build your career and you want to build your reputation.

Onboarding in the Video Game Industry: Harry Mack

Harry Mack is a Canadian sound designer and composer for video games. Harry discusses what he learned (and some of the mistakes he made) onboarding into the game industry.

> When I started my career, I started freelance, which is to say I didn't have a job. I didn't know what I was doing. I just wanted to be a music composer. I didn't want to do sound effects. That's how I approached clients – I said, I could do the music, and you'll have to hire someone for the sound effects. Every single one was like, "No, we usually just get one [person]." If I had experience from the beginning working in a company, I would have seen that the audio team does it all. After a year of not really getting anything to click, I knew I needed more experience. Sure, I've played games all my life and I've had music as part of my life. But [it's different] to put it all together, see how it works in a project, learn the jargon, and know who to talk to . . .
>
> There's a huge advantage of working in a company [in-person]. Like, if you need to talk to a programmer or an artist, they're just right there. You walk over and you say, "Hey, let's get this implemented." What does this look like as a freelancer? Often times, it could be a couple of days to see something implemented or to get an answer to a question, and that can be kind of frustrating . . .
>
> Some of the biggest mistakes I've made are not quite knowing when to keep my mouth shut (in regards to being so excited to be a part of a project), and just going out and saying, "I'm working on this

> project." Something as simple as confirming [a project exists or title] is a big no-no . . .
>
> I would [recommend] getting some experience in-house first, if you can. Once you're comfortable learning all that stuff, then you can feel free to go independent . . . Even if you're not the audio designer, get a testing job, something [entry-level] even like a secretary position or whatever . . . so you can see the project cycle from beginning to end. That's extremely valuable.
>
> *From Sound Design Live podcast*

After the preparation phase is the **establishment phase**. An emerging professional becomes an established professional when their income sustains their lifestyle on an ongoing basis. Opportunities can come to an established professional, but it still requires effort to maintain relationships and seek out work. "At the intermediate level where you're making par, it suddenly is no longer about skill. It's 100% a people fit at this point," says Megan Frazier.

This phase is also when specialization can occur. Rob Jaczko says:

> It's fairly evident to people that, at least at the beginning of your career, you can't be a super-specialist in doing one thing. Now, as your career gets on, you become a mixer, you're known as a vocal producer, you're just a general musical producer, you're a sound designer for shows. You define what your identity is a little bit more narrowly later in your career.

For Canadian mixing/mastering engineer Mariana Hutten, being established means she can be more selective about who she works with, and also niche into the job roles she would prefer to do. Mariana shares:

> I've got the bill-paying stuff down. It's now about how much your project interests me. If I worked with the person before, is that person pleasing to work with? My decisions now are more based on: What's going to further my career? Do I want to have this in my portfolio?

The Reality of Audio as a Job

Turning an interest into work (and eventually a career) comes down to being paid by other people. In the modern audio industry, this can require starting a small business versus finding a full-time job with a consistent paycheck. Podcast editor/consultant Britany Felix shares:

It's like every other small business. There's a grind. It's hard to get clients at first and there's competition. You have to treat it as [a business] rather than just, "Oh, I have these audio skills. I'm just going to go do this thing now."

For some entering the industry, living independently, finding work, and managing money are completely new experiences. Canadian recording engineer James Clemens-Seely says of coming into the field:

I had never done any math, like, how much would it cost to pay rent or how much debt I would accrue versus whether or not there were any income-generating avenues for me the first few years. You read these books about the old white guys who the business is built on and it's just like, "I dropped out of high school and knocked on the door to Abbey Road and they said make some Beatles records." It's totally unrealistic. That's not how it is anymore. There aren't really [full-time] jobs, but there's maybe more work than there ever has been out there before.

Michelle Sabolchick Pettinato says of live sound as a job:

As far as hanging out and partying with the band . . . the band is your boss. While yes, at times, you can and will develop great friendships with the people you work for, there is a fine line between employee and friend. You need to know your place as crew. It's not backstage partying with the band when load out is going on. You are there to do a job. This is a business like any other and as much as we don't want to believe it, it's about making money – money for the artist, their management, the promoter; all of which depends on each person doing their job.

Sal Ojeda shares of work in the music industry:

There are a lot of people that go into music thinking it's going to be fun all the time. As soon as the real world hits them, they realize it is a job like anything else. They struggle because it's like, "I didn't know it was going to be so tedious." You're going to be listening to snare samples for half a day, or organizing a library for a composer. It's a job at the end of the day. Even if you're on top, there are going to be days when you don't want to mix.

Doing audio as a job may entail looking for work opportunities regularly. Nathan Lively, creator of *Sound Design Live*, explains:

In 2016, I got a gig touring full-time with the Ringling Bros. Circus for US$1,500/week. That was the moment when I was finally convinced that I could really have a career and make a living in audio. Four months later, I lost the job and had no home. There are no promises in pro audio. The only thing you can rely on is change.

What Is a Business?

- **Business** is the exchange of goods or services for money

- A **business owner** is a person (or entity) who owns and operates a business

- A **client** is a person (or business) who pays for specialized services, goods, or advice from a business

---

Soft skills, such as how you communicate with others, are crucial in many areas of the industry. "You're not going to get fired for having the wrong EQ on your snare drum. You're going to get fired because you're treating someone poorly," says Scott Adamson, touring front of house mixer (Haim, Khalid).

Chris Leonard, Director of Audio at IMS Technology Services (a concerts and live events company), adds:

> I don't care if you can mix your butt off. I need to know when the CEO of this Fortune 500 company comes up, whether you could speak to them and put a lavalier microphone through [their] shirt – and can do it with confidence, speak to them and carry a conversation, and not look completely disheveled.
>
> (Potter)

Kevin McCoy says of soft skills:

> I am far less interested in the specific gear that someone knows than whether they're pleasant to work with, and whether they're able to learn new things. Presenting yourself in that way is a lot more important to me than telling me that you've used a specific console, or that you've worked with all these different speakers or whatever. I care a little bit about that, but I'd rather know that you're a pleasant person.

It is common to work with others when doing audio as a job. David Bondelevitch, an Emmy-winning re-recording mixer and sound editor who teaches sound for picture, shares:

> One of the biggest criticisms I get from students is that I make them work in groups. Students rarely understand that they will never be working alone. At the very least, they must please their clients. If they work at a facility, they don't get to choose who does their Foley or who the music editor is, yet they have to create a working environment with them. I have seen brilliant students turn in terrible work because they were unable to work collaboratively.
>
> (38)

Devyn Nicholson, Manager of AV for McGill University, says of working on teams, "You could be the world's best DSP programmer but if you're going to wreck the culture of my team, I'm not interested. That's the bottom line."

A **client** is a person or business who receives a specialized service or professional advice. For example, if a music mixer is hired by an artist to mix a song, the artist is the client, and mixing is the service. A client is hiring you for your expertise (skills and ability), but they are also seeking professional support and communication, or **customer service**. Even when a job is to contribute to a product (such as a film, music, video game, or podcast), it can require customer service. Recording engineer/studio owner Catherine Vericolli explains:

> A band comes in and they say they want their record to sound a certain way. That's what you do because it's a service – just like if I was an electrician coming to your house to fix your kitchen outlet. I'm not going to rewire your whole house because I think it should be wired that way, then charge you a ton of money for a bunch of work you didn't ask me to do.

Customer service is also providing what your client wants even if their tastes and preferences are different from your own. Mastering engineer Sarah Register explains, "This job boils down to client service. Cook the burger how the client wants. Sincerely give opinions and feedback when asked, but they have to eat the burger. Don't f— up the burger" (qtd. in Campbell).

Karol Urban adds of post-production sound (for film and television):

> This is [the producer and filmmaker]'s project that they have put so much into. They're putting into your hands and in your ears to manifest for them . . . I'm literally trying to put on the glasses (or the hat) of another person's taste. Whereas I go, "I don't feel emotionally connected at all when we do that," that person does. I need to change my mind when I'm in the chair working for that person.

Customer service applies outside of entertainment industries, as well. AV manager Devyn Nicholson shares:

> Every single role on my team has an element of needing to manage somebody's expectations (clients, professionals, vendors). Communication is a hugely important piece to that, whether it's setting the right kind of tone for a meeting, or whether it's letting them know that you can't deliver on something that they request . . . Every person on my team needs people skills.

Audio engineer Samantha Potter adds, "Being able to earn respect, being trustworthy, and having a customer service mindset will take you really, really far in this industry, regardless of your discipline."

The Reality of Working in Live Sound: Michael Lawrence

Michael Lawrence is a live sound engineer and technical editor (ProSoundWeb and Live Sound International). Michael tells a story of a "shadow" (someone observing him to learn) on a touring live sound gig.

> A company I work with had a fresh audio graduate from a program that I was not familiar with [shadowing]. As soon as we showed up, it was all about, when are we opening the console? I said, "Look, that's nine hours from now. We're going into an empty theater. Let me show you how to pin the boxes, and how to run these." She had no interest at all . . . until the console flipped, and then she was all about it . . . then she showed up 20 minutes before the show and said, "Can I mix?" I was like, "No, you can't mix. What are you talking about?"
>
> You don't just walk in and put your hands on a console. On those show days, I might be in the venue for 16–18 hours and 90 minutes of that is mixing. If that's the only 90 minutes that you're interested in, you're going to have a really rough time because that's really not what the job is. There are a couple of people out there who do the "white-glove" walk up, mix, and go back to the bus. But for the vast majority of us working in the field, that's just not the reality . . . If you just sit in a classroom for four years and you don't go push a road case, you're in for a really rude awakening.

Freelance Work and the Gig Economy

Full-time jobs do not exist for any business unless they have a full-time need *and* can afford that employee. For example, a restaurant or church may only need a live sound engineer on the weekends, or a studio may only hire a mixer when there are projects to work on. Game sound designer/composer Harry Mack explains:

> When I started my career, I was single, and I was happy enough to work freelance. I thought as soon as I had a family, I would want [a stable job] . . . know my hours and know what my salary was going to be . . . But game companies are very tenuous. They can fold, you can get laid off, and projects get shifted around. You could work a couple of years on a game and have it canceled because of funding.

Working for yourself, you can have many different clients so that if one project fails, you can move on to another one. You can choose, hopefully, the clients that are more stable. The real trick is to just always have a backup. You can't just assume that everything is going to work out.

Kate Finan, co-owner of Boom Box Post, says of working in post-production sound in Los Angeles (USA):

> In our industry, you're hired on a per-project basis. Sometimes you're [on staff] somewhere, but a lot of times you're in for just one episode, or the run of a series, or just one movie . . . I spent a lot of time at the beginning of my career hustling, having a lot of hiatuses, spending nine months with no work and just trying to pinch pennies, and then getting another project that lasted for a couple of years and having the same thing happen again.

There are two primary classifications for workers: **contractor** (or freelancer, independent contractor) and **employee**. What a "boss" looks like is different between a contractor and an employee. Beth O'Leary (freelance live sound engineer/tech, Sheffield, England) explains:

> These people who give you work (or give other people work instead of you) are not your bosses, and definitely not your friends. They're your clients. They don't owe you fairness, help, or career progression . . . You don't owe your clients loyalty or unpaid hours unless they deserve it . . . You are free to do what you want. There's no point in complaining about your boss keeping you in a bad work situation because your boss is you.

A contractor can do identical tasks as an employee. For example, a band could hire a recording engineer (contractor) or pay a studio that hires a recording engineer (employee). In general, a contractor will have responsibilities for their work (related to business, financial, and taxes) that an employee does not. The legal definition of "employee" and "contractor" vary by country, and it is important to familiarize yourself with the differences where you live.

A **gig** is a single-occasion music performance, but the term has been adapted to mean any temporary employment. A **gig economy** is a workforce where temporary, short-term, and non-employee (i.e., freelance or contractor) work arrangements are the norm. For a worker in the gig economy, it takes a combination of gigs to have income similar to full-time employment. Live sound engineer/tour manager Claudia Engelhart explains:

> In this day and age, you've got to have more than one skill or you've got to work for more than one person. If you have a couple of different gigs

simultaneously, then it generates more work. You're constantly trying to hustle another gig so that you can stay alive. The more irons in the fire, the better off you're going to be. You have to be willing to try different things until you're established enough. You're going to have to maybe do lights, or be a driver – whatever it is to keep going.

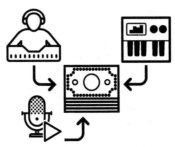

Figure 1.1 Gigs across the audio industry can be combined for an income equivalent to full-time employment.

Live sound engineer Aleš Štefančič says of his work in Slovenia, "My freelancing activities include audio education, operating a small studio for music and voice-over jobs, and mixing or teching live sound gigs for bands or rental companies. Touring is about 10% of my total income" (17).

The advantages of working in a gig economy:

- Choice over whether or not to take a gig
- Choice of rate, ability to raise rates, and by how much
- Freedom to pursue a variety of work (if you want to)
- More flexibility to take time off or work preferred hours (for some projects/gigs)
- Owning your equipment and choosing what to buy

The disadvantages of working in a gig economy:

- Inconsistent income. Less work means less pay, and no work means no pay
- A need to take gigs or work you may not like to generate income
- Unpaid responsibilities like bookkeeping/accounting, networking, sales/marketing
- No guaranteed benefits (retirement, healthcare in some countries, paid vacation, sick days)
- Paying for all your own equipment and repairs, and responsible for troubleshooting

The challenge of working in a gig economy is the lack of stability. Jeff Dudzick, who worked with dozens of contractor audiobook editors for Audible, says of relationships with contractors, "There are no guarantees ever [for] work . . . We're always upfront about that. You can't make promises to people. You can say it's been a steady stream of work for years, and it looks good, but no promises."

The Modern Audio Industry

The **modern audio industry** refers to a time period when the internet and social media have been widely used as a business tool. While there is not an exact year to its beginning, social media and smartphones exploded in popularity in the early 2010s. Social media usage went from 970 million worldwide in 2010 to 3.81 billion users in 2020 (Dean). New features at the time (like direct/private messaging and groups) made it possible for an emerging professional to communicate with professionals who may not have been as easily accessible before.

There are differences between those who started their careers in the modern audio industry and those who came into it beforehand. John McLucas uses the terms "old guard" and "new guard" to describe professionals of each era. The old guard may not recognize how important technology is to build a career in the modern audio industry (through social media and content creation, virtual methods of networking, and online learning). The new guard may not connect with the values of the old guard, such as in-person networking, onboarding through an entry-level job, or spending years learning the craft from other professionals (versus teaching yourself). There are many lessons to be learned from both.

Business in the "Old Guard" Audio Industry

The audio industry began in the 1880s with the growth of music production as a business. The 20th century brought new entertainment industries in film, television, video games, and amplified performances. Outside of entertainment, career opportunities developed such as education, preservation/restoration, manufacturing (products for professionals and consumers), AV, and more. For most of the 20th century, professional audio equipment was not widely available (due to size and cost), but consumer-level audio equipment sparked an interest for future audio engineers. Audio engineer/educator Agnes "Aji" Manalo (Manila, Philippines) says of the technology she used growing up in the 1970s and 1980s:

[I was] recording broadcasts via line out of the TV sets . . . creating mixtapes from the radio, recording my piano performances using a karaoke mic and machine. [I had] a Walkman with a 5 band EQ . . . If it were not for those exciting playtime discoveries and exposures, I would not be in this kind of business. In hindsight, what I am doing now is an extension of my audio playtime when I was a lot younger.

In the 20th century, the primary barrier to working in the audio industry was finding a work opportunity, or "getting a foot in the door." This route was necessary to access professional audio equipment and skilled professionals who knew the craft. Brian Schmidt, composer/sound designer and founder of GameSoundCon, shares:

When I started in the industry [in 1987], and through the early 1990s, game companies hired composers and sound designers as regular, full-time employees. The bulk of game music and sound were created by these in-house employees. Part of the reason for that was creating music and sound was highly technical, with very unique, specialized skill sets needed. You couldn't just go into a studio and start recording your soundtrack or record proper Foley and put it in the game – the technology didn't allow that. Each company had their own particular quirky systems and tools and required specific technical skills.

Job seekers typically had to "start from the bottom," or take an entry-level position to onboard. For example, in the recording studio world, a runner was (and still is) an entry-level position with responsibilities like cleaning and running errands. Tina Morris, studio manager of the Village Studios, says of the facility, "Once you get in, it doesn't matter if you can get a really great kick sound. It matters if you know how to make the coffee and get a food order right."

The reward of this path could be building relationships and learning from professionals on the job, stable employment, growth opportunities with the company, and possibly working on high-profile projects. However, those opportunities could take time. Front of house mixer Dave Swallow started his career in 1998 as a recording studio intern and says of the experience:

I suppose it all seems pretty straightforward. Get a job sweeping floors, learn how to use a mixing console, get asked to go on tour. Well, the gap between starting my unpaid job and getting paid was about six months and the gap between getting paid and going on tour was about three years.

(45)

The number of audio education programs grew in the latter part of the 20th century. In 1979, the Audio Engineering Society's first *Directory of Educational*

Programs listed 101 programs worldwide, and in 1996, the *Directory* had 211 (Pritts). In 2021, the *Directory* (on the AES website) showed 572.[1] Tommy Edwards, who attended Berklee College of Music (Boston, Massachusetts, USA), says of audio education in the late 1980s:

> The only way you could get your hands on that equipment was really to go to one of the recording schools. For me, it was Berklee, Miami, USC being the three big traditional schools, like, "Mom and Dad wanted me to go to University or college" . . . I also remember (in maybe 1987-88) the staff at Berklee sitting everybody down and telling us, you're not going to walk out of here and be making a lot of money. At first, you're going to get minimum wage jobs and have to work your way up . . . but in the 1990s, the record industry was still intact. You still had studios. You could still get a [full-time engineering job] back then.

By the year 2000, the audio industry was passing down the craft to the third generation of professionals in its most established locations (like the USA and the UK). For example, Grammy-nominated recording engineer Steve Genewick was mentored by legendary engineer/producer Al Schmitt; Al Schmitt's mentors were his uncle, Harry Smith, and industry pioneer Tom Dowd (Schmitt et al.). At the start of the 21st century, emerging professionals still relied on personal relationships for career information and guidance. For example, Kevin McCoy decided to pursue a career in touring musical theater after hearing stories about it while working on a small regional production:

> We would load up a 15 passenger van . . . The stage manager had toured on larger shows, and she told me all of her stories. All we had to do is drive, and there was time to tell each other stories. It seemed like the thing that I wanted to do. This was in the early 2000s, so there was little data access.

Catherine Vericolli (who attended audio school in the early 2000s) adds, "We couldn't watch YouTube videos to teach us how to make records. We thought our professors were Gods and magicians because they made records."

The 21st century brought a rise to home studios and project studios worldwide. Audio technology became more widely available, accessible, and affordable in the 1980s and 1990s with the shift to digital audio and computer-based recording. Brian Schmidt says of game sound:

> [The year] 2001 was the real inflection point in game audio with the introduction of the DVD as the game delivery format for the original Xbox and PlayStation 2. Now, a composer or sound designer could just go into the studio and record the music just like any other media and put in the game. That really changed the dynamic of working in the industry; those specific technical skills [of the past] weren't as necessary.

In 2022, some entertainment companies and professionals still function with an "old guard" mentality: Everyone must start from the bottom and work their way up on the job; careers are built on face-to-face relationships; learning from established professionals (in-person) is the proper way to learn the craft. While it is still possible to build a career with this mentality, it no longer is a necessity. Dave Swallow explains:

> We are now in a time when access to education, jobs, and careers in creative industries is far greater than it ever was for me and probably most of the people teaching [educational] courses. Many among my peers believe education in this industry to be a bad thing. These opinions tend to be grounded in conservative fear more than anything else – Fear that our industry will be flooded by an influx of "kids" who know all the tricks and can do our jobs for half the price . . . It's only through education that we can do better. The sharing of knowledge is beneficial for everyone.
>
> (50)

Business in the Modern Audio Industry

In the modern audio industry, many of the "old guard" barriers have been removed due to the growth of social media and technology. Access to information and equipment are no longer major barriers, which means emerging professionals can come into the field with skills (and experience) well beyond those of past generations of emerging professionals. Beginner music and sound creation apps cost as little as a cup of coffee and are easy enough for a three-year-old to use.[2]

The amount of content has exploded and continues to grow. For example, Nigeria's film industry ("Nollywood") produces nearly 50 movies a week (ITA). In 2020, the countries with the highest percentage of adults who played video games were Vietnam, Nigeria, the Philippines, and China (Buchholz). In 2019, four of the top five countries showing the most growth in podcasting were Spanish-speaking: Chile, Argentina, Peru, and Mexico; China was the fifth (Grey).

Work opportunities are abundant in the modern audio industry, and there are more global opportunities than ever. Someone from anywhere in the world could hire an orchestra to record in Prague or a sound team to record Foley in New Zealand. The COVID-19 pandemic (beginning in 2020) accelerated audio-related freelance opportunities worldwide. Agnes "Aji" Manalo observed in the Philippines, "Because of the recent pandemic, freelance audio engineering/

work-at-home setups seem to have proliferated because of the surge in pod-cast, events streaming, music video content materials and vlog requirements online."

There are more uses of audio in the world than ever before, which is leading to opportunities for people with a background in audio *and* other areas of expertise (such as networking, IT, acoustics, business, video, system design, radio frequency/RF, and more). AV professionals may be included as part of a building construction team, and businesses from retail stores to hospitals now have audio needs. Nathan Lively says of having varying skills (in live sound):

> In the next few years, the highest demand in touring will be for generalists. Everyone wants to specialize in mixing [front of house] or sound system optimization, but the highest demand is for people who can mix monitors, tour manage, operate a lift, and drive a truck. It's the same with local gigs. If you can set up a projector and operate a video switcher, you'll get a lot more work than if you can only take audio gigs.
>
> (25–26)

In the modern audio industry, it is possible for an emerging professional to move away from the "old guard" process of onboarding into the industry and instead learn and work on their own terms. This is possible because of social media, which has become a common tool for networking and connecting with potential clients. "We legit can lay in bed and make shit happen," says Alesia Hendley. John McLucas says of a "traditional" recording studio internship experience:

> I spent hours cleaning caked dirt off this guy's fridge left outside. I [thought], this is just dumb. I could go on Facebook and talk to people and maybe figure it out instead of just taking crap off of a fridge . . . I understand the technical principle of hard work, patience, and showing that you can do tasks. Why would they trust you with the session when you can't be trusted to do this baseline thing? I understand conceptually what it is, but for myself as a child of the Internet, I was like, I can do more things with [my phone].

In disciplines like music production, video games, podcasting (and more), a professional could perform all of their work from a location of their choosing (including at home). It is possible (but not the norm) to acquire a client online, perform all the work remotely, receive payment remotely, and not have a single verbal interaction. At the same time, for those working independently, some learning is lost by not having in-person learning (and listening) oppor-tunities. Mentorship has become more informal (and brief) with many of these interactions happening online. Dutch post-production sound designer Aline Bruijns explains:

> [Working alone], you're missing the complete picture . . . I sometimes find it really hard to keep up with all the technology and all the new ways of doing things because you don't really have a coffee corner, or you cannot walk into someone's office or in someone's studio say, "Hey, what are you doing?" Just looking over their shoulders and thinking, "This sounds cool. How did they do it?" I'm lucky that I have multiple groups right now so I can exchange a lot of ideas and learn from other people's ideas, and I can ask questions. Organizations are so important like SoundGirls, AES, and MPSE. All those kinds of organizations open up the world a little bit more because we're all situated in our own little caves working.

The paths used to begin a career in the 20th century still exist, such as working full-time for a studio or company in an entry-level role. However, the cost of living can be so high (compared to wages) that it may not be feasible to take these opportunities without outside financial support. Luisa Pinzon (Bogota, Colombia) explains:

> Part of this industry I've been fighting a lot is this old sound engineering thought that you need to be an assistant in a studio for five years getting coffee in order to be able to plug a cable. It's fair – I know the experience is there, and I know there are a lot of credits, and I know this is how it used to be. But how can a human being survive and eat and pay rent by not getting paid anything? That's how it was, but why does it need to keep being like that? You need to pay your assistant, not because your old boss didn't pay you . . . There are a lot of people with this [old] way of thinking that you need to work forever without getting paid in order to one day get a Grammy.

It is possible to work for yourself and be in control of your business in the 21st century, but it also has created a barrier: visibility, or standing out from your competition. "We've moved from a castle with a moat mentality to anything goes/anybody can do this. That's a great thing for art, but it's not good for people working right now," says mastering engineer Piper Payne.

Rob Jaczko says of the industry in the 1990s versus the 2020s:

> When I was a staff engineer at A&M (now Henson Recording Studios) in Hollywood, I didn't have to worry about many things. I wasn't out networking because I went to work every day at a five-room studio and I met people by virtue of that. I didn't have to worry about my taxes because I got a W-2 [tax form], and my accountant did my taxes. I didn't have to grapple with certain things that I would if I were starting out in [the 2020s] . . . Now we have to grapple with our brand, our social media

presence, and our business acumen. We need to get paid – probably through a series of different streams of income that we generate.

In the modern audio industry, it is increasingly difficult to have a financially stable career by relying on audio skills alone. In the interviews for this book, professionals (emerging and established) across all disciplines talked about evolving job responsibilities: Touring front of house mixers who double as a tour manager or production manager; aspiring producers and recording engineers who assist with studio management; podcast editors who handle marketing tasks. Additionally, independent audio professionals are expected to operate their own business on top of doing audio work. Re-recording mixer Mehrnaz Mohabati says of audio work, "It's a hard route. It's not an easy thing. We study and we work hard as a doctor, but we never have their security."

Emerging professionals come into the field with or without formal education, but formal education is still intended to be a starting point. Audio engineer/podcast editor Jonathan Hubel explains:

> My broadcasting degree was helpful in that I did gain some knowledge and understanding of the audio processes, but it only helped at the very beginning of my career before I had much experience on my resume. Everything else since then has really been gained on the job, growing through practice. A degree is a good starting point, but the audio industry is honestly almost entirely about who you know and how connected you are.

One difference in the modern music industry is the term "music production" has become synonymous with music creation. Some emerging professionals today seek out audio education to enhance their *own* creations whereas, in the 20th century, music production education was intended to work on other people's art. This massive industry shift has led to a crisis in formal education: Are music production programs (particularly university degrees) intended to teach an art form or help students be employable? The balance between being a creator – versus being *in service* of creators – was brought up by nearly every educator interviewed for this book. Susan Rogers, professor at Berklee College of Music (in the music production and engineering program), says of this challenge:

> In our department, we debate this all the time, and we have not arrived at a consensus. We've asked other audio tech institutions what their perspective is on it. The question is, what are the foundational principles? What's the core curriculum that every audio student must know? We simply cannot agree on it. I think, because I'm from the older generation, every audio student who gets a degree in music production and engineering should understand engineering in terms of the basics. They should be able to explain the principles of electromagnetism, they should be able to

describe gain staging, and then those simple "old school" analog principles. But, our younger teachers don't necessarily agree with that. The younger teachers think that knowing analog audio is just for the people who repair the gear, and that the engineer should focus on being more of an artist with sound – a sound sculptor. I wouldn't technically call that engineering because engineering is a proper word that assumes [a good deal of] theoretical knowledge.

I was holding a hard old-school line but in these recent years I have kind of given up. I feel like maybe that ship has sailed, and maybe today's audio youth need a different theoretical knowledge . . . I don't see how you can have a career without knowing the basics. I guess people do it.

In the modern audio industry, it is rare to find formal education in many specialized audio niches, especially those that require cross-disciplinary skills. For example, Avery Moore Kloss has built a career around audio storytelling (for radio, podcasts, and more), which requires journalism and audio skills. "A lot of the audio engineering you take is going to be about music. I like that and I appreciate that, but at the same time, I want to know about audio engineering with voice, too," says Avery.

Jeanne Montalvo Lucar has a degree in music production and also works in spoken word audio, where she learned radio producing on the job (at NPR's *Latino USA*). "I don't have a journalism degree, but somehow I managed to move my way into producing," says Jeanne. "I have definitely found it an asset because it makes me more marketable to people . . . it's very rare to find someone who can engineer and produce."

Audio preservation requires audio skills and detailed file and database management, but audio professionals typically have to learn the latter on the job. Mastering engineer Anna Frick explains:

I'm fortunate my mom was a librarian and so I learned those skills of just being methodical and cataloging and wanting to know where everything goes, and where everything exists. That has been hugely helpful to me in this world. I don't think a formal education necessarily prepares you for that – unless you're doing a library sciences degree. I don't think actual preservation of audio is a focus in library science programs. So, it's almost like a marriage of the two is what you need for this kind of work.

Tommy Edwards, Vice President of Product Development at Warm Audio, says of the skills needed for product design of audio products:

[You have to] understand the physics of basic hearing. Then, you need to understand audio. Then you need to understand things like ROI [return on investment] and business metrics and positioning and market research and how you get a product in Sweetwater Sound. [Having] all those things in a single person is wildly rare . . . If you are going to acquire those skills in a traditional college pathway, that's like two or three degrees.

The New England School of Communications (NESCom) at Husson University in Bangor, Maine (USA) has a rare degree program: a bachelor's degree with a concentration in live sound production. Assistant Professor Eric Ferguson says one advantage of their coursework is the hands-on technical experience students get that they likely would not get in a music production program:

> In the studio, they don't get behind the gear. They don't plug in the console. The tipping, the lifting, the plugging in – the signal flow you learn by actually plugging things in – that never happens in the studio . . . I see my students graduate, go get a job, and within one year work as a PA tech for [a major tour] . . . In our program, the goal is to help them get a job.[3]

Online education programs and informal communities are thriving, but online learning has its limitations. Utility sound technician (for production sound) Patrushkha Mierzwa explains:

> Many people want to help you become successful, and you'll see a multitude of YouTube videos and tutorials claiming to teach you techniques used in mainstream industry work. I've watched several of them myself, and they can be informative. But if you're just starting out, it can be impossible to know what is a valid technique and current practice, and what will get you passed over for the next job.
>
> (19)

Video game sound designer Javier Zúmer adds:

> If you try to get a sense of people working in the video game industry by reading the people who talk on Twitter, that's a very small proportion [of the industry]. Most people working in video game audio are working, they don't talk much [online], and they do their thing. So, if you only listen to the loudest people, you're going to get a distorted sense of the community . . . Be careful with survival bias because we only see the survivors – the "successes." . . . That can give you a wrong sense of what works and doesn't work. It happens also in science.

Fundamentally, the modern audio industry is the same as it always has been: It takes time and effort to turn an interest into a career. In the 20th century, a career was built through meeting people, building relationships, gaining credits, and skills. In the 21st century, all of this still exists plus the "internet mountain" – the need to build a web (or social media) presence to find work opportunities (Hendley). This leaves aspiring professionals to ask: Should I listen to the advice of the old guard and follow their playbook to build a career, or can I do it my own way and on my own? Needless to say, the answer to this is complicated. Foley mixer Jeff Gross shares:

> The way to learn is any way that you can. Ideally, the hybrid method (working for others and learning on your own) is the best because you learn what things are supposed to sound like. It's kind of like learning to play guitar. You can learn by ear and not learn the theory (and all of that side of the experience). That's one way, but you have no real foundation. Your foundation is only what your fingers are doing and your brain is hearing. It's the same thing with Foley. In a perfect world, you try to at least hear and see what people are doing who have been doing it [a long time] because you're not necessarily going to know what sounds right. The worst thing is if you do a show, send it to the stage and then the mixers say, "What's [wrong with] this? This doesn't sound [right]."

The ideal career for you (and the potential paths to get there) will depend on many factors – from what kind of work you are trying to pursue, where you live, your learning style, and your barriers. As Javier Zúmer says, "[Use opportunities] to learn . . . but finding a job is a job itself."

The Reality of a Competitive Industry: Tommy Edwards

Tommy Edwards started in the music industry pursuing a career as an artist and musician. He pivoted into product management, where he has overseen the creation of music technology and audio products for industry companies like Warm Audio, Blue Microphones, and Line6.

> Music is a lot like sports. When we're younger, a lot of people want to do it. It's a funnel and it just keeps getting smaller and smaller and smaller the higher up you go. Pick your favorite sport. The set of people who play women's volleyball in eighth grade is large, and then it gets whittled down in high school, and then you go to college, and now it really gets constrained. How many women end up playing volleyball in the Olympics?
>
> I hate that sounds so pessimistic . . . but how many people in my class at Berklee were going to win a Grammy? It's not all of us, for sure. I happened to be roommates with someone who won seven of them . . .
>
> A lot of [audio programs] didn't exist when I was going to school [in the late 1980s/early 90s]. Now your competition is even higher. We certainly have less traditional jobs, like, I'm going to go work at

Ocean Way and be a staff engineer. Recording studios barely exist. A lot of them have closed. That's the really big shift in what it means to be a pro audio engineer in 2021.

What I do like about the new era is there are less gatekeepers. Because there are fewer gatekeepers, a lot more people have access to this arena. That said, there's a lot more noise. So, then it becomes the challenge of how do you get above the fray above the noise, and really stand out?

The industry has become more friendly to generalists . . . being wide versus being a specialist in one area . . . It is probably harder to be [specialized such as] a mastering engineer because you're just doing one thing. You're focused on one thing, there's a lot of really good people out there, there's a lot of really good tools to do that, and you're battling AI . . .

Jack of all trades is not necessarily a bad thing . . . In our arena, it can be somebody who knows how to record a band [or] orchestra in a studio or at a home studio, who also knows video, and who also knows social media. Maybe they're not the expert in any one of those disciplines but think about how powerful that is to have that broad skill set.

You hear the adage "renaissance woman" or "renaissance man." I really resisted that for a long, long time, because I associated it with "jack of all trades but master of none" . . . But then I realized how rare that actually is [in my work] where you've got somebody who can understand things on a very technical level and immediately switch gears and go to marketing, and then switch gears back to human factors, behavioral stuff, and then go to DSP. The set of people who can do that rapid switching is actually small.

Notes

1 The AES's *Directory* only includes educational programs which have voluntarily participated. These numbers are not reflective of all education programs in existence.
2 At age three, the author's son was able to learn (on his own) how to record and create music using software intended for adults.
3 More on Husson University's program in Chapter 13.

References

Adamson, Scott. "Getting Started in Small Clubs – The Pandemic Sessions Pt. 2." *YouTube*, uploaded by The Production Academy, 24 March 2020, youtu.be/Q0NLn6GetNI.

"AES Education Directory." *Audio Engineering Society*. www.aes.org/education/directory/. Accessed 13 May 2021.

Bondelevitch, David. "Teaching Sound Mixing: Post." CAS *Quarterly*, Winter 2021, p. 38.

Bruijns, Aline. Personal interview. 21 June 2021.

Buchholz, Katharina. "Where Video Games Are Popular Among Adults." *Statista*. 12 Nov. 2020, www.statista.com/chart/18914/adults-video-game-playing-behavior-selected-countries/. Accessed 20 Aug. 2020.

Bushong, Chris. Personal interview. 27 Sept. 2020.

Campbell, Madeleine. "Sarah Register, Mastering Engineer." *Women in Sound*, no. 2, 7 May 2016. www.womeninsound.com/issue-2/sarah-register/. Accessed 9 May 2021.

Clemens-Seely, James. Personal interview. 20 Dec. 2020.

Dean, Brian. "Social Network Usage & Growth Statistics: How Many People Use Social Media in 2021?" *Backlinko*, 10 Aug. 2021, backlinko.com/social-media-users. Accessed 20 Aug. 2021.

Dudzick, Jeff. Personal interview. 1 Oct. 2021.

Edwards, Tommy. Personal interview. 29 Oct. 2021.

Engelhart, Claudia. Personal interview. 8 June 2020.

Farley, Shaun. Email interview. Conducted by April Tucker, 15 July 2020.

Felix, Britany. Personal interview. 22 June 2020.

Ferguson, Eric. Personal interview. 2 Dec. 2020.

Finan, Kate, panelist. "SoundGirls Career Paths in Film & TV Panel at Sony Studios." *YouTube*, uploaded by SoundGirls, 26 Oct. 2018, youtu.be/Y_g3drC2yyA.

Fisher, Gabrielle. Personal interview. 18 Aug. 2020.

Frazier, Megan. Personal interview. 11 and 21 Aug. 2020.

Frick, Anna. "Ask the Experts – Audio Restoration & Archiving." *YouTube*, uploaded by SoundGirls, 1 May 2021, youtu.be/_Mn1Y0F8mbs.

Gaston-Bird, Leslie. "31 Women In Audio: Eveanna Manley." *Mix Messiah Productions*, 31 March 2017, mixmessiahproductions.blogspot.com/2017/03/31-women-in-audio-eveanna-manley.html. Accessed 30 March 2020.

Greco, Giosuè. Interview with April Tucker and Ryan Tucker. 28 Sept. 2020.

Grey, Georgia. "Top Growing Podcasting Countries – March 2019." *Voxnest.* blog.voxnest.com/top-growing-podcasting-countries-march-2019/. Accessed 20 Aug. 2021.

Gross, Jeff. Interview with April Tucker and Ryan Tucker. 14 Jan. 2021 and 30 Dec. 2021.

Hendley, Alesia. Personal interview. 17 Dec. 2020.

Hubel, Jonathan. Email interview. Conducted by April Tucker, 22 Oct. 2021.

Hutten, Mariana. Personal interview. 19 Nov. 2020.

International Trade Administration (ITA). U.S. Department of Commerce, "Media and Entertainment Industry (Nollywood and Nigerian music)." 9 Sept. 2020, www. trade.gov/country-commercial-guides/nigeria-media-and-entertainment-industry-nollywood-and-nigerian-music. Accessed 20 Aug. 2021.

Jaczko, Rob. Personal interview. 15 Oct. 2020.

Lawrence, Michael. "Should You Go To School For Audio?" *Signal to Noise Podcast,* ep. 44, ProSoundWeb, 29 April 2020, www.prosoundweb.com/signal-to-noise-episode-44-should-you-go-to-school-for-audio/.

Lively, Nathan. "Nathan Lively." *Get On Tour: A Sound Engineer's Guide,* edited by Nathan Lively, self-published, 2018. pp. 23–26.

Mack, Harry, guest. "Career Advice For Freelance Designers." *Sound Design Live,* 16 July 2013. *SoundCloud,* soundcloud.com/sounddesignlive/career-advice-for-freelance.

Manalo, Agnes "Aji". Email interview. Conducted by April Tucker, 3 Oct. 2021.

McCoy, Kevin. Personal interview. 22 Dec. 2020.

McLucas, John. Personal interview. 25 Jan. and 8 Feb. 2021.

Mierzwa, Patrushkha. *Behind the Sound Cart: A Veteran's Guide to Sound on the Set,* Ulano Sound Services, Inc., 2021. p. 19.

Mohabati, Mehrnaz. Personal interview. 29 Jan. 2021.

Montalvo Lucar, Jeanne. Personal interview. 12 Oct. 2021.

Moore Kloss, Avery. Personal interview. 21 Oct. 2021.

Morris, Tina. Personal interview. *Once You Have the Gig – What Makes You Stand Out,* SoundGirls, 20 July 2020, youtu.be/QJhdm86kIlM. Accessed 21 July 2020.

Nicholson, Devyn. Personal interview. 23 Nov. 2020.

Ojeda, Sal. Personal interview. 27 Sept. 2020.

O'Leary, Beth. "Where Are You Going?" *SoundGirls,* soundgirls.org/where-are-you-going/. Accessed 30 March 2020.

Payne, Piper. Personal interview. 29 May 2020.

Pinzon, Luisa. Personal interview. 10 Nov. 2020.

Potter, Samantha. "Veteran Audio Professional (And STN Co-Host) Chris Leonard." *Church Sound Podcast*, ep. 11, ProSoundWeb, 16 July 2020. www.prosoundweb. com/podcasts/church-sound-podcast/.

Pritts, Roy. "Education and the AES." *Audio Engineering Society*, vol. 46, no. 1/2, Jan./ Feb. 1998, pp. 88–90, 92.

Rogers, Susan. Personal interview. 12 Oct. 2020.

Sabolchick Pettinato, Michelle. "Good Questions." *Mixing Music Live*, 26 June 2019, www.mixingmusiclive.com/blog/goodquestions. Accessed 22 May 2021.

Schmidt, Brian. "Re: Seeking permission to use quote in textbook." Received by the author. 16 Aug. 2021.

Schmitt, Al, et al. *Al Schmitt on the Record: The Magic Behind the Music (Music Pro Guides)*. Rowman & Littlefield Publishers, 2018. Chapter 2, Appendix A.

Štefančič, Aleš. "Aleš Štefančič." *Get On Tour: A Sound Engineer's Guide*, edited by Nathan Lively, self-published, 2018. p. 17.

Swallow, Dave. "Dave Swallow." *Get On Tour: A Sound Engineer's Guide*, edited by Nathan Lively, self-published, 2018. pp. 45–50.

Urban, Karol. Personal interview. 17 July 2020.

Vericolli, Catherine. Personal interview. 29 May 2020.

Yu, Fei. Personal interview. 9 Nov. 2020.

Zúmer, Javier. Personal interview. 24 April 2021.

2
Defining Your Dream and Goals

A career starts with an interest and a dream, but a career grows by bringing a dream into reality. What does a dream job look like in the real world? How long will it take to get there? What are the not-so-fun aspects of a career you might not see from the outside? Are your goals in line with the sacrifices you're willing to make for your career?

Front of house mixer Michelle Sabolchick Pettinato says of touring:

> You love music . . . That's a good start, but is it your passion? Is it something you are willing to bust your butt for, to work 16 hour days getting dirty and sweaty? Are you willing to slug it out in smelly bars and clubs if you are trying to cut it as a mixer? Or, working as a grunt on the audio crew for tour after tour until you are experienced enough to be the system tech? Having a sincere passion and strong desire is a great start. Getting your expectations in check is also helpful . . . If your motivation is to make a lot of money, hang out and party with your favorite band, or just because you like music, you probably won't make it. I'm not trying to burst your bubble, but realistically, you can make a lot of money in live sound . . . but it can take a long time to get to that point.

Theatrical sound designer/composer Elisabeth Weidner adds:

> You can learn the technical stuff, you can learn the software, you can learn all of those things – but if you aren't passionate about telling stories, and you aren't passionate about using music and sound to make that happen, the hours are killer, and it's a lot of work. I joke that getting the next job is a full-time job – on top of the full-time job of designing the show I'm designing. It's a lot of work, so if you're not excited about it, if you don't love it, it's a hard thing to do.

Susan Rogers defines **personal priorities** as "what you want out of life, what matters to you, and what you're willing to give up to get what you want." Your personal priorities can help guide you toward realistic career interests and goals.

DOI: 10.4324/9781003050346-3

Career Path Guidance and Personal Priorities: Susan Rogers

Susan Rogers is a recording engineer/mixer (Prince) and producer (David Byrne, Barenaked Ladies) who pivoted from the audio industry in 2000 and earned a Ph.D. in psychology from McGill University (Canada). She is a professor at Berklee College of Music (USA) in the music production and engineering department and director of the Berklee Music Perception and Cognition Laboratory. Susan's course curriculum helps students define their personal priorities.

Personal priorities . . . start with knowing yourself. I talk with students about the power of daydreaming. Put your phone in the other room, get away from that screen, go for a walk in the woods. Don't give your mind any problem to solve and don't give it any entertainment for a little while . . . where it goes over and over and over again, that's you. That's your deepest desire . . . I think a good life is when you can get your body to be in the same place where your mind is. Your mind is going there anyway. Why not go over there and do that?

Ever since I was a very young child, I have fantasized about the recording studio. For years before I ever saw one, I pictured them, and I imagined being there. But I did not picture myself playing or singing . . .

With students, I get them to list what they want. I ask them individually to describe their future, to describe themselves 20 years from now, how they'd like to live and what feels right. I ask them questions like, when you wake up in the morning, are you at work? In other words, do you work at home, or do you drive to work? I know so many people who built home studios and hate them. For me personally . . . I loved it. Fine for me, not fine for other record makers.

Are you in a city? Are you in the suburbs? Are you in the country? When you work, do your clients come to you? If your studio is at home, do you have strangers coming to your house to work with you? When you lay down your head on your pillow at night with your sweetheart, are you talking shop? It's a big one – Some of us love that kind of thing, others hate that sort of thing. In other words, do you want a spouse who's in the same field as you who at least knows what you do for a living? Or do you want to separate that personal life from your professional life?

Do you have children, and how old are they? Do they go to public or private school? Then I ask – and this is a tough one for folks in the

> music business – in the evening, the family is sitting down to dinner.
> Are you there? Are you there to tuck your children in and say good
> night? Are you in the studio?

Personal Priorities Exercise

The questions in the following sections are to help define your personal priorities (at this time – they may change in the future). The questions generally do not reflect the industry as a whole and only apply to specific disciplines.

Location

- Would you relocate to pursue or advance your career (if possible)?
- Is your location more important than your career aspirations (or vice versa)?
- Is it important to stay in a specific location (such as near family)?
- How far away are you willing to move, and for how long?
- Would you be open to relocating for a new job every few months or years?
- Would you like living in a major entertainment city or metropolitan area?
- Are you ok with having a lot of local competition ("a small fish in a big pond")? Or, would you prefer a smaller and less competitive environment ("big fish in a small pond")?

Eric Ferguson says of picking a location:

> If you're going to be unhappy in a place, don't go there. I grew up in Northern California and I grew up hating Los Angeles. So, when I moved to Los Angeles, I just hated it. It took a lot of years to get over that. Had I just gone to San Francisco like I wanted to (and grin and bear it through a much harder market), I would have been a much happier person.

Kelly Kramarik says of living in Denver, Colorado (USA):

> I am a big fish in a little pond here and I like that. I don't have to fight with people. I'm one of the only [women] . . . I'd welcome [others], obviously, and I'll help anybody that wants to do it. But at the moment, if someone's looking for [a woman in production or post-production sound], it's me every time.

Location and Musical Theater: Clare Hibberd

Clare Hibberd built a career in sound for musical theater in the West End (UK), as well as touring with shows nationally and internationally. Clare spent three years in Oman as the senior sound technician at the Royal Opera House Muscat. She currently teaches at the Royal Conservatoire of Scotland.

I moved over 20 times in the last 10 years. That has a real effect. It has been an incredible career. Of course, I've enjoyed every part of it, but I've had losses along the way. I've lost a lot of friends because I just haven't been able to see people. Maybe if I'd been living in one place, then that would have been a lot easier. But I was very focused on wanting to work on musicals, like a lot of people are.

The reason I went into teaching was I wanted to work with people, and I wanted to be part of a community. You just can't do that when you're changing contracts every year and living all over the place. You get other benefits, obviously, but it has a very lasting effect on you . . .

I tell students, if you want to work on a musical, then you need to go down to London and you could go on tour. But, it's very, very difficult for students to imagine that life, even though they've dreamt it since they were [younger] . . . when you actually point out that that involves living away from home probably for the rest of your career, it's very difficult. There's a lot of very good theater made in Scotland, but it's not big musicals.

Career Advising on Location: Rob Jaczko

Rob Jaczko is Chairman of the music production and engineering department at Berklee College of Music (USA), where he advises the program's students on their career plans.

Part of [career] exploration is geography. When students say to me, "I'll go anywhere that there's an opportunity," that's not very helpful. Let's examine what your life would look like. What are the day-to-day mechanics of being an adult and living on your own? How do you get around? You [work] in Manhattan and you don't want a car. You need to grapple with public transportation. Does that agree with you? We need to be flexible to try a bunch of things on to see what feels right and what fits.

I'm from New England [US], born and bred. I lived in Los Angeles for seven years and worked in Hollywood. It was tremendous. I couldn't have had the career that I've subsequently had without those years in the studio in Hollywood doing what I did. But ultimately, I like four real seasons. I love living in New England.

Leaving LA or New York is not a defeat, necessarily. Sometimes you may feel like, "I put in this time, and it didn't really hook up. I never worked on great records. I'm struggling to pay the rent. This seems to me like an abject failure." In retrospect, you will probably not analyze it that way. But the beauty is you can leave . . .

If you were to move to a secondary market and be a bigger fish in a smaller pond . . . there's a ceiling to your growth. If you move to Ohio and you're doing local bands out of Cleveland, that may be immediately gratifying . . . But it may not be the end game of where you want to go. Now, you may gain skills in a smaller market, and mature as a young adult (just functioning on your own), which better prepares you to move [later] . . . There's no script.

Work Environment

- Could you work over 40 hours a week? 60+ hours? 12 hour or more workdays?
- Could you work evenings or night shifts? Do you mind working weekends?
- How well do you function on little sleep? Could you work less than 12 or 10 hours after you stopped for the day?
- Can you do heavy lifting or other physical labor for a long period, or would you prefer more stationary work, like sitting at a computer or mixing console?
- Could you work in a studio all day (not getting sunlight most daytime hours)?
- Could you do a job where you would be expected to work outdoors, possibly in extreme cold, heat, rain, or snow?
- If you're considering touring work, could you be gone from home for weeks or months at a time?

- Would you prefer constant/ongoing employment, or could you handle instability (inconsistent schedule, seeking out future work)?
- Could you do a job where you're on-call 24 hours a day, seven days a week, year-round?
- Could you do a job where someone else dictates your work schedule or do you prefer to be in control of your own schedule?
- Do you like jobs where you will be busy/have a task continually, or where you may have downtime (such as waiting on clients to make decisions) while on the clock?
- Could you be in busy, loud, or social work environments during work hours, or do you need some space and privacy?
- Could you work with the same group of people gig after gig (or year after year), or prefer seeing new people on a regular basis?
- How much variety do you like to have day to day? Do you want to do something completely different and unusual every day?
- Would you enjoy (or tolerate) doing client services for high-profile people (making coffee or taking a meal order, making small talk)?
- Could you sustain your choices for five, 10, or 20 years? Or, do you foresee changing at some point in the future, and how?

Eric Ferguson (music engineer/live sound engineer/educator):

> I did 24 hours [straight] in the recording studio. I've also done 24 hours on tour and on a cruise ship. I've done 40-hour flight days and then gotten there and done a show.

Camille Kennedy (TV/film boom operator/sound technician):

> We work long days. We only get 10 hours turnaround from the time to wrap to when we have to be back at work. [With an hour and a half commute], you only have seven hours to exist, have a shower, eat, and try to get some sleep because the next day is going to be another 14–15 hour day. Five days a week is typical . . . But every day is a little different. You have no idea what you're going to be doing. I might be putting a wire on a goat [to mic them]. Like, I've wired up animals before.

John McLucas (music producer/content creator):

> I've tried to work for other companies, and I've hated it. I would be so unhappy and miserable if I had to wake up and have a fixed agenda. Even if I was in a [great] studio every day for 12 hours, six days a week, I'd be so unhappy. I'd be fundamentally depressed, and I would probably quit music.

Karrie Keyes (touring monitor mixer):

> I've met people who come out of school and they think they're going to do studio and live. Then, they find out pretty fast that they're suited for [only] one. It becomes what your personality is suited for. I couldn't sit in a studio. I would lose my mind.

Sound in Space: Alexandra Perryman

Alexandra Perryman has a unique job running audio for the astronauts on the International Space Station (ISS), monitoring day-to-day operation recordings, and communication between the ISS and NASA (among other audio work for NASA). Alexandra explains her schedule, which is dictated by the events she is covering.

> If I'm doing a live event with the space station that's going out on NASA TV, I come in an hour before that to set up. Sometimes that could be 3 am. That could be 11 am. That could be 11 at night. I just structure my day around that. There are days where if it's a space-walk, I'll come in at 3:30 in the morning, and I work until about 2:30 in the afternoon. On another day where I have multiple shows, I can come in, say around 7 am, work that first show, then I have to record a podcast, do podcast editing, and then I might have a field shoot that same day across the NASA Center. So, it really just depends on what I have going on that day.

Chris Bushong (on recording studio work vs. touring live sound):

> You don't have to set up the studio every day. On the road, you're literally building your office and tearing it down every day.

Jonathan Hubel (audio engineer/podcast editor):

> I personally struggle with live mixing, because there's more at stake if I mess something up, and that causes stress which then causes more mistakes. However, if I'm sitting in an editing booth, I can really focus on what I'm doing, rather than having the pressure of thousands of fans/listeners weighing on me. On the flip side, some people thrive under pressure and love to be surrounded by the glamor of the live world. These people would often really struggle with the isolation of a small studio.

Scott Adamson (touring front of house mixer):

> It's incredible if you have the personality for [touring] and you're excited about music. You get the satisfaction every night of mixing shows all around the world for people . . . You get a lot of adrenaline. If you're drawn to that sort of thing, there's nothing else like it . . . I have a travel bug – I always want to get on a plane and go places. That's fun for me. Some people hate that.

Patrushkha Mierzwa (utility sound technician for film/TV):

> I can't tell you how many hours of my life were wasted on sets while a committee of producers discussed all the different ideas that they wanted the director to take into consideration – while the crew waited. All the waiting would honestly add up to months . . . You will be kept away from your family, sleep, and regular meals, tearing your hair out to create good work for shots that will never be in the movie. And you'll do it over and over and over.

(243)

Working in the Elements: Camille Kennedy

Camille Kennedy is a production sound boom operator/sound technician (for film and TV) based in Toronto (Canada).

> We call it the traveling circus because we're like carnies. We're on the road, then you're staying in hotels, but then you're loading your gear in and out every day. You're in buildings or schools, and you're not in a proper soundstage where they actually have proper ventilation that they could turn on and off. Obviously, for sound, we're turning off all the air-conditioning or the heat while working. The summer in Toronto [can get hot]. People are passing out.
>
> My first big show in the sound department we did night exteriors for a month and a half . . . in the coldest winter in Toronto in years. We were outside in the middle of the night filming . . . It was so cold the propane turned back into gel.
>
> Or, we're in a studio. I have a blanket and a heater that I keep next to me beside my cart because even though we're inside, it's still cold. You can't run [a heater while shooting]. So, you have to be ok with minus 35°C temperature, plus 45°C temperature [−31°F to

113°F]. You have to have all the appropriate gear for everything in-between. You could have all four seasons in one day, and you have to be prepared for that.

That's one big thing for new people: make sure you're prepared. Check the weather. Is it going to rain? Do you need rain gear? Do you need to bring a fresh pair of socks? Do you have that stuff in your kit? You never know. You could be outside for 15–16 hours and you have no idea when you're going home . . . With sound, you're hauling all this stuff yourself every day to work . . . If you're taking [public] transit, you're carrying a good amount plus maybe carrying your boom pole.

Lifestyle

- If you're considering a major entertainment city (or a city with a high cost of living), what would your lifestyle look like? What would your diet look like, your social outings, etc., if you only have a small budget to work with?
- How important are future goals that involve saving money (owning a home, vacations, car, and money for hobbies/interests)?
- How much separation do you want between your work and personal life?
- Is it important to have free time and energy to do something else (performer/artist, be available to family, etc.)?
- Do you mind living with roommates? Living at home? Or, is living alone a priority?
- Could you live and work with your co-workers in a small space (and out of a suitcase) for weeks or months at a time?
- Are you ok with working holidays or missing social events for work (such as birthdays, family gatherings, etc.)?
- Do you enjoy (or could you learn how to tolerate) spending your free time networking and meeting strangers (with similar interests)?
- Would you enjoy going to audio meetups, events, conferences, etc. for fun or vacation?

Jack Trifiro (front of house mixer/tour manager):

If you want a stable home life with a retirement plan, paid time off, and vacation days, this is not the career for you. If you like working 20 hours a day, seven days a week, missing kids' birthdays and holidays, not seeing your family for extended periods of time, it's for you.

Britany Felix (podcast editor and consultant who has traveled/worked from headphones in the USA, Europe, and Asia): "The whole 'why' behind my business is freedom and travel. If I couldn't do that, then I wouldn't be able to do what I'm doing . . . As long as there is WiFi, location is absolutely irrelevant."

Working for Prince: Susan Rogers

Susan Rogers worked as a recording engineer for Prince on many of his successful albums in the 1980s, which required relocating from Los Angeles to Minneapolis. Susan worked on-call to Prince, which often included sleeping only a few hours a night.

> It didn't feel like I was making personal sacrifices. I felt like the luckiest woman in the world. I was getting what I wanted. Of course, there's sacrifice there. I sacrificed a personal life. I had no personal life or family life for all those years, but it was a fair trade – a more than fair trade. That cost-benefit analysis is a really important thing that I encourage always doing throughout your life.

James Clemens-Seely (recording engineer):

> For my whole career . . . the paychecks for classical projects were bigger than the paychecks for cool friends making great music in a storage locker after hours, even if I was more passionate about that . . . I'm not always working in the genre that I'm passionate about, but it's easier to be clinical and it's easier to separate work life from home life because of that.

Kevin McCoy (head audio engineer for musical theater):

> The way to make money in theater is often to tour. Sometimes people find it worth it to trade off some of the family stuff in order to make more money to make the family that they have happier. It's tricky. That's never been a priority for me, so it's not been a huge issue.

Nick Tipp (music engineer/producer):

> I had somebody paying to live in my living room in a one-bedroom apartment in Hollywood . . . then I was just getting an apartment with some friends or moving people into the house. Houses are better because then you can make studios.

Kelly Kramarik (television audio editor): "I don't want the stress of [freelancing] all the time because it's not fun. I like going to work and coming home from work, and then having the choice to take on more work if I want to."

Balancing Family with Career: Devyn Nicholson

Devyn Nicholson is Manager of Audiovisual Services at McGill University in Montreal (Canada). Devyn has had a successful career in professional AV that stemmed from his education in music production and interest in system design. His wife, Shannon Simpson, is a musician who has a bachelor's degree in music technology.

One of the things that enabled me to focus was my personal life. Not everyone's going to have their personal life sorted out to the point where I did when I entered the workforce, or ever, necessarily. Meeting my future wife and knowing how she felt about moving elsewhere was definitely a feeder into how I felt about where I wanted to see myself. For me, I've always been a big family person. I've always known that I wanted to be a family guy and have kids, and I intuitively understood that. Being a freelancer in music and leading that rock star/recording guy life does not equal out with a family and kids . . .

I definitely had an interest in pursuing studio stuff, but I wasn't really willing to make no money, burn all hours of my day, and put my life on hold, which is really what it takes at first. I recognized as well, being a native English-speaking person [in a French-speaking province], that I would probably have to go somewhere other than Montreal if I were going to try and make a real run at it . . .

I still do have my business on the side, so I am still doing recording and mixing, but now I'm selective about what I want to do . . . Shannon sees me mixing and she'll give her two cents. It's cool. Or when I say stuff, she understands to the point where she can give me feedback . . . I did do some of the bigger gigs that would come along at McGill, like, I was mixing front of house when the Dalai Lama came.

I have things that I am not willing to just drop and walk away from for anything. It's hugely important. It's not even knowing what you're willing to do or not in terms of a career. It's about knowing what your priorities are at the macro level.

Personality

- Are you introverted or extroverted?
- Is your best creative work (or most productive work) done alone or with others?
- Do you prefer to work on other people's products and art (versus your own)?
- Do you like working on something over and over to its highest level of quality, or do you prefer work where imperfections may be inherent in the work?
- Do you thrive under pressure and tight deadlines, or does it exhaust you?
- Can you take constructive criticism, and do you like getting feedback for improvement?
- Do you enjoy collaborating in person with a team toward a goal? If not, can you tolerate it?
- Could you tolerate working with difficult or unique personalities or people behaving poorly (a demanding artist, actor, person in a position of authority)?
- Could you stay calm and keep working in situations where it may not be appropriate to leave or defend yourself at that moment? (Getting yelled at, being blamed without merit)

Susan Rogers (recording engineer, mixer):

> An introvert typically gets their battery charged up being alone, and that's where their most creative ideas happen. The extrovert is the opposite. They get their batteries charged by being with other people, and that's where their creativity comes. I'm the kind of person who's going to do her best work when you put me alone in a room and shut the door, which is why I was a good mixer. I'm a good scientist because when I'm alone, creativity flows. For other people, that's a choking, stultifying environment. They need to be with other people in order for their ideas to flow.

Pete Reed (video game sound designer):

> Video games is an industry where you need thick skin and to be able to take feedback – positive as well as negative. If you don't like someone telling you your work could be better, it may not be the best job for you.

Andrew King (broadcast mixer):

> We're left to do our thing. Like, producers and directors don't know what I do. They have no clue how my stuff works. They'll say, "Hey, it sounded great tonight," and I appreciate that, but I want to be better. There is no

feedback unless you happen to know someone at home that is also an A1 [mixer] that is listening to your show and can tell you how it sounded.

Becky Pell (monitor engineer):

> I liked mixing front of house, but I like mixing monitors more. I feel more at home on monitors. I like the camaraderie and the sense of being a part of it and having a very interactive experience with the band . . . It's very rewarding to earn a musician's trust.

Jeff Gross (music producer/engineer and Foley mixer): "I'm a huge believer in respecting everybody . . . but [entertainment] is not a politically correct industry. You're going to be around a lot of people who speak very freely and creatively, and there might be unique personalities."

Jonathan Hubel (audio engineer/podcast editor):

> Are you generally flexible, spontaneous, and outgoing? You might thrive as a freelancer! Are you like me and love consistency, structure, and working by yourself or with a small team? Then you'll probably be better suited for traditional employment . . . I personally prefer structured environments, where the work is brought to me. I'm not a super outgoing person, so putting myself out there, constantly meeting new people, and making new connections is stressful and difficult. I also like the guaranteed income and predictable schedule of a 9–5 job, even if I could possibly make even more as a highly motivated freelancer. It all depends on your general personality.

Professional Interests

- How interested are you in learning about (and learning to use) new technology?
- Do you feel strongly about working on art or with people focused on their art (filmmakers, musicians, etc.)?
- Do you want to work on media (movies, games, podcasts, etc.) or content creation?
- Could you be happy in a role that supports those doing that work (like tech support, system design, or developing tools for the industry to use)?
- Is there anything you would refuse to work on (based on ethics, religious or personal beliefs, etc.)?
- Do you have any other interests that you would like to incorporate into your work, if possible (such as educating, religious beliefs, advocacy work)?
- Is there anything you feel especially drawn toward (or away from) already?

Michael Lawrence (live sound system engineer):

> The more I learned about mixing and about system engineering, the less
> attractive I found mixing, and the more interested I became in the engin-
> eering side of things . . . I'm really interested in the side of engineering
> where something just broke and we have to fix it. If the show's happening,
> you have to fix it quickly. That really engages my technical mind.

Andrew King (mixer for broadcast sports): "I couldn't do studio work. You're
going to mix this song and listen to it 400 times? I can't handle that."

Kevin McCoy, who was present for 400 shows and mixed over 250 musical theater performances of *Hamilton* ("And Peggy" Company) in a year:

> I love doing the same show every night. It feels like I'm more finely crafting
> it each time, which is really fun. If you're going to be angry or frustrated
> about having to do the same show over and over, don't do this [path].

Avery Moore Kloss (podcaster/audio storyteller):

> I feel prouder of the work I do when I have the chance to sit with audio,
> edit it, re-edit it, and catch something I didn't catch the last time. Then,
> I listen to it with my eyes closed and tap my foot and make sure it fits the
> cadence I want. I just like that. What I hated in daily news [for radio] was
> just sending out stuff I wasn't super proud of because I had a deadline. If I
> have a choice, long-form is the choice I make.

Working in Music Production vs. Post-Production: Sal Ojeda

*Sal Ojeda worked for recording studios in Los Angeles before pivoting into post-
production (sound for picture) as his primary career path.*

> I'm more sensitive to bad music than to bad shows. I cannot stand
> working on a bad [music] track, but I don't mind working on a bad
> show. Dealing with the characters in the music business – I'd rather
> be in post.
>
> There's dealing with all these egos in music. If they feel like doing
> anything at the moment, they feel entitled to just do it because
> they're famous, I guess. They always have people around who can
> encourage them. [One session with a well-known artist], I was the
> recording engineer, and it was really hard to just get one song down
> because they brought tattoo artists to the session.

If you're unsure, try both. I know friends of mine that tried to work in post for a little bit and hated it. As for me, I like it. You're building a puzzle. This piece of dialog doesn't work, so let me try to make it work. For me, that's fun. If I didn't enjoy it, I wouldn't be doing it.

In music, at least on the engineering side, you're either one of the greats of all time (Tony Maserati, Michael Brauer, CLA), or you're the flavor of the month. In post, there's a reward as you progress little by little. Most of the heavy hitters are older. I like that because I aspire to do that. I'm not saying post is an old guy's game, but it is a craft that you get better with experience and with age. A lot of it is how you interact with clients and your networking skills.

I don't like it when music people disrespect the craft of post and think that they can do it just because they know how to work in music. It's completely different. You have to respect the craft. It's not just EQs and compressors. If you don't have the right mindset, you're not going to make it. I wouldn't lead anyone to think that working post is easy (or is the easy way out) because it's not. It just requires a different mindset. It's not for everyone.

Applying Personal Priorities to Work

It takes some hands-on professional experience (even for those with a degree or formal education) to find a good career fit. Kristen Quinn found her path in game audio after exploring music production and live sound:

I really did not like live sound at all. I found it very stressful. It's about the number of things I have to focus on at once. Even though I have to pay attention to a lot of things [in games], I still get to go and be head down in the studio, and just tweak and really focus on something. Whereas in live sound, I felt like everything needed my attention all the time. It also gave me anxiety around if something goes wrong . . . I interviewed with the audio department at [video game company] Monolith, and they ended up bringing me on as an intern. From that second on I knew this is what I wanted to be doing. I was in love.

Mariana Hutten stepped away from an opportunity in game audio to pursue music production:

I love the gaming industry – how cool it is, how much money there is. There is so much passion there from people. But I don't like gaming. In interviews, there are always the questions: What's your favorite game? What games have you been playing lately? I haven't played a game in five years. They're not going to hire me because I don't have that crazy passion.

Theatrical audio engineer Becca Stoll knew post-production would not be a good fit for her:

I [visited] someone who worked in post-production . . . I watched him edit a clip of someone saying "what's a frittata" for an hour. I was like, this is not my game. I get that to you, it has genuinely grown better. I'm like, what's the next line? I want to keep moving. If you really like to finesse and do artistry and prefer your work to be more project-based, you might find a lot of success in film.

Rachel Cruz's first job was at a jingle house, but the bulk of her career has been in product development:

If I had stayed in my first studio job, I would have eventually gotten very tired of doing some of the things that you really have to enjoy doing, like managing some of the personnel, clients, big personalities, and doing it on a daily clip. At first, it was like, so many buttons, so cool. But now it's moved beyond the technology and into systems. What I found out was that my personality is also into these long strategic arcs. I like having more complication in problem-solving in doing what I do.

Javier Zúmer started his career in Spain where he took any opportunity he could find. He pivoted into post-production sound after moving to Ireland, and now works in game audio in the UK. Javier says of his experience:

I think first, you take everything. But with time, for example, I noticed that I was better with post-production than location audio. That was something I realized in my [early] years because I'm more of a shy person, and also I didn't like working nights, like music mixing. Slowly, I realized post-production was more my thing and it fits my personality and my skills. My love for cinema also helped. So at that moment, it was like cinema or video games, 50/50 preference. But with time, I started to learn about interactive sound design and middleware, and I realized how interesting it was compared to post-production. Production is linear – you make the perfect thing tailored to that moment on screen . . . You don't create the mix for video games. You throw the ingredients under rules, and the player mixes the game when they play . . . it's more of an open thing. It's even more creative. That's why I wanted to jump to that area.

Creating a Faith-Based Career in Audio: Jonathan Hubel

Jonathan Hubel has incorporated his faith into his work for churches (AV and live sound), radio, and podcast companies. While pursuing a bachelor's degree at John Brown University, Jonathan did an internship in Guatemala where he helped record an audio version of the Bible in native Mayan.

I had already had ideas that I wanted to pursue faith-based work in my career, and my experience over those few months [internship] helped even further solidify my desire to do so.

The majority of opportunities that combine religion and audio skills will more than likely be connected with a church to some extent. However, there are other opportunities, albeit much fewer and far between. Some Christian bands and musicians do live concert tours, requiring an AV team, though your mileage may vary whether the band has their own techs as opposed to using the services of whatever venues they stop at. Christian radio stations are also an excellent opportunity if you're more interested in the broadcasting side of things, whether on-air talent, programming/scheduling, or editing and production. There are also other small organizations and ministries that will produce syndicated radio shows or podcasts . . .

There are quite a few faith-based podcasts . . . A lot of faith-based podcasts are being produced by non-profits, churches, and people who have little to no budget, so it's going to take a lot of work to make it financially viable, especially as a freelancer. It's definitely a calling based on shared beliefs, not just a career and money-maker.

References

Adamson, Scott. Personal interview. 14 May 2020.

Bushong, Chris. Personal interview. 27 Sept. 2020.

Clemens-Seely, James. Personal interview. 20 Dec. 2020.

Cruz, Rachel. Personal interview. 9 Oct. 2020.

Felix, Britany. Personal interview. 22 June 2020.

Ferguson, Eric. Personal interview. 2 Dec. 2020.

Gross, Jeff. Interview with April Tucker and Ryan Tucker. 14 Jan. 2021.

Hibberd, Clare. Personal interview. 7 May 2020.

Hubel, Jonathan. Email interview. Conducted by April Tucker, 22 Oct. 2021.

Hutten, Mariana. Personal interview. 19 Nov. 2020.

Jaczko, Rob. Personal interview. 15 Oct. 2020.

Kennedy, Camille. Personal interview. 21 Feb. 2021.

Keyes, Karrie. Personal interview. 19 June 2020.

King, Andrew. Personal interview. 15 Dec. 2020.

Kramarik, Kelly. Personal interview. 18 June 2020.

Lawrence, Michael. "Systems Engineer Arica Rust." *Signal to Noise Podcast,* ep. 48, ProSoundWeb, 20 May 2020, www.prosoundweb.com/category/podcasts/signal-to-noise/.

McCoy, Kevin. Personal interview. 22 Dec. 2020.

McLucas, John. Personal interview. 25 Jan. and 8 Feb. 2021.

Mierzwa, Patrushkha. *Behind the Sound Cart: A Veteran's Guide to Sound on the Set,* Ulano Sound Services, Inc., 2021. p. 243.

Moore Kloss, Avery. Personal interview. 21 Oct. 2021.

Nicholson, Devyn. Personal interview. 23 Nov. 2020.

Ojeda, Sal. Personal interview. 27 Sept. 2020.

Pell, Becky. "Becky Pell, Monitor Engineer & Yoga Therapist." *Signal to Noise Podcast,* ep. 41, ProSoundWeb, 15 April 2020, www.prosoundweb.com/category/podcasts/signal-to-noise/.

Perryman, Alexandra, guest. "NASA's live broadcast engineer, Alexandria Perryman." *SoundGirls Podcast,* 30 June 2020.

Quinn, Kristen. Personal interview. 30 Sept. 2020.

Reed, Pete. Email interview. Conducted by April Tucker, 24 Nov. 2020.

Rogers, Susan. Personal interview. 12 Oct. 2020.

Sabolchick Pettinato, Michelle. "Good Questions." *Mixing Music Live,* 26 June 2019, www.mixingmusiclive.com/blog/goodquestions. Accessed 22 May 2021.

Stoll, Becca. "Ask the Experts – Mixing for Broadway and Theatre." *YouTube,* uploaded by SoundGirls, 3 Feb. 2021. youtu.be/ifBC_ErEDCs.

Tipp, Nick. Personal interview. 22 Sept. 2020.

Trifiro, Jack. Personal interview. 4 June 2020.

Weidner, Elisabeth. "Ask the Experts – Sound Design for Theatre." *YouTube,* uploaded by SoundGirls, 5 June 2021. youtu.be/rFMLCU1g19k.

Zúmer, Javier. Personal interview. 24 April 2021.

3
Defining Your Barriers

Make audio work for you – don't work for audio.

– Alesia Hendley

A dream is something you envision regardless of your real-life circumstances. In a dream, you can pursue anything that interests you, anything is possible, and you can be whatever you want. In reality, anything (and everything) can get in the way of that. Rob Jaczko explains:

> Starting out, we need to balance the abstract goal of "I want to be a music producer for records" against the fact that two weeks from now is the first of November and the rent is due . . . I think we need to have goals, of course, because that helps us navigate what our choices are . . . We need to look around and figure out how we're going to put a patchwork of events together that in hindsight we call a career.

A **barrier** is something that limits your ability to freely do what you want to do. Music producer/composer Brian M. Jackson discusses some common barriers to working in the audio industry:

> Finances: This is the most obvious and common limitation at the onset of a career path. If you don't have the money to do what you need to do, then figuring out how to earn income so you can save up what you need to get started will be a big part of your near-term plan. You might have to work at a few jobs that really suck for a while. You will have to make sacrifices.

> Obligations: Are you married? Do you have kids? Do you help run a family business? Having honest discussions with significant people in your life is imperative . . . Their level of support will have huge ramifications for your plan, especially when it comes to a significant other . . .

> Geography: Even with the Internet, where you live will influence your plans . . . It might be necessary to move from a small town to a nearby city, college town, or big city, or to move from a big city to somewhere more

DOI: 10.4324/9781003050346-4

affordable. Moves can also be small, and might include simply moving across town . . .

Physical/psychological: ALS didn't stop Steven Hawking from achieving greatness as a physicist, and blindness didn't stop Ray Charles or Stevie Wonder from becoming great musicians . . . If you have some sort of physical or psychological limitation, you will have to tailor your plan accordingly, but there is no reason not to go for it.

(Jackson, ch. 2)

A Dream Meets Reality: James Clemens-Seely

James Clemens-Seely is a Canadian recording engineer who has worked with garage bands, electronic artists, award-winning artists and producers, and nearly every professional orchestra in Canada. Growing up in Montreal, Quebec, James dreamed of working in the music industry in the USA, but found career success by remaining in Canada.

I was thinking (naively at the time) I'll get good at studio stuff while I was in school or volunteering in studios, and then I'll move to New York or LA (where it's raining Grammys) and make rock records and be famous. It was pretty much just the dream. I assumed it would be hard work, but I was also at that stage (especially where you're a young white guy and nothing hard has ever happened to you) where I thought I'd walk in and they'll say, "Here's a Grammy. What do you want to do?" It seemed like it would be hard work but just a question of perseverance, and not really a question of luck or suffering . . .

As a Canadian, getting to New York or LA to start working was implausible in terms of the hurdles that you have to go through. My increasing realism merged with the path of least resistance. I had been making recordings in my parents' basement in high school (in Montreal). I was riding my bicycle around town to churches to record concerts and using that money to buy more gear to do more basement recordings. It just turned into a growing network of work there. Montreal is a great city for cheap rent and lots of vibrant artists. So, what ended up being the most sensible path was to stay there, rock the part-time gigs (freelance or steady gigs), and either teach at a trade school or work backstage as a stage manager . . .

As a freelancer, it always feels like moving to a new location will be a big setback no matter how prosperous the new area is, since

> so much of what we do is based on our networks. Time is one of the strongest ingredients in a solid professional network. Staying in Montreal let me build on a head start I didn't even realize I had and probably jumped my career several years ahead of where I'd be in any other city.

Everyone will face a unique set of barriers, which is why there is no single (or simple) answer of how to successfully build a career in the audio industry. Career building may appear as easy as searching for job listings, applying, and landing a job, but in reality, it is rarely that simple. To give an example of how few barriers have to exist to *easily* land work: During the COVID-19 pandemic in 2020, New Zealand's film industry was booming. New Zealand was one of the few countries in the world where film production work was not shut down and the country's borders were also closed to non-residents, which eliminated global competition. Production sound assistant Eliza Zolnai (originally from Budapest, Hungary) was living in New Zealand in 2020, and says of seeking work:

> There was no COVID here and there was a lot of work . . . I looked up some names on iMDB.com about shoots that I knew happened here. They are shooting *Avatar* here. They shot quite a lot of *Lord of the Rings* here. So, I just looked up some names and sent out my CV to people and I got work. I just looked online for contact information and sent out a few emails but really not a lot – I think three. I met three guys and then things just started happening . . .
>
> Budapest is so much busier than New Zealand. I think compared to others my age, I'm lucky that I arrived with a lot busier CV than a lot of people [in New Zealand], and just the fact that I have a diploma for what I'm doing. It was really easy for me, and I feel pretty lucky about it. Now that people are getting vaccinated in countries, it's sort of swinging back a little bit, so now there's not a lot of work in New Zealand and it's getting very busy in Budapest again.

Barrier: Getting a Foot in the Door

For emerging professionals, one major barrier is getting a foot in the door – hearing about work, being considered for work, and also landing that work. Video game audio director Kristen Quinn says of the competition, "One job we had 500 applicants within like a week. That doesn't always happen in every company, but that was a sought-after company."

Mehrnaz Mohabati worked for seven years as a music engineer in Tehran, Iran, before moving to Los Angeles, where she decided to pursue post-production. Finding opportunities was a barrier for Mehrnaz:

> For me, the first and the biggest problem [in the US] was I never looked for work [before]. Work always found me back home. But here, I had to find work . . . I had a rough year [after school]. I was hearing just crickets. I was applying everywhere. Nobody was returning my calls. People [told] me, "you need to change your name" . . . Why do I need to change my name to get an interview? They [also] said, "You're overqualified," and "You are under-qualified." It was so crazy. It was a year of very tough living. But in the end, I made it through.

Finding gigs and opportunities that are not formally posted takes effort. Music producer Jeff Gross shares a story of an up-and-comer whose career was cut short after moving to Los Angeles:

> He came out here with three or four months' worth of savings and sat on a couch every day practicing [skills] versus networking and getting out there and meeting people. He ran out of money . . . I always tell everybody when they're coming to LA to come with a plan. Come to LA, get a job at Starbucks, at FedEx, anywhere with flexibility. Just get *a* job. If you think you're going to be famous inside of six months – it's going to be more like six years.

It can also take effort to access equipment to learn specialized audio skills. David Bondelevitch, an educator with over 30 years of experience as an award-winning sound editor and re-recording mixer, says:

> I learned the most by mixing student films and film scores . . . I had keys to the [school] and spent most of my time in the studios . . . One of my pet peeves is that students have access to millions of dollars of equipment, but only use it when they have an assignment due, rather than taking advantage of all the studio time they can get to practice. The ones who take the time are the ones who, typically, become successful.

Barrier: Unrealistic Expectations

One barrier to building a career is having unrealistic expectations – particularly when it comes to landing opportunities beyond your experience level. Boom Box Post (Burbank, California, USA) is a post-production sound studio with a formal internship program. The program advertises responsibilities including completion of their internship program curriculum, studio errands, and light cleaning. Their internship curriculum includes one-on-one lessons in each post-production sound job with their highly-experienced team. It culminates in creating pieces for the intern's personal demo reels. Co-owner and supervising sound editor Kate Finan says:

> We do get a lot of applicants who come in and they just graduated college or a short program somewhere, which is great. They say, "I'm here to interview to be the supervising sound editor on all your shows." You would be surprised how often that happens. That's not realistic.

Radio/podcast producer Sarah Stacey, who studied in the UK, shares a story of an emerging professional who had unrealistic expectations about how long it takes to build a career:

> I knew someone who came out of college at the same time as me. She gave it six weeks. She didn't have a job and she just gave up and said, "I'm going to go do something else." I don't think that's long enough. If you're really serious about this kind of career, it [takes time]. You're going to deal with a lot of rejection, and you're going to feel quite disheartened, but you just have to persevere. Then once you get that foot in the door, the rest will follow.

Recording engineer/music studio owner Catherine Vericolli adds:

> If you do not have realistic expectations from the get-go, you will be very frustrated. You're not fun to be around if you have some big giant expectation that you're not meeting because you think that something is owed to you. I have fired engineers for that – because they thought they deserved something.

Barrier: Finances

Finances are a common barrier for emerging professionals. Fela Davis is front of house mixer for Christian McBride and co-owner of 23dB Productions (a New York City-based audio production company). As an emerging professional, Fela attended Full Sail University (USA) to learn audio before returning home to South Carolina. She took jobs like deboning chickens and folding airbags for Honda to save money to move to a larger city with theaters, where she found stagehand work.

Karol Urban, re-recording mixer and sound editor for high-profile scripted television shows, had five part-time jobs while being a full-time college student. Karol shares:

I had two jobs that weren't audio-related that made more money. I sold makeup and formal dresses at the [local] JCPenney, and I worked at a computer lab. I was very aware of the fact that it was going to get a lot harder a lot quicker financially [after college].

Akash Thakkar says of one job:

I used to clean toilets when I first started. I was a public pool janitor. I don't recommend that as your side job. It was the worst. So, it's okay to have a side job. It's good, actually, to have some sort of financial stability while you're studying and while you're working on your craft. That is so, so crucial.

Andrea Espinoza used a side job as a way to access opportunities that would not be available otherwise:

I was about to graduate high school and I wanted to do the New York City Fringe Festival. They weren't paying, and my mom wasn't giving me money. So, I was working at McDonald's during the day, and I would buy my MetroCard to go to the city. I was working for free just to have that experience under my belt. It later paid off.

Low pay for audio work can be an ongoing challenge for professionals in some disciplines and locations. Mabel Leong says of working in audio in Singapore:

When your local poll [about] artists being the top non-essential job makes waves worldwide,[1] you know where the industry stands in Singapore's society . . . The media industry is not a good-paying one here, with full-time positions earning 50% or less of the expected median income for employed "professionals" in Singapore. However, this is the job I am passionate about, so I keep on keeping on with my full-time and freelance non-audio work to remain involved. Passion doesn't always pay, but it makes one fulfilled.

Taking Available Opportunities: Jeanne Montalvo Lucar

Because of financial barriers, Jeanne Montalvo Lucar had to seek out paid opportunities as an emerging professional. This took her in an unexpected direction: Classical music. Jeanne's experience at the Banff Centre for Arts and Creativity (Canada), Tanglewood Music Festival (USA), and restoring the Metropolitan Opera broadcast archive prepared her for later pursuits: radio, podcasting, and Latin music production in New York City.

I couldn't do free internships because [in Dominican culture] you don't work for free . . . My parents wouldn't have paid for me to live in New York with no income. That's part of a group of people who are privileged enough to have that, and that wasn't me. So, classical music was paid. It was on the table, and I took it . . . I kept telling myself that the thing that I was going to be good at would reveal itself to me and I would end up there . . .

[It took] putting in extra time and effort so that I could step up my game. If you're not willing to do that, in whatever part of audio it is – nobody's going to spoon-feed you. You shouldn't expect to come in and be like, "I'm only going to work eight hours. If I don't learn what I need to learn in eight hours, then this isn't for me." That's not how it is. If your learning curve requires extra work, then you have to put in the extra work and you have to put in the extra time. Maybe that means listening [or practicing] on the weekends.

A lot of people want the sexy job, and a lot of times classical music is not the sexy job . . . It does take a little bit of [thinking], all right, this isn't exactly what I envisioned for myself, but I am working in audio. I find this work fascinating. Eventually, it can lead to doing what you might want to do later on in life. It's not like if you're in classical music, you're tied to classical music for the rest of your life.

Do whatever job gets offered to you . . . I actually think it's better to be a jack of all trades and be *good* at all of those things. Maybe you are better and excel at one of those things over the other ones, but I don't necessarily think you need to be a master of one thing. Being a master of one thing means that you're not going to get all the other jobs that are a little bit tangentially off – a couple of degrees off of that one thing. If you're good at many different things, then you can do many different things. I did classical, and I now do radio and podcasting. I can do a lot of different things.

Barrier: Overvaluing Formal Education

Formal audio education and degree programs can be an excellent way to build foundational skills (for those who can attend). However, having a degree or certificate does not ensure career success, advancement, or job prospects. Most disciplines still expect onboarding, or learning the norms, etiquette, terminology, and technology used by established professionals on the job.

Assistant Professor Eric Ferguson says of coming into the field:

> Realize that you're getting education part two. It's about the journey. It's not about the destination of becoming the big mixer. You've got to realize that you're not good – You're not going to mix right away, and that's ok.

Kate Finan, who works in post-production sound, adds:

> Coming out of school or a technical program or a different career, expect to have an entry-level position. That doesn't mean it needs to be a crap position. It just means that it's appropriate for the amount of experience that you have. [It's] working in the machine room, being a mix tech, being an assistant, making the coffee, doing the scheduling. Those are all things that put you in a position to learn so much about all the inner workings of a studio, all the inner workings of a production, how to handle people, who's who, and also it gives you the opportunity to see all of the jobs.

> You're not going to want all the jobs. I hated every single Foley session that I attended but I loved doing dialog editorial. Those are things that you learn when you're in those entry-level positions, because you're exposed to such a wider swath of what's going on. Rather than pigeonholing yourself into like, "I'm going to be the next premier sound effects editor" when you just got out of school, that's really not the best place for you to start your career either. So, keep an open mind.

Associate Professor David Bondelevitch says of his education (as a grad student at the University of Southern California):

> Faculty at USC were surprisingly unhelpful in finding work for me. As a faculty member now, I understand why. A university does not exist to be a job placement service. It exists to give you the knowledge, skills, and disposition to work as a professional. I do try to help students get a foot in the door whenever possible, but it's a rarity when I can have direct influence like that. (My successful former students have earned everything on their own.) What really helped me was the network of USC [graduates] working in Los Angeles. I can trace almost every mixing job I have ever had back to someone who knew me at USC.

Associate Professor Leslie Gaston-Bird says being open-minded is crucial, especially for students:

> If they've got tunnel vision and they want to do [one] thing, then they're not going to see this cornucopia of other cool [stuff]. When I was in school, I was open-minded about it. I wanted to be in a studio and a rock star, but then I actually had to eat, and I ended up in public radio. There are other people who ended up at Capitol Studios. I think you can be wherever you want to be, but if you're confused and not listening, you're doomed.

Navigating Barriers Starting at the Bottom: Camille Kennedy

Canadian boom operator and sound utility Camille Kennedy had to push a broom to start her professional career – even with two undergraduate diplomas and a graduate certificate. By starting at the bottom, Camille learned about filmmaking, professional sets, and production sound in a way she could not have learned in school.

After I graduated, I got an internship at a mastering studio in Toronto. It was five days a week, 40 hours. You can't make a living working full-time and not getting paid in Toronto. So, I started doing guitar restoration and repair with this gentleman who's a special effects member at IATSE 873 [a labor union]. He said, "F— working for free. Come make some money and get on set. You can do sound here, as well." He took me under his wing, and I ended up getting my [union] permit status as a construction laborer. I honestly started by pushing a broom around the studio cleaning up . . . Then I started scenic painting, and I did a few cool shows . . .

I then got a name hire from the guys that I was working for. They wanted me to come up [on set] and they wanted to teach me gripping . . . A grip is the muscle of the crew. They do all the rigging . . . On the first day, I walked onto set and everyone was moving around and there was equipment flying back and forth. I was just taken aback. I didn't know what to do. I felt that anywhere I went, I was in the way. That was the biggest thing – to know where to be and how to stay out of the way. It was so overwhelming . . .

They really threw me into the fire. One guy sat all day with me pointing and saying, "What's this piece of equipment called? We call it this, but we also call it this. Why would we need this one?" By the end of the day, I didn't remember everything, but I had a good grasp of the equipment I was using. That's what people will do for you. You have such a great opportunity to learn on set if you want to learn. If you can take criticism and you want to learn it, they'll teach you everything you need to know.

Once I finally made my way on sets, I would go up to the mixers and say, "I know you don't know me, but I want to do sound." They could see I was getting on set, and I was doing every job possible just to get there. Eventually, if someone didn't show up, I'd get a call last minute. "Hey, come in. Do you want to do sound utility for us?" That's how I broke into that.

I was fortunate enough to work with some senior boom operators that really, really helped me along – not just how to boom and techniques but how to be a better person on set, how to handle situations, and the personal aspect of it. You have to keep a cool head. A thousand things could be flying at you, the world could be on fire, and you have to wire this actor and just keep cool and not be rude . . .

Being on set really shows them your work ethic and the fact you can actually hack it on set. They watch you. They want to see if you're going to be able to meet the demands of the department. Sound is such a unique department. Everyone has their own job and you're just expected to know it. You're not told what to do. That's why it's so hard to get into it because there really isn't an entry-level to get in . . . It is definitely beneficial to get your skills on a professional set because you learn how it's done the right way.

Barriers and Privilege

When you look at the career path of someone you respect or compare your career progress to others, parts of their stories may be hidden. Privilege is an advantage, benefit, or special opportunity available to an individual or group. While everyone will face barriers and challenges, privilege may allow some to move faster in their careers or have advantages not available to everyone.

A financial or economic advantage can include having financial support (or access to loans) to be used for school/education, an unpaid internship, or to work a low-paying job. Free or reduced-cost housing is an advantage that can offer protection during the emerging years. Kevin McCoy says of getting started in theater in Minneapolis (USA):

I was lucky. I had a roommate who had a really good regular job, and he was able to cover for me when I couldn't quite make rent, which was very generous, and something that no one should ever expect to run into.

Theatrical audio engineer Becca Stoll adds:

I have to acknowledge the unique privilege that I have that has made my career path different – the access to free housing in Manhattan. That one really can't be overstated. If I'm home in Connecticut but someone needed me in the city tomorrow, I could be there.

Where someone lives can create barriers. For example, not every country has a well-established entertainment industry or formal education opportunities. Audio education materials (including textbooks and manuals) are not in every language. Even basic utilities like stable electricity or internet (used to learn and practice skills) can be a barrier in some countries. Phebean AdeDamola Oluwagbemi says of this challenge in Nigeria:

> Sadly, for any business in Nigeria right now, you have to consider power supply. It's a big issue. While some parts of the country have been able to stabilize the power supply . . . you still have to try and get a power generator into your budget. That takes a lot of money from a lot of startups.

Moving to Another Country: Mehrnaz Mohabati

Mehrnaz Mohabati worked as a music engineer in Tehran, Iran, for seven years before moving to Los Angeles in 2010. Mehrnaz pivoted into post-production sound for TV and film and has since built a successful career, but it took overcoming an extra set of barriers.

> When I came [to the US], I needed to know more about the artistic culture (not American culture, because I think that's two different cultures). In Iran, I knew my musician culture. Their way of living is very different than what other people are doing in [Iran]. [The US] is the same.
>
> The hardest part was knowing the sense of humor here because it's totally different . . . Sometimes it's hard for me to understand [British] sense of humor, as well. For me, working on a documentary was way easier than a comedic movie because my way of telling a joke was very different than [in the US].
>
> In the beginning, when I started doing [post], my accent was a little bit weird for some people, and I didn't know what to do for them. I couldn't change my accent. It is what it is. So, at some point, if it was a good time, I'd just joke about it. [We're] laughing and it breaks the ice.
>
> But you learn. That's the beauty of our job. You always learn something, and you always discover something that you don't know. It's not routine at any time by any means.

Nepotism is special treatment (especially in business) given to someone because of a relationship (regardless of their ability or qualifications). One professional shared this story of a boss whose career was built on nepotism: "He did bad work, and he did bad work because he wasn't really doing the work. He was networking instead of doing the work, and then he hired people who were competent to do the work."

Worldwide, our industry contains a diverse and unique variety of professionals (by age, race, gender identity, religion, and much more). At times, these differences can, unfortunately, create barriers and challenges. Some professionals, businesses, and industry organizations are actively working toward change – from the Recording Academy and the Audio Engineering Society to advocacy organizations such as Audio Girl Africa, Black Sound Society, Color of Music Collective, SoundGirls, and Women's Audio Mission, to name a few.

A discourse community is a group that shares common values, goals, and ways of communicating. Susan Rogers says of her experience in the music production community in the 1980s–1990s:

> I definitely was aware of the gender differences. I couldn't quite fit in . . . It's not like there's no room for me, or Sylvia Massey, or any of the other great young female engineers. But part of what we learn when we're getting into a profession is the discourse community. When your discourse communities engage in dialogue that is a little bit foreign to you, you have to learn how to speak that language rather than having it come naturally.

For Langston Masingale, hitting the "ceiling" of the companies he worked for led him to start his own businesses, including a recording studio and Handsome Audio, where he designs and manufactures music industry products (more of Langston's story is in Chapter 20). Langston shares:

> In terms of my barriers to success, it was my race. I'm a firm believer in that . . . I kept dealing with these situations where I would only get so far, or opportunities were not coming to me that were going to others. I kept looking in avenues and in directions that made more sense for me . . . I [decided], "I'll open my own business." So, I did. I did that numerous times over the years.

One key to building a career within a discourse community is finding people who can offer support – anyone from other emerging professionals or professionals, educators, advocacy organizations, or support on social media. Langston Masingale explains:

> Part of your armor is your support network. You need to find a mentor, you need to find an advocate, and you need to find a supporter even if

they don't work in the same company as you . . . Because if you don't have those people to talk to, to console you, to encourage you, to invigorate you, you're toast. If you don't have those people behind you, then it doesn't matter . . . Everybody needs that person in their corner that can give them that jolt of re-encouragement.

Navigating Barriers

Barriers are like an approaching storm. If you're lucky, you have a choice whether to go through it or to go around it, but you are also prepared if you need to face the storm. Knowing where the storm is (or recognizing your barriers) can help you decide where to put your efforts – what skills to learn, who to network with, where to find your support communities, and what type of opportunities to pursue.

Not all barriers need to be overcome. For example, James Clemens-Seely considered moving after graduation but recognized it would be a difficult barrier to overcome. He would be starting over versus building on his existing relationships and clients in Montreal (Canada), where he grew up. James explains:

> There's a transition leaving school and all of a sudden . . . you've got student debt to repay, you've got no gear, you have startup costs for your business, and you need to charge money now. You didn't have to worry before because you had all these school facilities. Client recruitment is a lot harder [now] because you have less, and it costs more. That's all brutal. I saw friends going through that and I didn't have to go through that because I'd been building a client base and building an equipment pool from high school.

Luisa Pinzon, who spent her emerging years in Bogota, Colombia, recognized she would have a barrier to building her career if she was too specialized (because of where she lived). Instead, Luisa was open-minded to opportunities, and found it paid off:

> When I graduated, I did assisting for live sound then I would jump into my teacher's post-production studio. Then, my friends would call me to record a music album and I would just go and do it. The world has evolved to something where you don't need to be good at just one thing . . . Besides the pandemic (a very different situation), I haven't been without a job at all since graduating three years ago. It's because back in the university when people were saying to me "you should choose one thing," I didn't. I was always into everything.

Some barriers can be avoided by knowing your personality or personal priorities. For example, someone who has difficulty focusing or holding attention for long periods may be better suited to certain types of work. Audio engineer/ radio producer Jeanne Montalvo Lucar explains:

> If you're mixing a podcast or something that you can stop and start, that's a good place if you have issues with focusing or attention. If you're going to do a live show, absolutely that won't work . . . If you're on a four-hour [radio] show live, absolutely 100% of your focus needs to be on that. There's no multitasking on what you might want to do personally – like you can't read an article online. My phone is upside down and I'm not even paying attention to it.

When a barrier arises, opportunities you may not have considered before might be of interest. Sarah Stacey shares an example:

> When I finished my master's, I intended to stay in the UK and try and find something there. But it got to the point where I had to come back to Ireland when my lease was up, and I literally had nowhere to live. But I thought . . . I can always go back if it works out that way. Then, I ended up falling into some work – a commercial radio station in Dublin. That's what I did for the next four years.

Podcast editor Tom Kelly found he encountered many barriers working in music production, but stepping away led him into podcasting:

> It was almost five years of not making any money in audio and actually totally giving up and then falling into podcasting by accident . . . I was a barista at a coffee shop and made way more money doing that than any job offer that I got in audio. After moving to Denver, I got obsessed with podcasts . . . I produced 200 podcast episodes in a year [for free]. At a conference, someone said, "I've heard your work. I'm an editor, and I have too much work. Can I subcontract some to you?" I was able to quit my job within six months.

It will take sacrifices for those who choose to overcome difficult barriers. Eric Ferguson, who has had a successful career in both live sound and music production, explains:

> Anybody that wants to be successful in this business [has] to make sacrifices. The locational sacrifice, your time – You're probably going to sacrifice your love life. You're going to be poor. It's not for the rest of your life – but you're going to be for a while.

Our barriers – and our tolerance to take on barriers – can change over time. Meegan Holmes (Global Sales Manager at Eighth Day Sound) shares:

It's okay to try something for a few years and [decide] "this is not for me. I want to do something else" . . . I think that so many of us start out thinking it needs to be something that's decided, and it doesn't. It's very fluid.

Navigating Barriers with Broad Goals: Kevin McCoy

Kevin McCoy is Head Audio for the Broadway musical Hamilton *("And Peggy" Company) in San Francisco (USA). For Kevin, having broad goals and being open-minded to opportunities has been key to his career. Even as an established professional working in sound, Kevin took an opportunity to work as a production manager.*

I would *not* suggest that anyone early in their career be picky about what kind of show they're doing. I did *Sesame Street Live* for three years, and I wouldn't normally choose to listen to *Sesame Street* in my life. Do the show and then listen to something else on headphones later, if you need to. It's not that you can't do the stuff that you enjoy, but you can't be picky about it when you're getting started.

[Production managing] was still a fun gig and I was glad to take it. I think it makes me a better touring person in general. The biggest thing I was thinking about as a production manager was trucking logistics. As a sound person, I can be the jerk sound person who's just angry that my truck hasn't shown up, or I can be the sound person who understands the logistics of getting nine trucks in and out of three docks . . . I can even plan my day to account for it, or I can help the person in charge of that by giving them information that I know will help them.

If you are entering into your career, don't stress out about making a plan. The only decisions I've made in my career are that I wanted to work in theater, that I wanted to tour, and that I wanted to work on *Hamilton*. Other than that, I just took the opportunities that seemed best at the time they were presented to me . . . If I was offered a couple of different jobs, I would take the one that seemed like it was the best fit for me at the time. I didn't look at it like, "Which one of these will let me work on Broadway when I'm 42?"

When I'm asked, "What's your 10-year plan? What's your five-year plan?" I have no idea. My plan is just to keep doing the show until I can't do the show anymore.

Building a Music Career in Toronto: Mariana Hutten

Mariana Hutten is a music producer, mixer, and mastering engineer in Toronto (Canada). For Mariana, the time from choosing to have a career in audio to being an established professional took between seven and eight years.

Mariana started exploring the audio industry through a bachelor's degree in electroacoustic studies from Concordia University in Montreal (Canada). While studying at Concordia, Mariana worked part-time doing AV at the school. In the summer, Mariana worked for local video game companies doing quality assurance testing (QA), or playing games to look and listen for errors and write reports. This led to an opportunity with the audio team at Ubisoft, a global game company with an office in Montreal. Mariana explains:

> They needed extra audio people to do audio testing on *Assassin's Creed* to essentially test the game, but only pay attention to the audio. I did that for Ubisoft for three months and it was super fun. The team really liked me and said, "When you finish school, you should apply here to work as a sound designer."

Mariana had considered video games as a career and recognized a good opportunity, but it didn't fit her personal priorities. Mariana didn't want to stay in Montreal, and while she enjoyed working on video games, she didn't have a passion for playing them. Mariana says of video game companies, "When you work for one of those companies, that's their life. They're the same as music people – they spend the day working on the game, then they go home and they play more games."

Mariana also felt strongly about not doing unpaid internships:

> It's partly cultural. I come from Latin America. There are fewer people who went to secondary school, so it's really valued when you did go to school. It's an insult to be asked to work for free when you invested time in education. It's a different mindset, but I understand the mindset in North America that it's a continued form of education.

> I also couldn't afford it . . . I couldn't afford a car, and there wasn't an available car for me . . . I always had to have a day job. Whenever I tried to go to a studio, it was always, "No, you have to be here full time." It was never, "Sure. You can have a day job and then come here in the evenings and learn with us."

After graduating, Mariana moved to Toronto, the hub of the Canadian music industry. Mariana says of the experience:

> When I moved to Toronto, it took me forever to find work, to be honest. It was really tough after graduating. When I did find something stable (that wasn't small gigs), it was a year after I had moved, and it was a gaming company. So, my day job for two years was game testing.

Mariana also found apprenticeship (informally) with Post Office Sound, a new studio at the time:

> I would just go there, hang out with them, and watch how they worked. They taught me how to edit. They taught me how to do Melodyne. Then they gave me a few old projects to practice on. They were starting out, so they didn't have many opportunities to give me. But once they had too many projects, they started giving me the stuff that they didn't have time for, or they would recommend me for smaller budget stuff. So, it started off organic like that. Once I felt I had learned enough from them, then I stopped going there. They would recommend me for projects once in a while.

In 2018, Mariana left Toronto when she was accepted as an Audio Recording Engineer Practicum at the Banff Centre for Arts and Creativity (a continuing education program in Canada). She explains:

> I decided to go to Banff because of a couple of engineer friends who are amazing at what they do (and work on big projects) who went there. I thought it would be beneficial. I went for the engineering training, but also to network.

When Mariana came back to Toronto a year later, she landed a job at *Artscape*, a not-for-profit that supports the creative community. The Digital Media Lab where she works has state-of-the-art recording, photography, and visual effects studios, but initially, Mariana was not hired as a recording engineer. "I was supposed to be there helping all six studios at the same time. I'm not a photographer, I'm not a videographer, and I had to answer technical questions for people," she says.

Mariana wanted to work in the studios and saw it as a barrier she could try to overcome. She recognized the recording studio was being underutilized because artists did not know how to use their high-end equipment. Mariana explains:

> My dad always said, if you want to make a change in a corporate environment, you make it by writing a really good report showing

data and showing numbers. So, that's what I did. I wrote reports showing what other studios do, how they serve people, and how much money they're making. I did all that research, gave it to my manager, and my manager showed it to the CEO.

Her pitch was successful. Mariana works as a recording engineer and instructor at *Artscape*, and the studio is booked out months in advance. She continues to run her freelance business (mixing and mastering music), and her goals are to pivot more into mastering, and work with bigger artists. Mariana says of her career so far:

> Everything in my career has been really weird. It hasn't been traditional in the sense of, "I knocked on the major studio's door and made coffee for two years." For me, it's been a very, very weird way of going around and getting to the same spot.

Mariana recommends:

> You have to be creative about ways of achieving something you want to achieve. There's never one single way of doing something. My strategy is to build your own world – your own reality. People in this industry will tell you that the way of doing it is you go to audio school, then you do an unpaid internship in these studios, and so on. That's not a lie – That's a way of getting there. But, that's not the only way of getting there.

> Most importantly: keep trying. Life is not easy for anybody, so you've got to keep going. When I was working at the gaming company and doing audio as well, I did 14-hour days all the time. I still have 14-hour days now, but it's 14 hours of doing work that I like. It's a completely different thing when eight of those hours are doing a job you don't like. Few people can withstand that over long periods of time. But if that's what you want to do, that's what you have to do.

Note

1   Article being referred to: "Sunday Times survey saying artist is topmost non-essential job sparks anger in community" mothership.sg/2020/06/sunday-times-survey-artist-non-essential/

References

AdeDamola Oluwagbemi, Phebean. Personal interview. 10 Jan. 2021.

Bondelevitch, David. "Teaching Sound Mixing: Post." *CAS Quarterly,* Winter 2021, pp. 28–38.

"Boom Box Post Internship Program." *Boom Box Post,* www.boomboxpost.com/internship-program. Accessed 13 Aug. 2020.

Clemens-Seely, James. Personal interview. 20 Dec. 2020 and 21 Oct. 2021.

Davis, Fela, guest. "Fela Davis." *Roadie Free Radio,* ep. 20, SoundCloud, 26 Sept. 2016, www.roadiefreeradio.com/podcast-1/2016/9/26/fela-davis.

Espinoza, Andrea. Personal interview. 28 May 2020.

Ferguson, Eric. Personal interview. 2 Dec. 2020.

Finan, Kate, panelist. "SoundGirls Career Paths in Film & TV Panel at Sony Studios." *YouTube,* uploaded by SoundGirls, 26 Oct. 2018, youtu.be/Y_g3drC2yyA.

Gaston-Bird, Leslie. Personal interview. 2 Oct. 2020.

Gross, Jeff. Interview with April Tucker and Ryan Tucker. 14 Jan. 2021.

Hendley, Alesia. Personal interview. 17 Dec. 2020.

Holmes, Meegan. Personal interview. *What Makes You Stand Out,* SoundGirls, 20 May 2020, youtu.be/wYamiOK8Y6A. Accessed 21 May 2020.

Hutten, Mariana. Personal interview. 19 Nov. 2020.

Jaczko, Rob. Personal interview. 15 Oct. 2020.

Jackson, Brian M. *The Music Producer's Survival Guide: Chaos, Creativity, and Career in Independent and Electronic Music.* Routledge, 2018, ch. 2. doi.org/10.4324/9781315519777.

Kelly, Tom. Personal interview. 22 June 2020.

Kennedy, Camille. Personal interview. 21 Feb. 2021.

Leong, Mabel. Email interview. Conducted by April Tucker, 29–30 Sept. 2021.

Masingale, Langston. Personal interview. 14 Sept. 2021.

McCoy, Kevin. Personal interview. 22 Dec. 2020.

Mohabati, Mehrnaz. Personal interview. 29 Jan. 2021.

Montalvo Lucar, Jeanne. Personal interview. 12 Oct. 2021.

Pinzon, Luisa. Personal interview. 10 Nov. 2020.

Quinn, Kristen. Personal interview. 30 Sept. 2020.

Rogers, Susan. Personal interview. 12 Oct. 2020.

Stacey, Sarah. Personal interview. 19 June 2021.

Stoll, Becca. Personal interview. 27 April 2020.

Thakkar, Akash. "Successful Freelancing in Game Audio." *YouTube*, uploaded by GDC. 9 Jan. 2019. youtu.be/93ggs7hwJeU.

Urban, Karol. Personal interview. 17 July 2020.

Vericolli, Catherine. Personal interview. 29 May 2020.

Zolnai, Eliza. Personal interview. 27 Sept. 2021.

4
Networking

No matter how much you know about a microphone or some type of crazy software plugin, this is a people business. It doesn't matter what part of the industry you're in – if you're in live sound, post-production, video game sound, if you're in the studio – this is a people business.

– Catherine Vericolli

Networking is the process of interacting with others for professional purposes. Your **network** is a group of people whom you have a mutually beneficial business relationship with. A person within your network is considered a **connection**. When you hear catchphrases like, "it's who you know," or "you have to connect with the right people," these refer to networking activities.

Networking is a way to:

- Meet potential clients (people who could hire you in the short term or the future)
- Meet others with similar interests or career pursuits, including potential mentors
- Begin the process of building relationships and building a reputation (so others can learn who you are and what you do)

Networking is a skill developed over time, no different from learning a musical instrument or a new sport. Networking is not always a formal activity. Ariel Gross, founder of Team Audio and the Audio Mentoring Project shares:

When I hear the term networking, I conjure up images of people in fancy clothes mingling in a beige convention space, repeating their "what they do" line over and over, handing out business cards, performing their best fake laugh, basically marketing themselves into a better job. And you know what? That totally happens. People totally do that, and maybe it can work sometimes. But in my experience, there's a better path, especially for

DOI: 10.4324/9781003050346-5

people who feel like they're covered in a thin veneer of slime after an event like that. It's called making friends!

Music producer Brian M. Jackson adds:

> Even when you're in full-on networking mode you will meet amazing, interesting people who aren't in the scene or industry, and who won't help further your career in any obvious way. A diverse assortment of friends and acquaintances is always a good thing.
>
> (ch. 2)

Networking can also happen naturally as you learn about your craft. Avril Martinez, who started their career in theater before pivoting to VR, explains:

> I went to meetups and I learned a lot of things about VR. I learned about how to program and what audio engines exist. In the process, I met people, I spoke to them about my passion for audio, why I did the things I did [in theater], and how it translated to what they're doing [in VR]. That's how connections started happening – I just really cared about it.

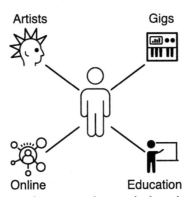

Figure 4.1 Different ways you can be connected to people through a network.

Your network may include:

- Relationships formed before starting your career (classmates, teachers, friends, relatives)
- People you have met through hobbies, interests, or volunteer work related to your career
- People you have worked with or met on a gig (co-workers, colleagues, clients)
- People met at industry events (meetups, conferences, seminars/training, parties)
- People you regularly interact with online (by following and commenting on social media, sending direct messages)

Music producer Nick Tipp says of networking:

> Everybody you know in your life has some connection to something that might eventually pay you [for work]. So just talk to your folks and your cousins and your friends and all these different people and just see: Is there some sound or editing or some tech or music thing that you know of that somebody's doing? Do that a few times and then other people are trying to connect you. If you say, "thank you" to them, and it works out and you make some money, then you make an earnest [statement] like, "I made rent because you sent me that dude," then they're in your corner.

While some networking happens naturally, it also requires effort to meet professionals and potential clients. The idea of contacting or meeting people you don't know may feel very foreign or invasive, but it is very normal in the audio industry (both online and in-person). Product manager Ryan Tucker explains:

> If you don't insert yourself into the industry machine, you're not going to be exposed to the people that can hire you. You're going to miss out on the right people knowing you, being able to recommend you to others, and so on. Your brand needs to spread like a virus. You need to spread it to the right people to infect them with your wares. If you don't network, [you're] out of sight, out of mind. You don't exist.

Michelle Sabolchick Pettinato shares the consequences of not networking:

> After I got my first tour, I just assumed the tours would keep coming. I didn't do a good job of making contacts and networking. There were a lot of opportunities where I could have, but I was kind of shy and not really good at approaching people I didn't know. For example, the band was playing a festival where I could have met a lot of other tour managers, but instead of approaching them I just wanted to stay out of their way.

> When that first tour came to an end, I no longer had a job. I actually had to take a job as a telemarketer (which I quit after ten days) to pay the bills. I had to learn very quickly how to network and make connections with people who could provide me with future work.

Effective networking is *not* about asking people for work. Karol Urban (re-recording mixer/sound editor for TV/film) explains:

> When I network, I never ask for a job . . . I have never come to [someone] and been like, "I need a job. Do you have any jobs?" Never done that . . . Often times, I'll just meet with people, and I'll ask them what their advice to me is, and what their leads are. I've always shared my very specific vision: I want to mix dialog and music for scripted narratives.

Music producer John McLucas says of being asked for work through social media:

> I probably get a canned message every week or two . . . I've had [the same] people send me five or six canned messages and I don't respond. I have not had a single client (nor heard from anybody else in my field) tell a story of cold messaging, "Let me work for you. Hire me," and had that turn into a career-pivoting movement.
>
> ("Quickest Way")

Re-recording mixer Sal Ojeda adds:

> Recent graduates email me now and then and say, "I'm moving to LA. I'm looking for work." Okay, good luck. I mean, I'm trying, too. But if they're nice and they ask me to be on a call or something, I do it. I'll tell them, "This is how [the industry] is and if I need help, I'll call you."

Quality interactions are more important than quantity. "Networking isn't how many connections someone has on LinkedIn – it's how many meaningful conversations you can have going at any given time," says film sound designer Shaun Farley.

"It's important that in your networking that you are genuine, and you aren't treating people as though they are stamps [to collect]," says Megan Frazier ("Interview"). Megan says networking in the audio industry is different from networking in other industries:

> Since I used to be in sales I know deeply how to "work the room" when going to networking events. [At GDC], I realized I needed to slow down if I wanted to make genuine connections with people. Ending up with lots of contacts wasn't what I was looking for; I needed to establish authentic and genuine connections, [which] take time and need to be nurtured.
>
> ("GDC")

Video game sound designer Akash Thakkar adds:

> People actually want to form some sort of connection with you . . . They'll realize, okay, you're easy to work with, you have a good vibe. [Working together] will probably go a lot better than someone who just handed me their business card I know nothing about and just ran away. It is worth it to form these relationships . . . Having quality [interactions] over quantity is really, really important.

Introverts vs. Extroverts and Networking

It is not necessary to be extroverted to be effective at networking. Video game sound designer Javier Zúmer shares:

A secret about shy people like introverts: we are introverts, but we like to talk about the things we like. I like to talk about audio, so I feel okay . . . Audio people don't like to do the business part and sell themselves, so, it's very hard for me to say those things about myself and sell myself. But somehow you need to play the game a little bit. So when you contact people, be humble, but also show cool stuff – because if the stuff is good, it's going to speak for itself.

Karol Urban says of the differences between herself (an extrovert) and her more introverted friends in the industry:

I actually get invigorated by talking to people. I work with many folks, however, who are more introverted at heart. Sound often attracts listeners who are more introspective. But, the best creative teams need all types of people and perspectives. I have seen people who prefer smaller groups and individual networking to larger groups be incredibly effective at forging deep-lasting connections. Their networking manifests itself in equally helpful ways. So, depending on your personality, your networking approach and comfortability may be different. But that doesn't mean that you will be any less effective.

John McLucas grew his social skills from an area of weakness to a major strength of his business:

I was not solid socially when I started out (as far as being a good human to talk to for an extended period of time). So, that was a really big friction point. The biggest thing that got me through it was going on tour because I was forced to talk to strangers every day . . . Talking to as many people as I could really helped me get through that. If you don't know how to approach somebody in a friendly, non-threatening, non-scuzzy way, you've already lost before you've even tried to make a [sales] pitch.

("Q&A")

Where to Network

Networking can take place at a variety of locations and events. For example:

- Courses and training programs; alumni events
- Industry organizations (international, national, regional/local chapters; in-person events, or online). These can be broad to the industry as a whole or specific to a discipline
- Places where other professionals may go socially such as venues/shows, open mics, film and music festivals, conferences
- Formal industry networking events (in-person or online)
- Social media

Other ideas:

- Hobbies or interests outside of audio where you might meet people who have ties to the industry or your work
- Aspects of your identity or life experiences (connecting with alumni, people who grew up where you did, other fans of your favorite sports team, etc.)
- Activities that could help you connect with professionals you respect but may not have access to otherwise (volunteering with an organization, interviewing a professional for a publication, podcast, blog, etc.)

John McLucas adds:

> Go to meetups, open mics, film screenings, low-key public events – whatever it is. Just go out and meet people because if you have that shared interest, you can always fall back on that for conversation. It's not like meeting somebody at the grocery store randomly and trying to spark something up. We're both at this film screening for this new indie film. We can talk about that, we can talk about what you are working on . . . people love talking about themselves. It's so easy when you start off with a shared activity.
>
> ("Reasons")

Chris Bushong gives this example of networking in Nashville (USA):

> There are industry nights at certain bars. Over in East Nashville, there will definitely be industry people hanging out, or go down to 3rd and Lindsley and Vince Gill plays down there all the time. There are always industry people hanging out there. During my first two years in Nashville, I would just hang out all the time trying to meet people . . . Sometimes networking was fun, depending on the people, and sometimes you're just like, I really just want to go home, put on my slippers and prop my feet up, but I really should go do this because it could be really helpful for my career.

Networking From School: Kelly Kramarik

Kelly Kramarik landed a full-time job in Denver, Colorado (USA) by following up with a guest speaker she met while in school. Matt Waters (re-recording mixer, *Game of Thrones*) gave a presentation at the University of Colorado-Denver, and Kelly introduced herself after the talk. Matt told Kelly to get in touch if she ever visited Los Angeles (where he lived and worked).

After graduating, Kelly wanted to stay in the Denver area, and she was interested in working for Starz, a Denver-based television network. When Starz had a job listing for an audio editor, Kelly applied but did not get a response. Kelly decided to attend an industry event in LA (*Mix Presents Sound for TV and Film*) and reached out to Matt about meeting up during her visit.

Matt invited Kelly and a couple of his colleagues for coffee. They asked Kelly what kind of work she wanted to pursue, and she mentioned the job at Starz. One of them had a friend at Starz and offered to reach out. Kelly tells the story:

> He talked to the manager at the time and said, "I think Kelly is a really good fit if you want to check out her resume." The manager did . . . [After interviewing me he said], "You're young, but you seem like you're really qualified, you can learn quickly, and you also are the only person that had a recommendation from the *Game of Thrones* mixers." And I got my job. So, it was a lot of networking and right place/right time, but they just liked me as a person. I went to LA to say hello.

Networking can also happen while in school. Video game sound designer Will Morton (Edinburgh, Scotland) explains:

> Your time at college is the ideal time to start networking. Make friends and work with people who are doing sound design. Make friends and work with people studying composition. Make friends and work with people on game development courses, with people studying TV or film production, [or] any kind of performing arts . . . That person studying film? After leaving college they may get a directing job on a film or TV show and need a sound designer. The people you helped out with their game development coursework? They may leave college to set up a studio, and you can imagine that they'd feel safer hiring a sound designer they already know and trust . . . Having friends is a lot more valuable than having acquaintances – people want to work with people they like.

> The time you have at college or university is very likely to be the last time you will have this amount of time to do this stuff . . . While you've got the opportunity to make friends with a bunch of like-minded creative people who also have a ton of time on their hands, make the most of it.

Networking can happen on the job, also. "It's a common freelancer thing to have drilled into your brain: Don't say no to anything. But, that's part of how you network because that's how you meet the most people, too," says Scott Adamson.

Networking with people who share an aspect of your identity can be a way to connect. Jeanne Montalvo Lucar, a Grammy-nominated audio engineer and radio producer, explains:

> I wish I could find more Latina engineers so that I could have more in my own network. It's nice to [find people who] understand certain issues you might have dealt with coming up – like not being able to take free internships . . . and groups of color are not going to have the same experiences as a lot of other ethnic groups.

Some industry advocacy groups offer support specific to certain identities. For example, Audio Girl Africa was created in Nigeria by Phebean AdeDamola Oluwagbemi and Bada Aramide to support women who are pursuing careers in the audio industry in Africa. Phebean says of the organization:

> [Our goals are] to create opportunities for women in the industry, open up doors (like networking doors) for young Africans who may want to create careers, make it easy for them to find their way into the industry, find mentors, find internship opportunities, and jobs . . . We're trying to build a safe space where every girl finds people like her that can say, "This person I can reach out to." Then they could reach out to a circle of friends for some kind of mentorship, just to guide them so they know it's a whole community.

Networking in Podcasting: Sarah Stacey

Sarah Stacey is a radio and podcast producer in Dublin, Ireland.

> Networking probably has been more important for me than anything. There's just so much to be said for people recommending you for jobs. If you're editing a podcast for someone, they'll recommend you to the next person they know who's doing a podcast and it just snowballs from there . . . Networking is a big thing for your first year because you just need contacts. I wouldn't be able to do any of this without the contacts that I've made . . .
>
> I don't really plan myself networking. I just keep an eye on opportunities that I find interesting, and I try to take on things that will allow me to do different sorts of work just to build up a good portfolio.
>
> I did more in-person networking when I was in the UK [going to school]. We always had different guest speakers come in from different areas of radio to talk to us – people would do masterclasses and stuff, so that was always a great way to get career advice.

The guy who hired me for my first radio job – I only worked with him for a few months. But then recently, he heard that I was freelancing, and he had a couple of opportunities (because he runs his own company now). So, he was able to get back in touch and ask if I would do some stuff for him. So, it's kind of come full circle.

Networking in Live Sound: Scott Adamson

Scott Adamson is a touring front of house mixer who has mixed at top venues worldwide. He is the creator of The Production Academy, an online education program for the live sound community.

In live sound, a lot of networking happens on the job, and there is no harm in networking with anyone and everyone. Meeting everyone from other crew to musicians could lead to future work. Almost always, your first touring job is going to be from going out and meeting musicians and talking to musicians and meeting those types of people. Or, if you're backstage at a bigger concert and you have access to talk to tour managers and production managers, that can be a great way, too, because those are the people that hire the touring staff for larger tours.

I help people more when they just come up and talk to me and engage. There's so much downtime on shows. Like, we might have an hour of downtime sitting around. Sometimes it's hectic, but there's time to chat and make connections.

[If you're just in the audience], a compliment goes a long way. "Hey, your mix tonight sounded great. Do you mind if we connect?" Just get someone's email address . . . Email is easy, and if someone doesn't respond to your email, it doesn't matter, right? Reach out, give a compliment, and then see what happens.

Online and in-person networking each have their advantages and disadvantages, and what works for one professional may not be ideal for another. Online networking has the advantage of connecting with people you might not reach otherwise (especially in different locations). Online networking also can reach more people faster – such as messaging multiple people in a matter of minutes versus attending an event for hours. Game audio director Kristen Quinn says:

> I used to write people that I really liked their work. I let them know I really love their work or ask them how they did something. I enjoy getting those, and I really love helping people who are trying to get their foot in the door. If someone writes me, I always try and write them back and find time to listen to their reel and give them feedback if they want feedback. If they want advice, I try and participate in any way I can.

One advantage of in-person networking is gathering a lot of information about someone quickly, which can lead to a higher quality interaction for the time invested. Megan Frazier explains:

> Online networking is a lot harder because when you're in person, people are talking about themselves. You can pick up probably 20 or [more] things that you can follow up with them over email. Online, they might mention something that you can follow up on, but you're not getting the holistic perspective. There's just a bunch of stuff that comes up in person you just wouldn't get online. Like, what are the names of their kids? . . . I would try and get as many video calls in as possible. Like, now I know you have a cat. It's all of those little things.
>
> ("Interview")

Networking Online: Britany Felix

Podcast consultant and editor Britany Felix has worked on over 2,000 podcast episodes, including 400 episodes of her own podcasts. Britany is a self-taught podcast editor and is making a living from her podcast work.

> I actually do very, very little in-person networking. It's just getting into these communities where your ideal clients are and constantly providing value. That's how I built my business in the beginning [in 2016], and that's how it still is now. Every day I was going into these certain active podcasting communities [on social media], searching for the questions that I could answer, and I was just providing a ton of free advice. Then, I became known as the "go-to" person. It got to the point where people were tagging me on posts or saying "you need to talk to Brittany" when somebody needed an editor . . . It can all be done online if you want it to be.

People have seen this in the online space in general, but once you get super active, you start to form real friendships with people you've never seen in person. That just comes from showing up every day and participating. It's no different from any other relationship. If you half-ass it, you're not really going to form any kind of a bond. But if you're constantly there and you're constantly nurturing those relationships, then it doesn't really matter whether it's virtual or in-person. In-person, obviously, can be nice. When I go to conferences, I do connect with people who I've met online and it's fun, but they've also already referred me business well before we ever met in person, so it's absolutely not required.

Basics of Networking at Events

Networking can be a byproduct of attending an industry event, or an event may be formally planned for networking purposes. For example, a conference or trade show might have an official "meet and greet" or "happy hour" designed for people to gather and network. There are networking meetups specifically for people in the industry or a specific discipline. Or, a group of people may informally chat after a seminar, guest lecture, or training event. Ariel Gross says of networking at events:

> There's always a sense of urgency when you're trying to find a job, whether you're fresh out of school and need to start paying off your debts or you're feeling stuck in your current job. It's why when I go to game [development] events, I'm inevitably cornered by a poor, desperate soul that assails me with endless questions that all amount to "Can you get me a job?" And I get it! I really do . . . One of the absolute best pieces of advice that I have been given and that I can pass along right now is to make friends. Not business associates. Not acquaintances that can be leveraged later. But real, honest-to-goodness friends.

When networking at an event, some basic preparation can help: have a business card, an "elevator pitch", and know the basics of social etiquette for networking. An **elevator pitch** is a concise introduction of yourself and what you do. The idea is your introduction should be short enough that you could say it on an elevator ride (as little as 30 seconds). Composer/music producer Catharine Wood explains:

> I actually have been in the elevator at BMI talking to the head of [a department] where I had like two seconds. You always have to have a bin of

things that you can pull from really quickly to be able to quickly connect and let people know who you are and what you do. When I was in engineering school, one of my instructors said this is going to be one of the most uncomfortable things you'll have to do as somebody who isn't a "braggy" person. It feels kind of gross to be like, "I just did this project" or whatever it is, [but] you have to let people know what you do.

("Importance")

Podcaster and audio storyteller Avery Kloss Moore says her elevator pitch is: "I can help you make a podcast that people will actually enjoy listening to." Avery explains:

That's where my specialty is – I can tell you in 30 minutes how to start a podcast. But if you have a podcast, no one's listening to it, you're not getting enough views, and you want someone to look through it and [suggest] ways to strengthen your storytelling in audio, that's what I do best.

Catherine Wood says an elevator pitch can change depending on who you are talking to:

No two "elevator pitches" are the same. It all depends on my audience and whatever I've got going on in the studio at the time. Much of it is dependent on a) knowing who your audience is, and b) knowing what will impress them. I think of my brand as an engineer/producer/studio owner like a stock [investment] . . . In addition to the quality work I'm doing, these encounters help to keep my stock up – and depending on *who else* is in the elevator can generate further relationships (and jobs).

("quote")

Business Card Etiquette: John McLucas

John McLucas is a music producer and content creator.

The business card is an incredible tool to make sure that you stay in touch. Don't give a business card too soon. Take your time, and make sure you've got a reason for them to have it directly, and then bring it out if you need to. The wrong way would be to approach any kind of a conversation where you already mentally or literally have your business card in hand. I've had it done to myself a lot. It's like a sub-20 second conversation and they've already whipped up the business card to put in your hand and then they try to dip out to get to the next person. That's the quickest way for somebody to look at the name and be like, "I will never hire or work with them ever because that was such a crappy interaction and very clearly a selfish perspective."

Before handing it out, make sure that they want it. Say, "What's the best way for me to follow up on this (discussion, project, opportunity, endorsement)?" Let them say, "Can I have your card?" Or, "Do you have a card?" Some people may not be into taking business cards. They might just say, "Take my card, you contact me." It could be a simple vetting way to see who's serious . . . Put a little note on the back of it so they know what you discussed.

From the Modern Music Creator Podcast, "How I Make The Most Out Of NAMM and Networking Events"

Networking at Events: Alexandra McLeod

Alexandra McLeod, a graduate of the University of Exeter (UK), started practicing networking at 18 years old at school events and media industry events.

I have found that the more events I attend, the more people I meet, the more professional opportunities I am offered.

Take a friend! Although most people attend networking events alone, there's nothing wrong with going in a group either. Maybe you'll feel more social when there's already someone else by your side.

Look up who the speakers are on LinkedIn and prepare some questions in advance in case there is an opportunity to speak to them, and you don't want to blank on the spot.

Have generic questions ready for other attendees including name, industry, interests, hometown, university, and city. You never know what you could have in common with someone.

Take a notebook, so if you really can't face speaking to anyone, you can look engaged and professional by writing things down. This looks better than being on your phone for sure!

From soundgirls.org

Networking with Strangers (In-Person)

One challenge when talking to strangers is getting in and out of the conversation. Approaching people you do not know can be incredibly awkward at first, but by the end of the conversation, no one will likely remember the first thing you said (yourself included). Some ideas to get into a conversation with a stranger:

- Start by introducing yourself. This is usually effective as long as you have something ready to say or ask after (to keep the conversation going).
- Look and listen for clues. If you see someone wearing a T-shirt of a favorite band or you overheard them talking about a piece of gear you also like, these can be conversation starters.
- Prepare a few questions (or think of a few topics) you would feel comfortable asking anyone (or answering yourself). You will encounter professionals who *love* to talk tech or sound and others who would rather discuss life outside of their work.

To keep the conversation going, Ariel Gross suggests:

> Don't just talk about yourself. It's hard to do because you want to let people know how great and employable you are, but it turns people off and can make you seem desperate and insecure. Wait for someone to ask you to talk about yourself before talking about yourself. When you do start opening up about yourself, keep it to a few sentences and then end with a question about the other person . . . Take some mental notes about stuff you have in common, and then ask follow-up questions about those things. In this way, you can steer the conversation to topics that you can contribute.

Jack Menhorn adds:

> If you go to a party and b-line for all the audio directors and hand them your business card as soon as possible, people will smell that and gravitate away from you. People want to hang out, swap war stories, workflow techniques, and nerd out with like-minded people. People do *not* want to hang out and get begged for a job . . .

> Make sure to not speak up in a group discussion just for the sake of it. Listen and absorb conversation rather than trying to make sure you are a part of it.

If a conversation naturally ends, falls flat, or it's time to move on, an easy way to get out is to say, "It was nice talking to you." That signals that the conversation is over and you can step away without needing to give a reason. It is better to step away too soon and feel successful than to drag out a conversation uncomfortably for no reason. This is also the ideal time to exchange contact information or ask for a business card, if you plan to do so. If you feel you need to give a reason to step away from a conversation, you can say, "Excuse me – I need to go" (get something to eat, a drink, make a call, meet someone waiting, etc.). A break can be a good opportunity to write down information like a name or details you were going to follow up on.

Meeting Strangers at Events: Karol Urban

Karol Urban has worked in television and film post-production sound in Los Angeles and Washington, DC (USA).

> I say "hi" to four people before we leave, because otherwise, it's totally giving up. I always look first for the person who's being left out of the action, because they're probably thinking like I am: I don't know anybody. How can I get into a conversation with someone new? Typically, they're receptive. They are at a networking event. So, unless they were dragged kicking and screaming, they also are looking to be social. So, I approach them and I say something authentic, empathizing with the situation, and I identify myself. Before you know it, you have met a new person and are deep in conversation. Sometimes you might even learn they are waiting for someone and then bonus – you get to meet three or four more people.
>
> Sometimes there are massive clusters . . . In this case, you will observe people joining those clusters and leaving those clusters. Why not be one of those people? Why not introduce yourself?
>
> People typically love to talk about themselves and their adventures. Curiosity about the other person goes a long way. But if you are feeling a lingering awkwardness, excuse yourself. Find a friendly way to thank them for their time and redirect your focus outside of the conversation. You can always exit that area and try the next group.
>
> Most people feel similar emotions in social situations. Don't assume you are the odd one out. There is likely at least a percentage of any given group that is feeling and thinking the same thing as you. So, I look around for someone who I can see on their face and in their body language is trying to find a way to connect.
>
> I don't normally ask people for contact information. It is a bit awkward when people do that to me. I find them later. I might explain I read their blog, checked out their site, or looked up their company and let them know what I found sincerely impressive. I will mention where we met and finish by saying how nice it was to meet them.

In LA, it's funny because some people don't go [to events] because they feel it is just a big group of people looking for work. But people who are in our industry are either looking for work or they have too much work. So, they may all be looking for work now. But now they have met me so when they have too much work, they can remember me. They might think, "I met a re-recording mixer the other day. She had a bunch of experience." That's how networking works.

I'm not above fangirling on somebody. I remember the first time I met Randy Thom [Director of Sound Design at Skywalker Sound]. It was at the CAS Awards cocktail hour and I was coming in from out of town. We didn't really know anybody. So, my husband and I approached Randy Thom and I said, "I'm sure people do this to you all the time, but we've got to do this, too. Your work was a massive inspiration as to why we started doing what we do and why we professionally do what we enjoy. We are major fans . . ." He was really, really friendly and really kind. We didn't make it awkward and began to exit ourselves as he is a very popular and busy man. He stopped us and recommended we meet someone else. He said, "Do you know so and so? You guys would get along. He's from Virginia, too – he's right over there." So it began – we went over to the next individual and opened with, "Hey there, Randy tells me you are also from Virginia." We were on a roll then, meeting new folks. It was such an enjoyable and exciting evening.

There are so many rabbit trails that you go through to find your place in our industry. For me, each new person offers the possibility of a new direction that could significantly enrich my career path or simply my experience as a member of the community.

Reading the Room when Networking: Jeff Gross

Jeff Gross has worked in the audio industry in Los Angeles for 30 years.

Every room is different, and I read every situation differently. What it boils down to is finding that common ground so that you can have a conversation . . . The whole idea is to make somebody feel comfortable. This is a people business. People = work. You get gigs through people. If people feel comfortable around you and you can do the gig, they're going to want to hire you.

Something in this business we can all talk about is gear. But, it's not like you're walking up to somebody for the first time and saying, "Hey, I've got this gear." It's not about you trying to impress somebody with your knowledge or your belongings. The whole point is just about relating to somebody on a human level.

Networking is not just, "What projects are you working on?" If you lead with that, you steer the conversation down a very sterile business-like conversation versus getting to know somebody. That's really how you ingratiate yourself into somebody else's world. It's much more about making somebody feel relaxed and comfortable speaking their mind. Not like, "Hey, I want to talk about business," and diving right into that kind of stuff.

The more versed you are in the ability to have a conversation about anything, the easier it's going to be. Toastmasters (www.toastmasters. org) is a good idea because it forces you to speak on a topic that you might not know about . . . The best education is experience.

The Importance of Following Up

If you meet someone through networking who gives their business card or contact information, it is good etiquette to send a message the next day. Jack Menhorn shares:

I am surprised by the number of times I have given my contact info to emerging professionals who then never reach out for a chat or advice. If you get my info, then please shoot me a message! I love talking! Let me talk to you about stuff!

Akash Thakkar adds:

Email them within 24 hours. Talk about what you talked about if you had a good conversation . . . Don't [send lengthy] emails, which is what I get 150 of every single week . . . Just say, "Hello, it was great to meet you."

Following up with your potential clients within 24 hours is the most powerful way that you'll actually move people from [a work] prospect to lead to opportunity to a client . . . Most people mess this up. They get a pile of business cards, they look at them and say they're not interested, and then throw them away. Or, they're just kind of languishing on their desk for months and months and months.

Michael Lawrence (live sound engineer, technical writer, and podcast host) says following up is a way to stand out from your peers:

> When I speak at a trade show or something like that, out of 50 people, maybe 10 will come up and ask for my contact information. I'm always like, "Here's my card, here's my number. Call me, text me, email me. We can [video chat]. Whatever I can do to help you get moving forward, I'm happy to do that." Out of those 10 people, four will send me an email and say, "I really want to meet up with you," and out of those four people, usually will get one email back and forth.

Networking at Trade Shows/Conventions

The audio industry has trade shows and conventions worldwide, which can be great networking opportunities. Lead audio designer Will Morton says of game conferences:

> It's easy to find events, and easy to get a feel for what the events are like (YouTube often has videos from previous events, so you can see what you are letting yourself in for). If you are new to game development conferences, make sure you do this research early – you will often need to schedule a few days to do a conference, including travel if you don't live near main cities, and there are often a lot of extra details to be sorted out when going to a conference.

Akash Thakkar (Seattle, Washington, USA) adds:

> Sending emails to people who might be there is very important before you go to an event. Like, "It'll be great meeting you. I'll be at your talk," or "I can't wait to see you at this event," whatever it may be. When I first started in Seattle, I messaged the [head of a meetup] and I said, "I just moved to Seattle. I can't wait to meet you. I'll be at this event." Nothing more than that. I didn't mention I was a composer, sound designer, none of that stuff. But when I showed up, I said, "It's so nice to meet you." [The response was], "Oh, you sent me that email? Let me introduce you to some people." They already had an idea of who I was just a little bit . . . the "stranger danger" was kind of gone a little bit.

For broadcast mixer Andrew King, networking at industry events eventually led to a major work opportunity:

> The biggest thing is meeting people and making connections. That's something I'm not very good at. I'm fairly introverted, so doing that kind of stuff is totally against everything inside my being. But I would

go to industry events, and I would just walk up to people and introduce myself and say, "Hey, I'm Andrew, I'm a truck A1. What do you do?" It was through a series of that happening which is how I got to work the Olympics [in 2018 in South Korea] . . . There's no way I would have gotten there [otherwise].

In addition to formal events, there are usually events outside of show hours, such as tours, parties, and meetups. Becca Stoll says of the USITT show (a popular show for theater), "There's seminars, workshops, and panels all day, but where the jobs happen is at the bar afterward."

Podcast editor Britany Felix says of podcasting conferences:

Multiple times at these conferences I've ended up going out to dinner after the event with some big names in the industry that I would have never been able to get hours of one-on-one time with any other way.

Some after-parties are advertised (including on social media) and others are invite-only. It's common to go to one party and hear about other parties or find groups who are leaving for a private party (and allow you to tag along). Andrew King shares how he was invited to one such party:

The night before the show opened, there was an industry networking thing. I went to that, and it was all the big names from all the networks, and they're just hanging out. So, I was just going around introducing myself to people. I met a guy, who introduced me to another guy, who introduced me to another guy that got me into another event. It was an invite-only event for [mixers].

Tips for trade shows:

- Ask around about getting a free or discount pass beforehand. Some shows give complimentary passes for basic events. Some shows give passes in exchange for volunteer time or have a student discount.
- Look at the event schedule for presenters or panel members who you may be interested in meeting or talking to.
- Pace yourself (taking breaks during the show, get enough sleep) if you plan to go to after-parties.
- Trade shows are also a great place to practice networking skills. If there are exhibit halls where companies display their equipment, there are employees at the booth whose job is to talk to people. Even if you have no intention of buying a product, this can be a great way to practice talking to strangers, asking questions, and getting in and out of a conversation.

Social Drinking

Going to a bar or drinking alcohol during networking is common in some disciplines and cultures. However, it is not required for career success. Kevin McCoy, who works in musical theater for Broadway shows, shares, "I don't drink at all. I'll go to bars and I'll just drink soda."

Music editor/supervisor Fei Yu (Beijing, China) focuses on doing her job well and connecting to clients through her work:

> I don't like drinking, and there is a drinking culture in [some parts of] China. I don't want to share my personal life. I [connect] with people through my expertise, from the music itself, how I share all of my knowledge with them, and being a professional, rather than doing other social things with them.

Jeff Gross, who has worked in the Los Angeles audio industry for 30 years, advises:

> Figure out how to be comfortable in your own skin. If you're comfortable in your own skin, you won't need alcohol or other mind-altering substances to be the life of the party. You don't need that stuff to have a good conversation, communicate, or just enjoy someone's company. I liken it to being around celebrities. People can be awkward around celebrities. I've been around enough people, celebrity or not, to just know how to be comfortable in my own skin. Other people will read that and feel comfortable or at ease.

References

Adamson, Scott. Personal interview. 14 May 2020.

AdeDamola Oluwagbemi, Phebean. Personal interview. 10 Jan. 2021.

Bushong, Chris. Personal interview. 27 Sept. 2020.

Farley, Shaun. Email interview. Conducted by April Tucker, 15 July 2020.

Felix, Britany. Personal interview. 22 June 2020.

Frazier, Megan. Personal interview. 11 and 21 Aug. 2020.

Frazier, Megan. "GDC is Like a Box of Chocolates: You Never Know What You're Going to Get." *A Sound Effect*, 25 April 2019, www.asoundeffect.com/gdc-game-audio-experiences-2019/. Accessed 2 Aug. 2020.

Gross, Ariel. "A Big Jumbled Blog About Joining Team Audio." *Ariel Gross*, 26 June 2012, arielgross.com/2012/06/26/a-big-jumbled-blog-about-joining-team-audio/. Accessed 8 Aug. 2021.

Gross, Jeff. Interview with April Tucker and Ryan Tucker. 14 Jan. 2021.

Jackson, Brian M. *The Music Producer's Survival Guide: Chaos, Creativity, and Career in Independent and Electronic Music*. Routledge, 2018. doi.org/10.4324/9781315519777.

King, Andrew. Personal interview. 15 Dec. 2020.

Kramarik, Kelly. Personal interview. 18 June 2020.

Lawrence, Michael. "New Co-Host Chris Leonard & Early Career Advice." *Signal to Noise Podcast*, ep. 16, ProSoundWeb, 9 Dec. 2019, www.prosoundweb.com/signal-to-noise-podcast-episode-16-new-co-host-chris-leonard-early-career-advice/.

Martinez, Avril. Personal interview. 28 Aug. 2020.

McCoy, Kevin. Personal interview. 22 Dec. 2020.

McLeod, Alexandra. "The Art of Networking." *SoundGirls*, 5 July 2019. soundgirls.org/the-art-of-networking-2/. Accessed 20 March 2020.

McLucas, John. "The Quickest Way To NOT Grow Your Network." *The Modern Music Creator Podcast*, ep. 9, 10 Sept. 2018, pod.co/the-modern-music-creator.

McLucas, John. "Q&A #2 – Why I Freelance, Realistic Vs BIG Goal Setting, Hardest Things When Starting." *The Modern Music Creator Podcast*, ep. 7, 27 Aug. 2018, pod.co/the-modern-music-creator.

McLucas, John. "5 Reasons You're Failing & How To Win." *The Modern Music Creator Podcast*, ep. 3, 30 July 2018, pod.co/the-modern-music-creator.

McLucas, John. "How I Make The Most Out Of NAMM and Networking Events." *The Modern Music Creator Podcast*, ep. 61, 8 Jan. 2020, pod.co/the-modern-music-creator.

Menhorn, Jack. Email interview. 30 June and 26 July 2020.

Montalvo Lucar, Jeanne. Personal interview. 12 Oct. 2021.

Moore Kloss, Avery. Personal interview. 21 Oct. 2021.

Morton, Will. "Make Some Noise Getting a Job Creating Sound and Music for Videogames." *Gamasutra*, 8 April 2015. www.gamasutra.com/blogs/WillMorton/20150408/234948/Make_Some_Noise_Getting_a_Job_Creating_Sound_and_Music_for_Videogames.php. Accessed 8 Aug. 2021.

Ojeda, Sal. Personal interview. 27 Sept. 2020.

Quinn, Kristen. Personal interview. 30 Sept. 2020.

Sabolchick Pettinato, Michelle. "Michelle Sabolchick Pettinato." *Get On Tour: A Sound Engineer's Guide*, edited by Nathan Lively, self-published, 2018. pp. 56–57.

Stacey, Sarah. Personal interview. 19 June 2021.

Stoll, Becca. Personal interview. 27 April 2020.

Thakkar, Akash. "Successful Freelancing in Game Audio." *YouTube*, uploaded by GDC. 9 Jan. 2019. youtu.be/93ggs7hwJeU.

Tipp, Nick. Personal interview. 22 Sept. 2020.

Tucker, Ryan. Personal interview. 27 Dec. 2021.

Urban, Karol. Personal interview. 17 July 2020.

Vericolli, Catherine. Personal interview. *What Makes You Stand Out*, SoundGirls, 20 May 2020, youtu.be/wYamiOK8Y6A. Accessed 21 May 2020.

Wood, Catharine. "The Importance of Branding and Social Media Webinar." *YouTube*, uploaded by SoundGirls, 2 July 2020. youtu.be/I-KfKVaXtFc.

Wood, Catharine. "Re: Using Quote in Textbook." Received by the author. 10 Aug. 2021.

Yu, Fei. Personal interview. 9 Nov. 2020.

Zúmer, Javier. Personal interview. 24 April 2021.

5
Business Relationships

Everything I've ever done is relationship-based. In fact, a friend of mine who's a tour manager referred me to another artist for a gig recently, and when their management called me, they asked if I had a resume. I was like, what? In the last six years, I honestly have never had to send out a resume for anything, really. It's all relationships. But obviously, you have to know how to do your job.
– Jason Reynolds

Networking is the process of meeting people, but business relationships take these interactions and turn it into more. Business relationships take time to build, and an established relationship could last for decades. Business relationships can (and do) lead to work – but not necessarily by asking for work. Live sound engineer/tour manager Andrea Espinoza explains:

I don't think in my career I've ever been like, "Hey, can I have a job?" to someone. Or, "Let's go out to dinner," and then at the end of the dinner, "Can I have a job?" . . . From my relationships with people and from me talking about what I'm doing, people have been like, "I recommended you for this," or "I put your name in for that."

Business relationships can be formed and developed in-person or online and are genuine relationships. Karol Urban (re-recording mixer/sound editor) advises:

Make contacts with people who are interesting to you, but seek to make sincere connections with people regardless of where they are in the industry. There is nothing wrong with seeking out a person you find fascinating, but actual friendships should be organic and sincere.

Meegan Holmes adds:

You don't know who is going to be able to hire you at some point . . . so working on your relationships and cultivating your relationships in everything you do – and on every job you do – is really important. Especially in the beginning because you're laying the groundwork for your future.

DOI: 10.4324/9781003050346-6

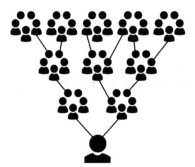

Figure 5.1 Business relationships can evolve into a team of people helping you look for work.

Developing Relationships with Watering Holes

A "**watering hole**" is a slang term for a gathering place (like a bar or pub) where people come together to socialize. John McLucas uses the term to explain how business relationships can form in a wide variety of environments, both in-person and online. "A watering hole is just a place where people can come together if they want to. That's literally what Facebook is, your Instagram page, your TikTok page," says John (McLucas, "interview").

The watering holes that will work best for you (in business) depend on the work you are seeking, where you live, how you like to socialize, and how much you like to be online. Initially, going to a watering hole is like a networking event where you are looking to meet people. If you continue to attend and participate, you will likely encounter the same people and start to build relationships. Over time, this can lead to friendships, introductions to other professionals, work leads, and possibly being recommended for work.

Kelly Kramarik participates with the Denver Audio Meetup group, which hosts in-person events for audio professionals in Denver, Colorado (USA), and also has a Facebook group. Kelly shares:

> When I was just getting into the industry, it was the best tool ever. Once a month, they would have meetups at different studios around the Denver/ Boulder area. I'd just go and hang out with engineers and talk about mixes and check out the studios. So, before I knew anything about what I was doing, I made sure that every single studio in Denver knew who I was. Every engineer knew every engineer, and they knew me.

Social media can be an incredibly powerful watering hole if you are an active participant in a community. Post-production studio owner Kate Finan explains:

It's easier than you might imagine to feel like you "know" someone after you've interacted a few times on social media. This is the virtual version of hanging out in the right room. If you're there long enough and you make your presence known (in a non-annoying way), you will inherently be seen as belonging. I would caution you to start small with commenting and liking, then move on to posting when you're more confident. Definitely don't just jump in and post every day. That's like crashing a wedding and then hopping on stage to give a speech.

John McLucas adds:

A social platform should always have that watering hole essence in it. It should always have that community aspect. When I post content on Instagram, I could just do that to build my reputation. But for me, it's more of a watering hole because I have a vested interest in interacting with the people who interact on my content. I also take time to be interested in other people and actually like the people that are in this circle. So, it's not necessarily just "look at me" flex for the sake of building reputations. Like, I actually am excited about Gabe's song coming out even though I didn't produce it . . . I'm building the watering hole by proxy of how I behave on the platform. If you have three people commenting, you should go back and care about them deeply because they're giving to you.

("Interview")

Megan Frazier suggests when participating in social media watering holes:

Ask questions that are open-ended, that everyone can contribute opinions about, and that there is no right or wrong answer. That is valuable because people do want to connect and talk. So, asking a question that a high-level person and a low-level person can both talk about – and still have valid points and have differing opinions – is actually valued and gives back to the community.

There are also significant numbers of what I call "private gardens" where people converse about audio in a professional context, but you may or may not hear about them . . . someone has to invite you. It's not a secret, but it definitely is an industry secret where only the people who are in the circle invite the new people.

Watering holes are also about joining a community (and being known by that community), so quality is preferable to quantity. Podcast editor Tom Kelly engages regularly in two social media groups to build relationships and also to offer his expertise. Tom explains:

I started very slowly just being absolutely everywhere. No matter where you scroll on these two forums, I wanted my name to be seen on every, every page length. So, when I messaged people, they've already seen me a bunch.

Developing Relationships: John McLucas

Music producer and content creator John McLucas is regularly contacted by people who are seeking his help. Some ask for work and others are trying to build a relationship.

> When people are curious about stuff in my world, it's such a different approach because it's disarming. But when you come in and say, "[Hire] me," it's like, why am I giving you my time? Why am I giving you my money? You have to show me your value . . .
>
> ("Interview")

> I'm more flexible with people who have been following my content or if they've been really active in my Facebook group, and then make an ask. I jumped on a call once to help a guy [who] was in my Instagram stories for months interacting on my content. I gave him an hour of my time with no ask in return. But if you just came out of the blue like, "Hey, John. I like your YouTube content. Can you listen to my song and give me a bunch of feedback?" No thanks.

> If you're being a proactive and helpful member of your community, in your niche, in your local scene, in your industry, you start to build that reciprocity for interacting with others. They're going to want to be more invested in what you're doing. [Online] this can be anything from posting your own statuses or photos, going live on [social], responding to comments, and interacting on other people's content. It also means interacting with other people's stuff, not just your own, because that starts to come off as selfish.

> In-person interactions and connections cannot be passed up. If there is one thing you cannot get away with, it's that. If you're not the kind of person that does well with crowds, that's fine, then it's going on a lot of coffee dates, a lot of lunch hangs. [Some] ask people to go on hikes. It's just something to connect with people on that deep one-to-one level and really solidify that relationship beyond just being "the dude that can mix [my] song." It's about creating that depth. That is a time-tested method that's lasted forever in business.

> Everybody's strengths and everybody's weaknesses are completely different. If you find social media disgusting and you want no part of the beast of the internet, that is totally fine. You do not have to do anything at all. But there's value in at least giving it a shot.
>
> ("Interaction")

From The Modern Music Creator Podcast

Developing Relationships in School

Universities and audio programs cannot guarantee job placement, but building a strong relationship as a student can increase the chances of a professor helping you post-graduation. Luisa Pinzon's career in Bogota, Colombia, started thanks to her professors. "When I came out of university, I was assisting – mainly my teachers. I was doing the music production of a TV show, and then I worked a lot at a post-production studio of one of my other teachers," says Luisa.

Burt Price, Chief Engineer at Berklee College of Music (USA) says of the students hired to work in the studios, "Every work study generation, there's one that I do what I can to get them a real job. Like, if I hear [about a job], I connect them, and they get a job and make money."

Over time, relationships that start in school can lead to professional work. Recording engineer James Clemens-Seely says (15 years into his career), "Now, work I'm getting is coming from people that I was classmates with or musicians that I grew up with. We spent more time drinking beer together than we spent working on things."

Tina Morris, studio manager and Berklee College of Music alumni, adds:

> I'm finding people that I went to school with – I'm seeing them being successful. It's kind of a cool thing, like once you all live life together and can help each other out, you could get somebody else a gig or somebody else can get you a gig. Developing your relationships is a really important part of being in this business. Finding people of like mind (even if they're not super successful or super famous) – If you all network, everybody starts to rise to the top together.

Build yourself a community within your profession. It's hard to talk about the vagaries of these industries with family and friends if they're not in it, as well; so, save them for the emotional (and maybe financial) support. It really does help to have other sound professionals to talk to. They may have ideas you haven't thought of before. Build that village and stay engaged with it, and it will help you deal with the issues.

– Shaun Farley

Relationships with Other Professionals

Relationships can be formed with anyone you encounter on the job – sound people and beyond – and can help you on the job and in the future. Patrushkha Mierzwa, who works in production sound for television and film, explains:

How does a person keep working in sound? *We talk!* We talk by text, by phone calls, by meeting for lunch, within Facebook groups, or Zoom panels and chats. We're a *very* welcoming and social group . . .

Knowledge is power: Who will likely know if an actor is delayed? Transportation, hair & makeup, on-set costumes. Who will likely know if the actor is in a bad mood? The second AD [assistant director], [second-second AD], their driver, hair, makeup, costumes, or personal caterer. Knowing these things is the difference between maneuvering around a potentially volatile situation or getting caught in the crosshairs.

(272)

The byproduct of having strong business relationships (and doing your job well) is that you may hear about opportunities that are not shared publicly, or you may be referred by others for work (with or without your knowledge). Boom operator Camille Kennedy was once hired for a project based on the recommendation of a camera operator she worked with prior. "I became friends with the operator [on the job], and he knew I don't mess up and I know what I'm doing. You get these opportunities because if you're pleasant [and capable], people want to work with you again," says Camille.

Luisa Pinzon says of music recording sessions:

It's a lot about just sitting down there with the musicians and talking to them, being like, "I do this and that work, and I work with these people." Then maybe you get their social media profile, then you talk to them online, then maybe at some point you start doing work for them.

Meegan Holmes, Global Manager for Eighth Day Sound (live sound for worldwide events) says of building relationships on the job:

Everybody you work with, people you work around, people you get to know, especially the people that you get along with, keep in touch with them. Tell them what your goals are. Don't be afraid to say to your friends, "I really want to work for this company," because their dad might own the place, or their best friend might be the person that does the hiring or whatever. You never know how you're going to create an opportunity. You just have to keep cultivating your relationships. It will absolutely pay off down the road.

These relationships can evolve naturally when working together. Fei Yu, who works for music for television and film, explains:

With people I really like, I would love to go out with them and talk to them more. I would love to know more about composers [I work with] because I really feel like personality shapes how you're writing music. If I know more about their personality, I know more about their music. So, I like to hear more of their story. If I know how to communicate with them and we're building a really good relationship, we become friends. Then, sometimes I'm willing to share more with them . . . Maybe they want to know me more. If we have some things in common, then we choose to work together.

Business relationships can be crucial for landing work opportunities across the industry. Ariel Gross (Founder of Team Audio, a full-service audio company for video games) says of job applicants:

> There are so many applicants qualified on paper . . . we need to keep costs down and rarely have wiggle room for hiring. Because of all this, we like to either directly solicit people we know and trust, or we reach out to those we trust for guidance . . . When I have a position open up and when I reach out to someone that I trust, you want your name to be on the list of people they suggest.

Meegan Holmes says of hiring:

> I've hired people not even seeing their resume on my friend's reference . . . We are a group of people that are all very close friends. I completely trust all the people that I work around to refer somebody that's super strong. They know what we're looking for and who they want to work with. It's your personality, too. If your personality lends itself to somebody liking you enough that they're willing to vouch for you, there you go.

Relationships are also important for finding information about working in the industry that may not be public, such as rates, potential job openings, or the reputation of a company you might work for. Rachel Cruz (Product Manager, Fender) shares:

> It's really important to really be friendly, be friends with, get along with, and keep in touch with your peers. It's not for smarmy political reasons. It's a tough place to get standard information about what the going rate is for anything, what's normal and not normal for working hours and pay rates – that kind of thing. Make sure your social support system and your social networks in the business are maintained. If you don't, it's going to be very hard to know your worth and to be able to calibrate yourself to what you know is reasonable.

> I'm at a place that is the best in the business for what we do. In a lot of ways, we set the benchmark, but in other ways, I'm always looking around and checking in and seeing what's normal for people and how things are going. It's important – that shop talk is still super, super important whether you're in a very small but highly specialized place, or if you're in a really big place.

> You never know where your next gig is going to come from, so don't burn any bridges. I know people who will leave a company or leave a client in flames. I find it crazy how I'm still running back into people all the time that I worked with 10 plus years ago. So, maintain relationships, reach out, and just stay connected.
>
> – Chris Leonard

Mastering Engineer Piper Payne adds:

> Be good to your neighbor because your network is the most important. I have seen people get shoved out of the industry entirely because they screwed somebody over, or because they were not paying attention to what the going rate of editing [was], or the going rate of a stay in the studio was, or they undercut someone who had helped them out previously, or taking someone's clients.

Touches

Chances are, you will be looking for work at different times from when your clients have work available. This is why it is important to build relationships *before* your clients think, "who should I hire for this?" Alesia Hendley explains, "Find genuine ways to connect because if someone's audio engineer has COVID, they know what you do because you've connected, and you've built a relationship."

Piper Payne adds:

> Eventually, someone will call you and they will say, "Hey, my recording engineer died, and I really need some help." That actually happened to me the other day. Like, the mastering engineer died right in the middle of a project, and they needed me to finish it. Legit happens.

Touches, or **touchpoints**, are a marketing term to describe interactions between a business and its customers. In the audio industry, this term applies to interactions between an individual and potential clients or professionals they would like to work for. The idea is to start building a relationship (even casually) without any expectation of getting anything in return. Karol Urban gives this example of a touch by going to a gym class, which was recommended by a person Karol was trying to build a business relationship with:

> [They] had to go after class, so I said, "It's cool to see you. Thanks for recommending this class." That was it. But, people have a tendency to

let their guard down after they've seen you two or three times. That was another time that person saw me where they got to know a little bit about me. I saw them probably three or four months later at a networking function, but then we had that in common to chat about. So, it was an opener. It did help, but you've got to have patience. People are not going to meet you and be like, "You're so awesome." It takes a development period to get to know people. I highly recommend that you have more than one person you're trying to get to know.

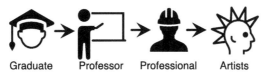

Graduate   Professor   Professional   Artists

Figure 5.2 Touches can help an emerging professional reach (and build relationships with) other professionals.

Touches can eventually lead to work. Megan Frazier explains:

> You usually don't make a sale between the first and third touches. You make a sale after 12 touches . . . These touches aren't something that you're just going to get money out of. You're going to get money in five years, but they gain knowledge of how you've developed over time if you've reached out to them 12 times. They've seen you grow up.

Alesia Hendley landed a full-time job by building touches online and in-person:

> I met a CEO on Twitter and we stayed connected just talking about random things. I drove three hours to a trade show where I knew they were going to be. Six years later, I worked for this company, not by chance but because I continued to build a relationship. I was in the right place at the right time, but they were willing to take a risk on me because the relationship was already there . . . I took the time out to *invest* in going to meet somebody. That goes so much further than people realize . . . [A relationship] might not transpire into something three months or six months down the line, but you never know. It could turn into something that you never imagined, and it'd be exactly what you were looking for.

Avril Martinez says of building touches, "You have to be at the same level . . . if this person is your friend or your friend's friend, that's fine to be direct. That's a two-degree separation. But not someone who's like seven or ten degrees away from you."

Fei Yu adds about trying to reach people:

> So many directors in China (especially if they are famous directors) have their own team. It's hard to break into their team because we're not in the same generation. But it's the same trying to reach an experienced director in Hollywood and you are a younger generation. You will need to work harder to break into that circle . . . If I know someone from their team I might say, "Could you introduce me to them?" Since that person knows me, maybe they will help make the introduction, and it may give us the chance to talk . . . If you were to reach out directly, it's not rude, it's just you don't have access to that. You sometimes don't know how to reach them. For me, that's not the industry. That's a relationship thing.

Getting Started with Touches: Megan Frazier

Megan Frazier is a VR sound designer in Seattle (USA). Megan completed a degree in international business and built a successful career as a banking industry headhunter before pursuing VR sound design as a career. Megan has worked at AltspaceVR (the social VR platform for Microsoft), and as a sound designer for Amazon.

> For someone who's starting out, reach out to people that you don't know in some way. If it's virtual, it doesn't have to be a specific platform. You just need to actually reach the person. So, send a direct email, a direct LinkedIn message, or a direct message (DM) via Twitter . . . In-person, being in a group is good for information gathering, but the actual direct reach out has to be one-on-one. It's not like, "I saw Cassie at that networking thing." That's not the same as "Cassie and I are having a conversation."

> There are three things that you need to do when you're reaching out to someone: Be concise, be genuine, and have some sort of call to action. When you're reaching out, you have to reach out for a purpose. It doesn't mean that your purpose must be "We need to discuss this business item." It means, "I want to check in with you to see how you're doing." . . . It can also be asking them a question.

Megan's Tips

- Don't reach out to a ton of people [at the same time]. Reach out to two people every single week. Sometimes you're going to go on vacation, or you miss your week, and that's okay.
- Try reaching out to people whom it would not crush your long-term dreams if you were to fail. There will be failure – lots of failure.

- Until you've gotten your smaller folk to respond to you – until you can consistently get someone to at least acknowledge you with a grunt, you probably shouldn't be chasing your big wigs.
- Avoid double touching [accidentally making first contact twice] . . . If you are starting out, email is pretty good at tracking (by looking at when you last sent it). If you need to take notes on something specifically interesting about that person (something you won't be able to easily remember), do that. There are [many] solutions and whatever you do, don't pay a lot of money for it.
- The biggest thing that newbies struggle with is not having deprecating language in their reach outs. They'll say, "I'm not that good." It's important to recognize [you are] not inferior by reaching out. [Think], "I am a valuable person and I could be valuable to your project."
- I always make it about the other person. They will be willing to talk about themselves unless they are seriously way too busy.
- If a person has indicated that they are interested in hiring in the next year, reach out to them every six weeks or so to say, "Hey, what's up?"
- If it's not as urgent, reach out to someone every two to six months. After a year it gets weird because they might not remember you.
- If someone is interested in working with you in any way, make sure that you're contacting them a couple of weeks before they said, "reach out to me in September." They'll know by September if they actually have that need now.
- I traditionally try not to ask about hiring too quickly because everyone is. Make the tone: "I am simply in this industry and being present around this person. I just want to get on the shortlist." If you're getting on the shortlist, that's all that you really need.

Two Common Mistakes by Emerging Professionals

1. Failure to reach out and network with people who aren't audio folks. It's actually helpful to have those audio connections, but they are support connections. They're [an important group], and maintaining those relationships is just as important as maintaining business connections.
2. Failure to follow up/not consistently touching the people that you are reaching out to. This mistake is more detrimental. You have to do networking to work in a network-centric industry. But, your networking will ultimately be useless if you aren't following up with the people that you're touching because relationships expire. You have to cultivate relationships just like plants.

Following Up and Maintaining Relationships

Live sound engineer Michael Lawrence explains the importance of maintaining relationships:

> When I get a call for a gig I can't make I'm often asked, "Can you recommend someone?" You're very likely to recommend someone you just dealt with. So, keep yourself touching bases with people. I have friends in the industry that I rarely see because they're off someplace crazy. I'll just send them a note and say, "Hey, what are you up to? What's going on in your life?" Just a couple of sentences to keep in touch. I think that helps both ways when they need somebody or when you need somebody.

Kevin McCoy (Head Audio, *Hamilton* "And Peggy" Company) adds:

> My general rule for a timeline of touching bases: Every six to nine months I like to keep in touch with important people in my network. I suggest a really brief email. I'm just saying, "Hey, how are you? I'm doing well over here in Minnesota. I just wanted to say hello. I've just finished my job as the production manager on *Paw Patrol* and wanted you to keep me in mind for any things you might have coming up. Have a great summer!" It's super quick because those people are always so busy and get a pile of emails. It's just a tiny little ping to remind them that you exist. Any more than that and they can feel like you're sort of obnoxious.

The standard etiquette for following up can vary based on where you live and the type of work you do. Karol Urban shares the differences between Washington, DC and Los Angeles (USA) working in sound for picture:

> What was too much in DC is just right here in LA. At first, I was waiting too long to contact people and I wasn't being as intense as I needed to be in Los Angeles. LA is a gig economy for the most part, whereas in DC, people are often salaried employees.

> In Washington, DC, [I would say], "Can I reach out in another four or five months? If you know of any opportunities, that would be great." Then I would keep my eye open for news for anything that they did that was kind of cool. That's when I would touch base again. I would have something relevant to say about something that they've worked on. I am showing them that I'm paying attention and reminding them of who I am. No further correspondence needed . . .

> In LA, I feel like everybody has either not enough work or too much work, and those are basically the two states of being . . . Which means that people

either need help, or they are so swamped they can't even talk to you right now . . . So, in LA it's a little bit more permissible to touch base a little more often.

Meetings

A business meeting can have many purposes from social/networking, information gathering, or an informal interview. If you apply for a gig or job and then receive an invitation to meet for a meal, it may be an informal interview, where someone is seeking to get to know you before making an offer. A business meeting could take place at a workplace, over a meal or drink, or even at an industry event. If you identify the purpose of a meeting, it can help you better communicate and prepare.

Social/networking meetings are casual meetings to get to know someone, their work, and find how each may be beneficial to the other. These meetings can be more like making friends than work-related, but it is still important to treat it professionally. If you are introverted or seeking to practice your social skills, social/networking meetings can be a great way to practice.

For emerging professionals, **informational meetings** are a great way to meet established professionals and let them get to know you. You will likely be the one asking for the meeting, so it is important to be clear about the purpose of your meeting when you ask ("Would you have time for a call or coffee to give some advice?"). Radio and podcast producer Sarah Stacey explains:

> I think people respond really well if you tell them that you're interested in what they do, and you'd like to find out more about it. People are more likely to offer you opportunities. I remember a lecturer told me once that people are always responsive if you say, "Would you mind if I came in to have a look at what you do?" rather than "Give me a job. Give me an internship."

Aaron Marks, author and owner of On Your Mark Music Productions (which specializes in music and sound for video games) says:

> Talk to any friends or acquaintances who are in the industry, find out how their companies work, who to talk with to get the jobs, and most importantly, how they do what they do. Talk to other game composers, sound designers, and voice talent. These people have plenty of knowledge and are willing to share much of it. Don't expect them to give you names and phone numbers of their clients. They have fought hard themselves for

them, but they will gladly discuss the process and the technicalities. Most are very open and friendly.

(32)

Standard etiquette for a business meeting can vary by location or culture, and entertainment industry norms may be different from those. Ask professionals in your community (who work in the same discipline) about appropriate dress, who customarily pays, if gifts are expected, and acceptable (or off-limit) topics of conversation. Some cultures are more open about personal questions than others. With all meetings, always follow up with a thank you message.

Mentors

Mentorship is a relationship where one person with more experience or expertise counsels or teaches someone with less experience. A **mentor** is someone willing to offer professional help and guidance to another, and a **mentee** is someone learning from a mentor. Mentors and mentorship can be helpful at all stages of a career. Karrie Keyes, who has been monitor engineer for Pearl Jam for over 25 years, says, "I still have mentors today. I couldn't do my job without them."

While "mentor" and "mentee" are terms used to describe the roles in the relationship, often these are not formal arrangements. Eric Ferguson says of a mentor at a recording studio early in his career, "I didn't realize it at the time I had a mentor. I just attached myself to this tech and he taught me all kinds of cool stuff."

In the modern audio industry, informal mentorship has become a key way of building relationships and learning skills. Large studios and major companies are no longer the "gatekeepers" of industry work, and the shift in many disciplines to working from home has made internship opportunities scarce. Aline Bruijns, who works in post-production sound in the Netherlands, explains:

> As much as I would love to help people, I cannot take people into my house, because I only have one room in my studio. I always say to them, "If you have anything – a project, if you're stuck on something, or you need fresh ears or a fresh point of view, or whatever – I'm always available to give feedback or help you out." I cannot offer a full-time internship . . . [Internships] are the toughest thing now to find because in the Netherlands, there aren't a lot of facilities around that are larger that can take on more interns.

The act of giving and receiving mentorship also may not be ongoing. Mentoring could be a brief learning opportunity, such as shadowing someone (watching them do their job) for a day. Or, mentorship could be guidance from a professional through a social media group or a course.

Andrea Espinoza says of mentors:

> We throw out the term of mentors now, but I would just recommend lifting your association. Find someone who's doing what you want to be doing and then ask [questions]. Figure out how they got into it and really see what you can glean from them, in that sense, but also surround yourself with people who are in positions you want to be in.

Ariel Gross, the founder of the Audio Mentoring Project (a game sound mentoring organization), adds:

> Examples of good interactions with potential mentors include asking for advice, seeking their feedback on works in progress, and bouncing ideas off them in a one-on-one setting. These interactions should all feel natural and authentic. You'll have to be observant and respectful of their time, and you'll have to show that you're taking their input to heart.

Mentorship is easier to find when you are seeking information versus a formal arrangement (relationship or program). Emerging professionals can also get off track expecting a mentor to be like a teacher who takes you "under their wing." In reality, a mentor may not be a teacher or a parent-type figure. Karol Urban explains:

> There have been a couple people who were dream mentors that I've gotten to know a little better and I've discovered that we aren't a super great match as friends. I don't think they dislike me, but we are not going to bond. We're never going to get ice cream, and it's okay. They're great. They're still mentors to me. I still watch everything they do and observe what they mix and am a massive fan but personality-wise, we are different animals, and that's okay. We respect each other. There's no negativity or anything.

Turning Relationships into Work

The ultimate goal of building relationships is to support finding work opportunities – whether it's being hired by people in your network, hearing about opportunities, being referred for opportunities, or being introduced to others. Avril Martinez, who has worked in theater and VR, explains, "If you do your job well and people like what you do, they recommend you, and they talk to other people for you."

Sometimes a relationship can turn into work quickly. For Kelly Kramarik, being active in a Facebook group for local professionals helped her land a gig in sound for picture:

> I posted, "I'm trying to get a 5.1 setup at my house. Does anybody have one that could give me some advice?" I met a guy who had a home studio [who] said, "Come check it out. I can meet you and we can go through my stuff." Just from meeting him that one time he ended up hiring me to do dialog on a feature film.

VR Sound Designer Megan Frazier landed a work contract at AltspaceVR which she traces back to a post for a free couch on Slack. Megan picked up the couch and stayed in touch with the couch's former owner (who worked for AltspaceVR). Over a year later, he told Megan a sound design position was available – a listing that wasn't public. Megan says, "They just hired me. I just talked to the person who was in charge of the project. There wasn't any other kind of competition . . . I interviewed and I got hired. So, that is networking."

Game sound designer Ashton Morris says of networking after relocating, "I found sound design work at the second meetup I went to – compared to having found remote work after emailing and applying to 62 different developers/ads. Those are actual numbers."

A **reputation** is the general beliefs others hold about you. If you have a good reputation, it means others have spoken highly of you (for a variety of reasons from your skill to personality).

Jason Reynolds says of his reputation:

> It doesn't matter if I'm doing something for free or doing it for the biggest paycheck – I treat them all the same. That's become my reputation: people know when they call me they're going get a certain level of effort and quality.

Having a good reputation can bring work to you over time. James Clemens-Seely explains:

> A friend asked my wife how I get clients, and she responded, "It seems like they appear out of midair." They sort of do at this point because I've built a network of people who know I'm reliable and who know other people. Even if someone doesn't know me, usually they find out about me because they know someone who they can rely on who says that I can be relied on.

Fei Yu is a music producer, music editor, and music supervisor in Beijing, China, who has worked with major Western composers and artists (including Andrea Bocelli and Hans Zimmer). Fei has been recommended for opportunities because of relationships with filmmakers she made early in her career:

> I already knew the directors, so they were already coming to me . . . because the quality was really good, the box office was really good . . . directors started to introduce me to other directors. Other directors might say, who helped you to do the music for your movie? They [would say] "Fei helped me" . . . So that's how I started building my career.

After 20 years working in theater, Kevin McCoy leveraged his relationships to land a job on *Hamilton* ("And Peggy" Company) in San Francisco (USA). Kevin explains:

> After I saw *Hamilton* on Broadway, I knew that I really wanted to work on it, so I just leveraged my network. I had a friend who was a stage manager on one of the other productions, and I had a friend who was the general manager. I had a bunch of different friends who knew the sound designer (Nevin Steinberg). So, I would call them and be like, "Hey, I'm really interested in at least being considered for a job on *Hamilton*. Will you please help me get in touch with Nevin?"

> Eventually, enough people talked to Nevin for him to be like, "maybe I should give this Kevin person a call." He called me and we had a really lovely chat. At the end of the call, he said, "I don't have any jobs to offer you right now. I just wanted to get a sense of who you are and talk to you." It was nine months or a year after that chat before there was even a sense that I was definitely going to work for him. Even beyond that, I had to wait a few more months for the actual job to materialize. So, it was quite a process. But it was just all about leveraging the existing network that I had.

Jeanne Montalvo Lucar landed a job at NPR's *Latino USA* in part due to groundwork laid through networking and building relationships:

> I had consulted for *Latino USA* because the engineer wanted to do more complex things in their day-to-day operation and he didn't know how to make that happen. Somebody had recommended me . . . I designed their studio system so they can do more complicated routing systems for their calls and their show.

> I had met the senior producer [for *Latino USA*] through another connection. I had been to a couple Ladio meetings [an organization for women in radio] to network and I'd met two of the producers there. So, several people on the staff already knew me.

Almost a year later, the engineer called me and said, "I'm putting in my resignation. You should submit your resume." I think on paper, it was also really good for them to have a female Dominican American engineer. *Latino USA* is a Latin American-based show and there are a lot of first gens there. It was also founded by a woman who's Mexican American. I think there was a strong desire and pull for them to want to make that work. Not to say that I wasn't the best candidate for the job, but I definitely think all of that helped . . . It all came together at that point and then I ended up getting a job there.

Reputations and Work: Claudia Engelhart

Claudia Engelhart is a live-sound engineer and tour manager who has worked with jazz guitarist Bill Frisell for over 30 years. Their relationship started because of her live sound work at the Knitting Factory in New York City (USA). Claudia tells the story, "Around the Knitting Factory, I was establishing a reputation just because I was around all the time mixing all these different bands. Every night was a different band. I never even knew who they were."

One night, Claudia mixed a show for John Zorn and Naked City because the sound person was sick. Claudia recalls:

> After that, I started working with Zorn. Within Naked City, every-body in that band had their own band and [they] started asking me to go on tour with them. Suddenly, I had five different bands I was touring with. Bill happened to be one of them.

Claudia says of building a reputation, "You're not going to [get in with a top industry artist] from day one, because who's going to hire you if they don't know what you're good at or what you do? You definitely have to establish your reputation."

References

Bruijns, Aline. Personal interview. 21 June 2021.

Clemens-Seely, James. Personal interview. 20 Dec. 2020.

Cruz, Rachel. Personal interview. 9 Oct. 2020.

Engelhart, Claudia. Personal interview. 8 June 2020.

Espinoza, Andrea. Personal interview. 28 May 2020.

Farley, Shaun. Email interview. Conducted by April Tucker, 15 July 2020.

Ferguson, Eric. Personal interview. 2 Dec. 2020.

Finan, Kate. "Creative Self-Marketing Ideas for the Audio Professional." *SoundGirls*. soundgirls.org/creative-self-marketing-ideas-for-the-audio-professional/. Accessed 11 July 2021.

Frazier, Megan. Personal interview. 11 and 21 Aug. 2020.

Gross, Ariel. "A Big Jumbled Blog About Joining Team Audio." *Ariel Gross*, 26 June 2012, arielgross.com/2012/06/26/a-big-jumbled-blog-about-joining-team-audio/. Accessed 8 Aug. 2021.

Hendley, Alesia. Personal interview. 17 Dec. 2020.

Holmes, Meegan. Personal interview. *What Makes You Stand Out*, SoundGirls, 20 May 2020, youtu.be/wYamiOK8Y6A. Accessed 21 May 2020.

Kelly, Tom. Personal interview. 22 June 2020.

Kennedy, Camille. Personal interview. 21 Feb. 2021.

Keyes, Karrie. "Career Paths in Live Sound." *YouTube*, uploaded by SoundGirls, 1 Oct. 2019, youtu.be/dbBxX9PddhI.

Kramarik, Kelly. Personal interview. 18 June 2020.

Lawrence, Michael. "Getting an Audio Education, Landing the First Gig, and More." *Signal to Noise Podcast*, ep. 12, ProSoundWeb, 6 Sept. 2019, www.prosoundweb. com/category/podcasts/signal-to-noise.

Leonard, Chris. "New Co-Host Chris Leonard & Early Career Advice." *Signal to Noise Podcast*, ep. 16, ProSoundWeb, 9 Dec. 2019, www.prosoundweb.com/category/ podcasts/signal-to-noise.

Lucar, Jeanne Montalvo. Personal interview. 12 Oct. 2021.

Marks, Aaron. *Aaron Marks' Complete Guide to Game Audio: For Composers, Sound Designers, Musicians, and Game Developers*. 3rd ed., A K Peters/CRC Press, 2017, p. 32.

Martinez, Avril. Personal interview. 28 Aug. 2020.

McCoy, Kevin. Personal interview. 22 Dec. 2020.

McLucas, John. Personal interview. 25 Jan. and 8 Feb. 2021.

McLucas, John. "3 Interaction & Connection Tactics That Doubled My Business in 2018." *The Modern Music Creator Podcast*, ep. 19, 19 Nov. 2018, pod.co/ the-modern-music-creator.

Mierzwa, Patrushkha. *Behind the Sound Cart: A Veteran's Guide to Sound on the Set*, Ulano Sound Services, Inc., 2021. p. 272.

Morris, Ashton. "Freelance Game Audio: Getting Started and Finding Work." *Ashton Morris*, www.ashtonmorris.com/freelance-game-audio-finding-work/. Accessed 1 Aug. 2021.

Morris, Tina. Personal interview. *What Makes You Stand Out*, SoundGirls, 20 May 2020, youtu.be/wYamiOK8Y6A. Accessed 21 May 2020.

Payne, Piper. Personal interview. 29 May 2020.

Pinzon, Luisa. Personal interview. 10 Nov. 2020.

Price, Burt. Interview with April Tucker and Ryan Tucker. 5 Oct. 2020.

Reynolds, Jason. Personal interview. 27 July 2020.

Stacey, Sarah. Personal interview. 19 June 2021.

Urban, Karol. Personal interview. 17 July 2020.

Yu, Fei. Personal interview. 9 Nov. 2020.

6
Preparing to Seek Out Opportunities

There are six reasons to consider a work opportunity at any phase of your career (in no particular order):

1. Money
2. Learning or improving a skill
3. Relationship/connection
4. Creative or personal fulfillment
5. Professional recognition/career advancement
6. Personal priorities

If an opportunity fulfills all six areas, it is a dream job. In reality, most gigs and jobs only fulfill a few at most. For example, the pay might be lower than you would like, and the content or topic may not be of interest to you. However, that could be outweighed by the chance to improve skills, build a relationship, and earn a credit (or something on your resume). If a gig only fulfills one or two of the six reasons, consider it carefully. If a project is low pay, has little to learn, has no relationship to gain from it, *and* it's not interesting to you, is it worth doing for the potential career advancement?

Some of these reasons may carry more weight than others. Some people will welcome unpaid opportunities, while others are strongly against working for free. A classical music engineer may love their work even if they never listen to classical music or attend concerts outside the job. A touring live sound engineer may find it more difficult to work with an artist for months without liking their music. Claudia Engelhart explains:

> The most important [part for me] is finding music that I like and that I can relate to. You've got to get something back from it and it's got to be with people or music that you love, otherwise why schlep around and work [long] days.

DOI: 10.4324/9781003050346-7

The reasons to accept a work opportunity can vary from gig to gig, as well. Mehrnaz Mohabati says of evaluating different gigs in post-production sound:

> If I'm working with a [re-recording] mixer, the personality of that mixer is important for me. If I can learn from them, if we can collaborate and have fun, those are important for me. If I'm working with a client directly from my own company, the first thing is the project . . . Even if somebody says, "I'll pay you a lot [to work on] my feature," I'll say, "Okay, let me watch it first." Working on features you don't like is the most painful thing.

Some entry-level opportunities will be low pay and involve menial tasks but can be beneficial for relationships and career advancement. Live sound engineer Kyle Chirnside explains:

> There can be a lot more things that happen at that show besides what you got paid. By being someone who pushed a cabinet [on stage] you just might be in the right place at the right time . . . "Do you want to come back and mix our next show?" Or, "Hey, can you mix the first band because I've got to go to the bus?" Or, "Our Front of House guy is sick. Do you want to step in for a bit?" I've got to mix some pretty sick shows just from showing up. If you just show up, half the battle is won.

The six reasons to accept a work opportunity combine like pieces of a pie. If you are feeling unfulfilled in your career or the type of work you're doing, the solution may not be to change career paths. It could be that you are missing out on some areas or too focused on others.

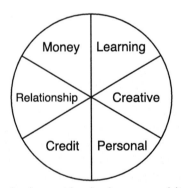

Figure 6.1 All six reasons need to be considered to have career fulfillment.

Money

Income, or the money you earn through your work, is one of the biggest challenges for emerging professionals. In addition to the need to earn income (for most), expenses need to be lower than income to not go into debt. "There are just exponentially easier ways to make money than game audio," says Megan Frazier, who began her career in business before pivoting into VR audio. "I'm not saying don't follow your dreams, but if you're in it for the money, go do something else."

Two common routes for emerging professionals where income is a challenge:

1. Pursue audio work (which may be low or no pay) and supplement with another source of income ("side hustle" and/or part-time job); or
2. Have a job (or another source of income outside of audio) as your primary income and transition into audio work when you can afford to switch.

Which route you take can be a matter of personal priorities. Music producer John McLucas says:

> How risk averse are you? Do you want the pressure of barely evading credit card debt, missing payments, and going down that rabbit hole? Are you okay with that stress? If you can go somewhere and make US$60,000 in six months – what would it be like to have a day without debt? Does that change how you create? . . . All of these [paths] work, but how much shit are you willing to eat?

Emerging professionals have done it all: working in a coffee shop, restaurant, supermarket, electronics store, bartending, and even selling dresses or doing solar panel installation. Video game audio lead Jack Menhorn says, "I worked at pizza places and produce departments before 'breaking in' to enough free-lance work in any way to sustain myself . . . I did not get a full-time job in game audio until I was 30 years old."

Video game composer and sound designer Akash Thakkar adds:

> I used to clean toilets when I first started. I was a public pool janitor. I don't recommend that as your side job. It was the worst. It's okay to have a side job. It's good to have actually some sort of financial stability while you're studying and while you're working on your craft . . . I do advocate for most people that have some sort of financial stability, because it can be a lot harder to so much as write a note when you're worried about where the next meal is coming from.

Jobs and gigs can strategically help your career even if they are not audio-related. Music engineer James Clemens-Seely says:

> The part-time job I had through my undergrad and master's degrees was assistant stage manager of a concert hall. It meant that I was in a concert hall all the time. I'm working concerts, I'm backstage helping musicians

who are having emotional or mechanical crises, and they're [thinking], "that guy was competent and kind." It led to working with a lot of musicians . . . I've always tried to find myself in situations where I can overlap with people that maybe I would have a reason to work with in the future.

Whitney Olpin (monitor engineer, Fitz and The Tantrums, Marian Hill) worked as a bartender as an emerging professional and quit once she was booking more sound gigs:

> I bartended for a long time . . . The reason it worked is they were flexible with my schedule enough to where I was like, "Hey, I can work Tuesday, Thursday, and Friday night," or whatever. I was open the rest of the week so I could take these little [shows] and mix it all together. So, if the shows dried up, I could still count on those bartending shifts.

> I was beginning to get more sound gigs and just couldn't handle the scheduling nightmare anymore. Plus, the more audio I did, the more I realized I had outgrown bartending. I didn't realize it then, but all the years I spent behind the bar would actually make me a better engineer. It taught me fundamental skills like multitasking, maintaining a sense of urgency, speed, and communication . . . I think it's really important to get your foot in the door with something, [even] stagehand, just so you can let the fire burn.
>
> (Keyes et al.)

Camille Kennedy found her way into production sound by taking opportunities in other departments on set:

> Even when I was working in props, if I had a puzzle box or something, I knew to take the pieces out so it doesn't make sounds when you move. Or if you have a beer bottle, you foam the bottom so when it gets put down, you're not going to knock it over dialog. Those are all little things you learn in other departments. So, when you get into sound, you can take all those little things and be able to tell the departments what you need, or you're able to just fix it yourself.

Live sound engineer Kyle Chirnside also suggests taking the opportunities available to you:

> There's a point in your career when you just can't pick and choose. You really have to take everything you get [offered]. If it is just pushing cases, you're still around it for the day. Even when I was out on tour with bands, I'd come home and if I got called for a stagehand [gig], I would still go because I was going to run into their front of house mixer, their monitor

mixer, [people] that work on stage. They were going to see my face, maybe know my name, and that all plays a part, as well. It's not always that you're going to be the front of house mixer . . . So, getting a job at the local church, at the local venue, getting a job with the local production company – it's all opportunities to reach that next stage, which is gigs all the time and perpetuating into an actual career.

Audio work outside your discipline can be pursued as a side job or side hustle, but what one person sees only as income could be a career aspiration for someone else. Karol Urban explains:

If it's falling in your lap and you've got to pay the rent, I get it. But honestly, you want to get out of any job like that as soon as you can. That's somebody's dream job. You need to find yours.

Guitar tech Claire Murphy saved money working stable jobs so she could take time off to work on tours without worrying about pay. Claire says of financial planning for tours:

You need to save up money to be able to take 6–12 months off . . . Once you make yourself available and you don't need to work for money, it opens a whole world of doors . . . A nine to five job isn't necessarily bad because it's going to put some money in your pocket and set you up. But, it is going to be soul destroying. You just have to remember to have the faith that if you keep going, it will happen. But it's not the worst thing because it's paying bills.

(Keyes et al.)

Transitioning into audio full-time (especially as a freelancer) takes time and patience. Podcast consultant and editor Britany Felix explains:

People just let it get to the point where they're burnt out and they're fed up. They have reached the end and they cannot take it for another second more. So, they make this big impulsive leap [into freelance work], which I did years before with another business, and wound up in the worst financial time period of my entire life. We start taking things out of desperation, and that's what I never wanted to do.

That's what I see so many new editors do – they decide they're going to jump into this headfirst. They just want that first client. They want to get the ball rolling. So, they take literally anything that comes along. They give discounts and it never ever, ever works out well. You do so much more damage by doing that than just taking your time and finding the right clients.

Side Hustles Outside of Audio: Phebean AdeDamola Oluwagbemi

Phebean AdeDamola Oluwagbemi is an audio engineer, production manager, and co-founder of Audio Girl Africa. She is based in Nigeria, where the lack of professional entertainment industry opportunities means most emerging professionals have to get creative with income.

> If you look at the larger population of young Africans, they are entrepreneurial people. What a lot of people do is they find other skills – something that you can do on the side that gives you the flexibility to focus on building your skills. There are over 1,000 legit ways to make money only if you're careful to look inward and just say, "What are the other skills that I can do?" Then develop the skill set.
>
> A lot of people don't know how to create designs for their business. I know how to do designs, so I can say, "I'll create a design for you at this amount" . . . a lot of people here do drop shipping, like buying and selling. There are a lot of churches in Nigeria, but not all of them can actually afford high-end equipment. You can come up with budget-friendly equipment packages for them, consult and get the equipment for them, and then make your money from it. Those are the ways a lot of audio engineers are doing it.
>
> Challenges either build you or you chicken out. So, when you face a problem, either you back out, or you find a solution to it. Find other ways to generate quick money that can help you.

Learning or Improving a Skill

Some opportunities have the benefit of skill-building, such as hands-on time with equipment, or observing other professionals on the job. Paula Fairfield (Emmy Award-winning sound designer for *Game of Thrones*) built her sound design skills partly on a film genre she does not watch. Paula explains:

> If you're a sound designer, horror is a place where you'll cut your chops. I don't like horror movies (I don't like violence). But, I have done more horror than I can shake a stick at because it's where you develop your ideas for creatures and for expressing sonically certain kinds of ideas.

As an established professional, Paula is still open to work of all budgets for skill-building or learning:

Maybe it's not the project of my dreams. But, you've got to find something that is for you, especially if you're working for free . . . I've done a lot of pro bono work over the years for emerging filmmakers. It's just something that I have done. But even in that, I try to find something that I will get out of it. Either I'm going to try a new technique, or I'm going to try this piece of software I've never tried before and see how that works. The person I'm working with will get what they need, and I get something out of it, as well. So, I'm always expanding my kit.

Film sound designer Shaun Farley adds:

One of the best things you can do is to always be working. Don't have a client to do work for? Do some work anyways. Make a list of areas where you're uncomfortable, or techniques (whether they be technical, aesthetic, storytelling, etc.) that you haven't tried before, and create a project for yourself. Finish it. Post it. Did you like how it came out? Awesome. No? Also awesome. Figure out why, and move on to the next thing on your list. Maybe one day you can revisit it. The more projects you get done, the more work you can point to for potential clients. The more projects you challenge yourself with, the more you'll grow.

Relationship/Connection

Some opportunities are worth taking to begin forming a relationship. Clare Hibberd says of working in sound for theater:

You could spend one week with one of the local sound designers who introduces you to a whole new concept of what sound design is and a whole load of new people. That could be the beginnings of you. If you are on the right show and you meet the right people, then suddenly every door is open because they will start offering you stuff . . . I've never had a job in theater through an ad. It's always been through word of mouth.

Tommy Edwards, who has worked for several well-known audio manufacturers, adds:

Every [major] job I've landed they actively recruited me. That's a real shorthand for the fine art of networking, which I know intimidates a lot of people, but it's really having to be your own advocate and being your own salesperson.

Chris Bushong says of the music industry:

Take any opportunity that presents itself, whether it be your friend's band that wants to record an EP and you're doing it in your bedroom, or mixing a show for them at a local venue. Those are the people that at some point might hit it big and they'll go, "We want our buddy to engineer the record."

Creative or Personal Fulfillment

Finding projects that fulfill you creatively or personally is important, especially in the emerging years when paid work may not always be satisfying. Sometimes a project comes along that you like so much or believe in so strongly that you will happily volunteer your time (or work for a reduced rate) to support it. A hobby project (or a passion project) is one done for enjoyment and not with the goal of getting paid. Even established professionals take on these projects.

Jason Reynolds' experience touring internationally with artists Stephen Marley and Shaggy can be traced back to the free gigs he did for another Jamaican artist, Papa San. Jason shares:

> Every time Papa San would come to Canada, I would go and run front of house for him. I knew a few members of the band, and they would call me and say, "We need someone to mix for free." I don't remember ever getting paid, to be honest. But for me, to be able to mix for Papa San as a young guy was just an honor. He was a big part of my upbringing, so I never really thought about money or anything like that.

Eliza Zolnai has worked in production sound for major Hollywood films and lower budget independent documentaries. She says of documentary work:

> It's that feeling that you're actually doing something that might matter – that's quite a nice feeling . . . I just love the intimacy of it. It's just a few of us. You can get a lot more involved – to be part of the story . . . In documentary, it's quite important what you're recording . . . If you don't record something on the spot, then that can be a huge problem. Maybe that whole emotional scene can't be used. That's a big difference.

> We shot *Terminator* 6 a few years ago. We were around a water tank for a month at night. There was not much sound there [and] obviously you're trying to do your best . . . but even if you recorded something, you knew for sure that they're going to throw the whole thing out at the end of the journey . . . Don't get me wrong – I love those jobs because I love to work on set, I love big crews, and being part of big technical things . . . When you're doing that for quite a long time on a studio movie like that, then I always feel that I need some inspiration again. That's when it's really good to do some documentaries.

Professional Recognition/Career Advancement

Some opportunities can be a stepping stone to better opportunities. For example, having some touring experience can open up doors to future tour work, or having one sound credit on a feature film can open up doors to more feature film work. Sometimes it is necessary to compromise on pay for opportunities that will advance a career. However, this should be considered with caution. While some opportunities are genuinely low/no pay, "exposure" can be a tactic to take advantage of job seekers. Phebean AdeDamola Oluwagbemi explains:

> Someone tells you to render a service and they'll pay you with exposure. How is exposure going to put food on the table? How is it going to help you solve bills and pay rent and all? A lot of creatives are [faced] with this problem of having to work for a certain number of years and not getting paid for it. I still go through it. It's still a challenge.

Music studio manager Tina Morris adds:

> I have runners that have met people [at the studio] and then they've worked for free because of the opportunity, and then that takes away from the community . . . I've called these clients and [said], "No, you are paying this engineer to engineer." I've had that a lot where people just kind of think . . . "I'm going to take advantage of the fact that they don't have any credits. They're worth enough to ask to work for me but not enough to pay them." I do not support that. I think that's total BS . . . Don't ever do that for free.

Evaluating Opportunities: Jeff Gross

Jeff Gross has spent 30 years working in the audio industry in Los Angeles. Jeff is a music producer, engineer, writer, and also a Foley mixer for television shows, films, and video games. Jeff gives examples of opportunities he took (or turned down) – and why.

Relationship and creative fulfillment: "I did a project with an incredible singer where he pulled the project from [another] studio and brought it to me here. At the time, I was working (music wasn't my only income), so I wasn't so concerned [about pay]. I basically said, 'Don't worry about it. We'll figure it out later.' I really just wanted the singer to get a record out there to start selling to be able to have more income. We did the whole record not talking about money, but he did say he wanted to take care of me and pay me when he could. He started writing checks to me after the record came out."

Relationship and career advancement: "I was working at a studio with three days booked [for a project], and all three days [had tech issues] and ate up hours . . . [My team asked me], 'Would you be up for doing a day or two on our own time to make this up for the client? The client is going to bring us more work.' I worked for free for like nine hours. It's a big client, and the team I was working with has a good relationship with the client. That client would be bringing another 150 days of work minimum."

Relationship and improving a skill: "My friend called me during the pandemic (ramping up to the [US] election) and needed audio clean up on a political ad he was producing. I said, 'Send it over.' There was no money in it, but I was more than happy to help out. It's a karmic thing. At some point, it's going to come full circle."

Money and personal priorities: "A studio where I did about 30 movies said, 'We still need to get the work [done], but we can't afford to pay you at this point.' Not wanting to burn a bridge, I responded, 'I love working with you all, but I don't know how that's going to go over with the gas company. They're not going to be too happy if I said I still want to use gas but I can't pay.' They were understanding in the end. I'm still friends with those guys to this day."

Personal Priorities

Personal priorities can influence whether or not to take an opportunity. These include location, work environment, lifestyle, your interests, and if the work fits your personality (more on personal priorities in Chapter 2). Could you tolerate a long commute or being away from home for periods of time? Could you adapt to the lifestyle of a low-paying position in a high cost of living location?

Opportunities will come up that are *not* a fit, which is why it is important to know your personal priorities (both for your career and your personal life). For example, Rachel Cruz says how her personal priorities determined the opportunities to pursue (as an emerging professional):

What I thought (at 18 years old) was that doing engineering and production was for me because it's something I always wanted to do. It ticked all the boxes in terms of creative expression, interest in technology, etc . . . I thought I was going to be working in a recording studio somewhere. I knew live sound wasn't the path for me, and I also knew that I probably didn't want to work in TV or film because the subject matter matters to me. All I cared about was music . . . I wasn't sure I wanted to own a studio, which was the way things were going. There were trends that were happening that I felt were not a good fit for me and the way that I like to work.

Rachel Cruz ultimately built a career in MI (creating products for the music industry), which fit her personal priorities. (More on Rachel's career is in Chapter 18.)

Sacrificing on Lifestyle: John McLucas

John McLucas is a pop music producer and content creator. After graduating from Musicians Institute (MI) in Hollywood, California (USA), John took a unique approach to his lifestyle, which helped him focus more on starting his audio career.

I didn't really have any skills beyond what I learned in music school. I can't sell. So, it was [choosing] either minimum wage jobs that would lock me into a time schedule (which I didn't want to do), or [odd jobs] like, somebody needs help cleaning this weekend on Craigslist . . . I made that decision to go fully into audio because the band [I was touring with] all had wives or families who were supporting them . . . But for myself, that wasn't the case. I could have asked for loans from friends or family, but that's not my style. It's not how my parents raised me.

I had gotten the idea from someone to live in a rehearsal studio . . . He said that'll keep your overhead super low, you get to put more time into music, and build yourself up. So, I thought it was a good idea. It's definitely illegal, like, it's not zoned for people to live in. There was a large community of people who all lived there illegally. It wasn't a great area. I remember my second or third night trying to go to sleep, and some people were having a dispute with a lot of shouting. I was questioning every decision I'd made at that point, but I'd signed the lease . . .

That was my life hack. I didn't see my friends – I disappeared for like 11 months straight. I would get up at 5:30 am, go to the gym, shower, come back, get groceries, work, try to see the sun, then go to sleep. It allowed me to still say "yes" to everything. It allowed me to have the studio (pretty much a desk with some speakers) and have it be presentable to have humans in it. My parents knew, and my mom was horrified. I think they knew that I was going to figure it out. They saw I was hustling . . .

That's when I started doing content. I didn't know what I was doing at all . . . I was just throwing spaghetti at the wall constantly and just trying stupid things or whatever to see what would work. I had all the time in the world. It was great. I had so much time to experiment and try new things. I could be on the phone with every single lead that was interested in even talking to me for half an hour. I was on-call similar to when you are at the studio. Anybody need something? Just give me that mix and I'll work on it. Don't pay me – just let me work on your stuff. That was the approach.

Steps to Getting Your Business in Order

There are many actionable steps to be better prepared for work opportunities (regardless of your pursuit). It is a lot easier to have business logistics sorted out *before* you are offered opportunities (or are busy with work). Your country may also have a legal definition for what is considered a hobby and what is considered a business. Depending where you live, there may be penalties and fees for not filing paperwork before starting to operate a business. There may also be fees for not filing taxes on time (or not filing at all).

Trade Names and Business Entities

A **trade name** is a fictitious name used to operate a business. Some countries have a term for this, such as "Doing Business As" (DBA) in the United States, "operating as" (o/a) in Canada, or "yagō" in Japan. In the UK, this is called a "business name." If you want to operate your business with a trade name (i.e., "Tuckermix"), you may be required to file paperwork and pay a fee. Consider if a name is easy to spell and pronounce and if it could be limiting in any way. Is it a name you would still like in 10 or 20 years?

Check if the website domain name is available (and reasonably priced) and if social media accounts are taken for the business name. Also, do a web search for the name to see if there are any companies with the same name or similar name or who might create confusion with people trying to find you. There can be legal implications when using a business name that is already taken by someone else.

A **business entity** is an organization created to operate a business. "Entity" refers only to the structure of the business (i.e., who owns it, who is liable for its activities). You may be required to choose a business entity when you start your business. There are hundreds of terms used for business entities world-wide (LLC, LLP, GmbH, corporation, partnership, to name a few), so it is necessary to look into the entities specific to where you live. Some business entities are designed for people working on their own. For example, in the United States, a self-employed individual is a "sole proprietor" and in the UK, a "sole trader".

To learn more about business entities, check with an accountant and your government's websites for taxes or business licenses. Your country or region may also have an official site or program to help small businesses. The types of questions to ask:

- What would be the best business structure for the amount of income I expect to earn?
- Is there any work I can or can't take under this business structure? For example, can a corporation (such as a record label or film studio) pay your business as that entity?
- Does my city have any requirements for a business license or certificate?

Setting Up Professional Communication/Contact Methods

A professionally named email address for work purposes sets the tone for professionals who may work with you (addresses such as "awesomeaudiodude" or "JTGames07" are not confidence-building email names). Social media accounts and LinkedIn are other methods people may seek you out to com-municate. Piper Payne adds, "Have a business card with a proper email address that is your first name and your last name at your business dot com, and your cell phone number available."

While clever and unique cards can be a topic of conversation, business cards are ultimately meant to be disposable. When you are getting started, it is not necessary to print a large number of cards. You may decide to change your dis-cipline, target work, or trade name.

For an audio professional, a website can act as a visual resume. A basic website for an audio professional should (at a minimum) include your primary area of interest, where you are located, samples of your work (if needed), and how to contact you. There are free website builders as well as low-cost options that include hosting (such as Squarespace).

Curating Your Online Presence

Any potential client (or future employer) could search for you online. Live sound engineer Beth O'Leary explains:

> Whether you agree with it or not, it is common practice to perform internet searches on people before inviting them for an interview. Plus, in an industry as tight-knit as ours, you'll inevitably see the same names popping up on discussion threads and groups, so we often form impressions of each other before we ever meet.

Mariana Hutten adds:

> If people already know you through social media, they're going to be looking at your work through social media itself. But, if somebody has never heard of you and they're just searching for a service, it's going to be through Google or word of mouth. Even if somebody recommends you word of mouth, that person's probably going to Google you after they've been told about you from somebody else. So, it's really important to see how you're represented when somebody types your name, and to see how far down the list you fall in when somebody searches for your service.

Search for your name online (in search engines and social media sites) to have an idea of what others may see (including photos). Delete (or make private) posts and information about you that could be seen as unfavorable or inappropriate for work. Some social media accounts show up in search engines and can be a simple way to start crafting a web presence. If you create content (such as a blog post, podcast, or YouTube video) these can also appear in searches.

By curating your online presence, employers and potential clients can get a sense of your personality, interests, creativity, professionalism, and your experience before meeting you. Potential employers will also look for signs of false information or anything that differs from your resume. Music studio owner Catherine Vericolli adds:

> Comb all of your social media for anything unprofessional or anything that is not related to what you are doing right now or are trying to do. This is essentially curating your brand and protecting your reputation. However, you don't want it to be stale because people now want to know who you are. They're not really interested in just your business. It's a really fine line to toe.
>
> ("Interview")

Potential clients or employers may be in a different generation than you, may favor different social media platforms, and may have very different social media habits. Some view lack of privacy, oversharing, posting too much/too often, too many selfies, or revealing/unprofessional personal pictures as unprofessional.

Pro Tip: Social Media

I remember being told, "If you're on Twitter, you have to be really professional." I think everybody in the room kind of panicked because they've been on Twitter since they were like 17 and just using it for personal rants or whatever. I think it's okay, but you just need to find the balance. Obviously, the rule that applies across the board is don't say anything that you wouldn't say to someone's face. You have to be mindful, as well, if you're working for a company. Most of them will have social media guidelines now, like things you should or shouldn't say. So, you have to be pretty careful. Use it for promotion as much as possible, and it's okay to have opinions, as long as they're not really offensive or damaging to who you're working for.

– Sarah Stacey

Some professionals will put a demo online. A **demo** is a brief collection of projects to show your work abilities. Some disciplines rely heavily on demos (such as game audio, podcast editing, and music production), while other disciplines will not require a demo. A demo will change over time, and it will be revised as you do more (and better quality) projects.

When sharing the creative work of others (both in-progress and complete), be aware of who owns the copyright and ask permission before posting online. If you are doing a re-design or re-mix (where you did not work on the released product), it should be clearly labeled so no one perceives you as taking credit for something you did not work on. Student projects or personal projects (such as recording and editing your podcast) can be used as part of a demo when you do not have professional projects to use yet.

Accounting Systems

Freelancers are generally expected to send invoices to clients to receive payment. An **invoice** is a document that states how much a client needs to pay for work you have performed. An invoice can be created through a free template (found online), software, or online service (some offer basic services for free). It is better to have an invoicing system in place before you land gigs so you can focus completely on your work.

Early in your career, you may not know how much time it will take you to edit a podcast, write a song, or do sound design for a project. Knowing this (through time tracking) can help you better gauge whether you are being paid fairly for a project. Time tracking is also necessary for billing and invoicing (if you are charging a client by the hour). A time tracking system could be as simple as a document, spreadsheet, or a time tracking app (see Chapter 12 for more on time tracking and invoices).

Resume/CV and Credit List

In the audio industry, a resume or CV (curriculum vitae) can be more of an introduction to other professionals than a means of landing work. Not every job or gig will require one. The length, format (whether you need a CV, resume, or both), and details included can vary from country to country, so it is important to learn what is the norm where you will be seeking work. "If you don't have a lot of [experience], it's okay," says Catherine Vericolli (Holmes et al.). "I would rather have a short resume with honesty and have it be concise than an actual full page full of filler that doesn't matter."

Ariel Gross, founder of Team Audio and the Audio Mentoring Project recommends:

> Always have someone proofread your resume. Ask them to point out any clunky phrasing or spelling errors. Ask them if it flows well and is easy to read. After they're done, tell them to turn away from it and ask them what they remember.

> Create a backup of your resume and then add their advice. You always have your previous version to fall back on, but experiment with changes that people suggest. Move sections around and try rephrasing things. Look at it like an experiment and try to have some fun with it.

For emerging professionals, relevant skills or experience stand out. For a resume for touring work, include if you have a valid passport, if you have any sort of criminal record (that would limit your ability to work or travel), or if you have a license that allows you to drive a truck. Tina Morris says of hiring for music studio runner positions:

> What actually is exciting for us is when we do see service industry jobs [on a resume], like if you worked at Starbucks or at a restaurant. It's not required, but it's [helpful] to know how to be of service, because that's really what we're aiming for. Even if you're one of our first engineers, it's still [about] service.
>
> (Holmes et al.)

Patrushkha Mierzwa adds of production sound:

> Because film can reflect anything in this world, you just don't know which
> skill becomes the deciding factor in choosing you over someone else. List
> any languages you speak, skills you have, sports you participate in (e.g.,
> several situations require booming in water).
>
> (274)

However, some skills (or other interests) should be shared with caution.
Patrushkha explains:

> You're applying for a job in the sound department; if you ultimately want to
> direct or produce, no one cares. In fact, it could hurt you by implying that
> you aren't serious about sound. If you're seeking only a particular kind of
> work, such as television or jobs in your area, then say so; otherwise, decide
> at the time a project comes to you.
>
> (273)

A **credit list** is a compiled list of the projects you have worked on and your
role on each. A credit list could be displayed on your website, or possibly
included with your resume/CV. Some websites maintain professional credits
like allmusic.com or discogs.com (for music) and iMDB.com (for film/televi-
sion). To some potential clients, your credit list may carry more weight than
your resume.

When including credits or past audio projects, be honest about your job title
and contribution. Being dishonest about your past work could cost you a job or
harm your reputation. Kristin Quinn shares:

> I've seen someone apply for a job who took credit for some of the work
> when actually the hiring person had done the work. That shows a lot about
> a person . . . Maybe I would have the opinion of that individual that they
> wouldn't maybe be the best collaborator.

Tips on Writing Your Resume: Michelle Sabolchick Pettinato

Michelle Sabolchick Pettinato has been a touring front of house mixer for
over 30 years. Michelle is the creator/owner of MixingMusicLive.com, and
co-founder of SoundGirls.

- Tailor your resume to the specific job you are seeking. For example: If you
  are applying for a job with a band as FOH (front of house) or monitor
  engineer, prioritize any live mixing experience you have. After that,
  list whatever other experience you have with live sound – any time

spent working for a sound company, on the sound crew at a live venue, as a stagehand working with audio, etc. If you are applying at a studio, prioritize any other studio jobs and experience you've had. If you are proficient in any digital audio workstations or software, include that. If you've got a knack for repairing vintage equipment, that would be an asset to a recording studio that specializes in vintage equipment. Mention it.

- Don't omit experience because it wasn't a paid position. If you volunteered to set up your church's PA system, or helped out at the local high school running sound for their musicals, or worked in audio in college, anything that you've done involving sound can be listed if you are shy on work experience. If you recorded and mixed your friends' band's demo in your home studio, it's experience that can be included on your resume.
- Do not list every single band whose show you've worked as "worked with" if you were not directly employed by the band. Just because you worked the main stage at Coachella doesn't mean you worked with Muse, Queens of the Stone Age, Foster the People, etc. Unless you were actually employed by the band, do not list them on your resume.
- Do not include a head shot or glamour shot selfie [in the United States].
- Do not lie about your experience. This business is very small and word travels fast. You will be found out.

From soundgirls.org

Where to Begin

Live sound engineer Jason Reynolds says where to begin:

> You've got to sit down and spend some time understanding who you are or what you want to do. I talk to so many people who hit me up and say, "I want to get into the music industry." To do what? It's a big industry. They're like, "I don't know, I just want to work in the music industry." Figure out what you want to do first and then go from there . . . There are so many ways to try to figure out what do you want to pursue, but you have to have something to chase. It can't just be a random thing. You have to have some sort of focus.

A **business plan** is a document laying out future goals of a business and how they will be achieved. Traditionally, a business plan involves an analysis of the market and a plan for operations, management, marketing, and finance. A formal and detailed business plan is excessive for most emerging professionals (in the audio industry), but the fundamental ideas are relevant:

- Target what work to seek out
- Have a plan on how to find potential clients and ways to communicate with them
- Identify professionals who can help you
- Have a plan on how to sustain yourself financially as you are getting started
- Identify how you can differentiate from your competition

A **specialization** is a specific area of expertise. Some emerging professionals will have a specialization in mind as a long-term goal (such as mixer or sound designer), but specializing too early can have negative consequences. For an emerging professional in the exploration phase, being open-minded to opportunities can help gain skills that could be beneficial later. Avril Martinez, who has worked in live sound, theater, and VR, says:

> There's so much crossover in audio skills as long as you have that strong foundation of what audio is, how it works, and some basic physics. You can use it in so many different avenues. Game audio uses the same foundation as live audio and theater.

Jessica Paz (Tony Award-winning sound designer for musical theater) advises:

> Take every job you can get . . . I hauled a 250-foot [76m] cable across a football field, pushed boxes, and unloaded trucks. I was an A2, an A1 [mixer], stage manager, designer – I did all of the jobs. It only makes you better able to understand what the other positions do. Now, as a designer, I can go back on that knowledge. I know what it is to mix a show, so I know what my A1 needs. I know what it is to be an A2 and the challenges that they face. I know how to support my team better because I've done all the jobs.

Grammy-nominated audio engineer Jeanne Montalvo Lucar adds:

> I actually think it's better to be a jack of all trades and be *good* at all of those things. Maybe you are better and excel at one of those things over the other ones. But I don't necessarily think you need to be a master of one thing. Being a master of one thing means that you're not going to get all the other jobs that are a little bit tangentially off – a couple of degrees off that one thing. If you're good at many different things, then you can do many different things. I did classical, I now do radio, and I do podcasts.

I can do a lot of different things . . . Eventually, you will find a niche that you're good at and then stick to that.

Music producer/music editor Fei Yu says of exploring different roles in post-production sound:

Maybe try to be a music editor, and then do the dialog editing or sound effects editing. After you try all the different jobs, then you will know what you really want to do. For sound, it's all about balance. It's about the balancing between music and sound effects, about the balance between dialog and music, or the balance of everything. If you're working with all the small details, you will understand more how to balance everything, rather than just concentrating on your own part.

In some niches, it takes a long time to build up clientele (even for established professionals). Mixing/mastering engineer Mariana Hutten explains:

The more specialized you are, the better you're going to be, of course. But, how likely is it that you can only master and make a living? It's very, very, very hard, if not close to impossible right now. I don't know hardly anyone under the age of 45 doing only mastering without doing anything else.

Rachel Cruz's job (as Program Manager at Fender) involves looking for emerging trends and ideas. She says of emerging opportunities in the audio industry:

There are jobs that are going to exist in the future (and even in the next five years) that don't exist now. There are opportunities that are starting to emerge that you couldn't have predicted and that you might not have thought, that's a good fit for me. Be open to that stuff and look for those emerging opportunities. Don't be stuck on the one place that everyone else is aiming for. The truth is, that one place looks one way on the surface but may not be so great. All you can do is work on yourself, work on the things that you know, and be ready to meet whatever shows up for you.

You have to future-proof your career. What doesn't change is the need to be good at your craft, the need to understand how to teach yourself things, and the need to practice. Then, no matter what changes in the industry, you can figure out a path for yourself. But, as long as you're tied to the idea that there's only one path, it's going to be a challenge for you. Be flexible because you also don't want to turn down an opportunity because you don't recognize it as an opportunity because it's not labeled the right label.

Specializing Too Early: Nathan Lively

Nathan Lively is a live sound engineer, educator, consultant, and creator of Sound Design Live.

There are a couple of times I remember when I tried to specialize too quickly in my career. One time I was offered a tour if I would be the front of house engineer *and* the tour manager. I would have liked to have done either by itself, but I said "no" because I thought that doing both would be overwhelming. Since then, I've met several sound engineers who do both and love it. Another time, I went into my first meeting at a new production company, and I said that I would only do sound system tuning. They said, "We really need people who can do anything – not just tuning." I missed two opportunities that might have worked out for me . . .

You are highly unlikely to just walk into a production company and say, "Hello, I'm Nathan Lively, and I only work on sound system design for large-scale outdoor theatrical productions," and to immediately get that highly specific, high-level position. Most of the time, what happens is that you start out as a stagehand or general tech and build up enough experience and credibility to get the more specialized positions that you're after in the first place . . .

It's really valuable to have a wide base of general skills so that if work in your specialty is unavailable at some point, you'll always have something to fall back on. Have a Plan B. If you move to a new city, for example, it might take you a while to build up your career capital again, but you don't go hungry because you can jump back into [other work].

References

AdeDamola Oluwagbemi, Phebean. Personal interview. 10 Jan. 2021.

Bushong, Chris. Personal interview. 27 Sept. 2020.

Clemens-Seely, James. Personal interview. 20 Dec. 2020.

Chirnside, Kyle. "Getting an Audio Education, Landing the First Gig, and More." *Signal to Noise Podcast*, ep. 12, ProSoundWeb, 6 Sept. 2019, www.prosoundweb.com/category/podcasts/signal-to-noise/.

Edwards, Tommy. Personal interview. 29 Oct. 2021.

Engelhart, Claudia. Personal interview. 8 June 2020.

Fairfield, Paula. "Mixing and Sound Design in Surround Sound with Ai-Ling Lee and Paula Fairfield." *YouTube*, uploaded by SoundGirls, 15 June 2020, youtu.be/Pdv2G2gbiOM.

Farley, Shaun. Email interview. Conducted by April Tucker, 15 July 2020.

Felix, Britany. Personal interview. 22 June 2020.

Frazier, Megan. Personal interview. 11 and 21 Aug. 2020.

Gross, Ariel. "A Big Jumbled Blog About Joining Team Audio." *Ariel Gross*, 26 June 2012, arielgross.com/2012/06/26/a-big-jumbled-blog-about-joining-team-audio/. Accessed 8 Aug. 2021.

Gross, Jeff. Interview with April Tucker and Ryan Tucker. 14 Jan. 2021.

Hibberd, Clare. Personal interview. 7 May 2020 and 15 May 2021.

Holmes, Meegan, et al. Personal interview. *What Makes You Stand Out*, SoundGirls, 20 May 2020, youtu.be/wYamiOK8Y6A. Accessed 21 May 2020.

Hutten, Mariana. Personal interview. 19 Nov. 2020.

Kennedy, Camille. Personal interview. 21 Feb. 2021.

Keyes, Karrie, et al. "Career Paths in Live Sound." *YouTube*, uploaded by SoundGirls, 1 Oct. 2019, youtu.be/dbBxX9PddhI.

Lively, Nathan. "Sound Engineer's Path Webinar." *Sound Design Live*, 16 Feb. 2015. www.sounddesignlive.com/sound-engineers-path-webinar/.

Lucar, Jeanne Montalvo. Personal interview. 12 Oct. 2021.

Martinez, Avril. Personal interview. 28 Aug. 2020.

McLucas, John. Personal interview. 25 Jan. and 8 Feb. 2021.

Menhorn, Jack. Email interview. 30 June and 26 July 2020.

Mierzwa, Patrushkha. *Behind the Sound Cart: A Veteran's Guide to Sound on the Set*, Ulano Sound Services, Inc., 2021. p. 274.

Mohabati, Mehrnaz. Personal interview. 29 Jan. 2021.

Morris, Tina. Personal interview. *Once You Have the Gig – What Makes You Stand Out*, SoundGirls, 20 July 2020, youtu.be/QJhdm86kIlM. Accessed 21 July 2020.

O'Leary, Beth. "Networking on Social Networks." *SoundGirls*, soundgirls.org/networking-on-social-networks/.

Payne, Piper. Personal interview. 29 May 2020.

Paz, Jessica. "Ask the Experts – Sound Design for Theatre." *YouTube*, uploaded by SoundGirls, 5 June 2021, youtu.be/rFMLCU1g19k.

Quinn, Kristen. Personal interview. 30 Sept. 2020.

Reynolds, Jason. Personal interview. 27 July 2020.

Sabolchick Pettinato, Michelle. "Tips on Writing Your Resume." *SoundGirls*, soundgirls.org/tips-on-writing-your-resume/.

Stacey, Sarah. Personal interview. 19 June 2021.

Thakkar, Akash. "Successful Freelancing in Game Audio." *YouTube*, uploaded by GDC. 9 Jan. 2019. youtu.be/93ggs7hwJeU.

Urban, Karol. Personal interview. 17 July 2020.

Vericolli, Catherine. Personal interview. 29 May 2020.

Zolnai, Eliza. Personal interview. 27 Sept. 2021.

7
The Search for Work

Getting a job is like a job. [Use opportunities] to learn, but if you really need a job soon, finding a job is a job itself.

– Javier Zúmer

As an emerging professional, you will have some control over your career, such as where you want to live and the type of work you want to pursue. Realistically, the trajectory of your career depends on receiving opportunities that are out of your control. Live sound engineer Kyle Chirnside explains:

I hear a lot, "I want to work for so and so." That's a reach because there are not many of us (who are mixers or system techs) that actually get to pick the artists that we're working for. It's really hard. It's kind of luck of the draw.

"You don't need to start at the top and you're not going to start at the top. Don't beat yourself up when you start at the bottom," says Kate Finan, co-owner of Boom Box Post. Eric Ferguson sums it up more simply: "Don't aim for your dream gig. Aim for just getting a gig."

Two Routes to Opportunities

There are two general routes to opportunities: working for other businesses and finding your own clients. Working for other businesses entails working under their terms, on their projects, and possibly with their clients. It could be a venue, studio, or production hiring you ("Can you work this show/session/shoot on this date?"). Or, it could be a professional hiring you for a task ("Can you tune vocals for me?"). This work can be freelance or employed (full-time or part-time). The other route is finding your own clients, where you are offering a service directly to them (such as producing a band's album, doing sound for a short film, or editing a podcast). See Chapter 8 for more about this route.

DOI: 10.4324/9781003050346-8

Some professionals find work through both paths. Jack Trifiro explains:

> The whole hustle part of it is the key to being able to fill your calendar with gigs that are making money – then keeping up with it to the point that if you start turning gigs down, you don't find yourself with an empty calendar. There's a fine line.

When working for other businesses, it is possible to find some audio jobs and gigs by searching online. Audio manufacturers, schools, corporations, video game companies, and AV companies tend to be public about job openings, including on their websites. Online classified ads (like Craigslist or Fiverr) can have job and gig listings depending on location. However, job boards and online job searching should not be your only means of seeking out opportunities.

Responding to Job Ads

Music producer Nick Tipp says attention to detail is crucial when responding to job ads:

> Every one of my job postings says, "Write a concise email that has no spelling or grammatical errors in it if you wish a response or an opportunity with this posting or internship." Somebody who reads the whole thing and was careful enough – that person is going to follow instructions [on the job]. The sorts of barriers that I create with the entry process will weed out a lot of people.

Educator Eric Ferguson adds:

> A common thing I've heard again and again and again is that no one will hire you if you're not already in town (Nashville, New York, Los Angeles). I've heard that [in music production] and live sound. There are so many people out there that want these gigs . . . You can't cold call Skywalker Sound from Texas and say, "I want a job."

Tina Morris, manager of the Village Studios in Los Angeles, agrees:

> Whenever we do have an opening, it's an immediate opening. It'll be between one and two weeks. I can't hire six months in advance. If you're in school and want to apply for a job, get out of school first, move to Los Angeles and then apply for a job.

While it is possible to be considered with no website or social media accounts, having *some* way for people to learn something about you online can be to your benefit. Theatrical sound mixer Becca Stoll says, "If I don't know you and I'm going to hire you for work, I want to find something when I Google you."

Tips for Responding to Job Ads

- If writing a cover letter, check names for accuracy. "A nice professional email is always the way to go . . . But [show] attention to detail. [Having the wrong studio name] will show me you do not have attention to detail and I will not want to hire you," says Tina Morris.
- Show how you are a fit for the job you are applying for, not what you are aspiring to do.
- Be honest (and direct) about your skillset and experience. This includes *not* giving yourself embellished job titles or claiming a title with too little experience.
- Don't give a rate or salary (especially when the employer is anonymous) unless absolutely necessary. It's better to wait for a call, interview, or meeting to discuss.
- Fake ads are possible. Be cautious about revealing personal information.
- Don't wait for a response (or stop seeking other work) after applying. It's normal to not get a response, especially when there are a large number of applicants.
- Set a limit for how much time you spend every day or week seeking job listings online. This should *not* be your only means of looking for work or trying to land gigs.

Responding to Opportunities for Freelance Work: Kirsty Gillmore

Kirsty Gillmore, originally from New Zealand, is the owner of Sounds Wilde, a sound design and voice directing business in London (UK).

Freelance opportunities aren't often formally advertised. They're passed on by word-of-mouth, email, and posted on social media. A casual approach to hiring may seem to encourage a casual response, but don't be fooled. Even the most laid-back "Hey we're looking for awesome peeps to join us" company will still be looking for a professional response.

Read the job description. When I've posted call-outs for freelancers in the past, it always surprises me how many emails I receive where the applicant either hasn't thoroughly read the application; or, they've forgotten to include any evidence that they have the skills and experience required for the role; or they seem to think that working in any area of sound for a few years is enough to be considered for a job that requires specific expertise.

Responding quickly to a job posting may increase your chances of the hirer reading your application, but it shouldn't be at the expense of the content. Similarly, responding to every freelance job advert, even if you're not qualified in the hope that someone might give you a chance, is not a winning technique. If a company is looking to expand their pool of freelance dialog editors and your background is solely in music production, they're probably not going to be interested. Freelance positions fulfill a professional requirement – the hiring company will want whoever they hire to be able to step in and do the job straight away. Avoid wasted effort on both sides, and make sure you understand what the hirer needs before you apply.

Do your research. It pays to do a bit of research before applying. If the job posting is on a company's website or social media account, it only takes a few clicks to get more information on what kind of work they do and for what they may be looking. If you can't find what you're looking for – ask. I'm always happy to answer people's queries about jobs I've posted, providing the answer isn't already in the job post itself!

All a hirer wants to know on a first quick pass of your application is: do you have the professional skills and experience needed for this job? If they can't see evidence that you could do the job, they're unlikely to follow up.

Application emails (general): From my experience as a hirer, I prefer a friendly, professional tone for application emails – not overly formal, but also not quirky. I don't need you to be creative in a cover letter to help you stand out – your portfolio or CV should do this for you . . . If it takes me more than a couple of minutes to get the information I need, I'll be inclined to delete it and move on.

If you've worked with the hirer or potential hiring company in the past, have met them in person, or have been recommended through a personal contact, this is always worth mentioning. I am more interested in working with people who I know to be reliable professionals, or who come recommended by someone I trust.

Application emails – what to leave out: Unless a job posting asks for it, the following has no place in a freelance application email (all of these come from real application emails that I've received):

You don't have the exact skills and experience, but you still think you'd be great for the job

A freelancer fills a professional need for a company, and they need to trust that you can do the job straight away. Unless expressly stated, you can assume you'll be expected to do the work as soon as it comes in with no training. If you can't provide evidence that you can do the job, then your application is likely to be discarded.

How much you love sound and want to work in the industry

I see this a lot from graduates and people new to the job market. If you're a working or trained sound professional, I'll take it as a given that you enjoy working with sound. You don't need to spell it out in an application email.

How you can only do the job if certain conditions are met, e.g., you can only do certain days per week

Your application letter is not the place to negotiate the day-to-day details of a job (unless specifically requested). If a hiring company decides to take your application to the next stage, you'll have the opportunity to ask questions and discuss requirements on both sides. Applications which include a list of unasked-for stipulations can make you seem inflexible, which isn't a desirable quality in a freelancer.

From soundgirls.org

Cold Contacting About Work

A **cold contact** is reaching out to someone you do not know to ask them about a work opportunity. Some companies receive cold contacts regularly. At the Village Studios (Los Angeles), studio manager Tina Morris receives 5–10 inquiries daily about jobs or resumes. Tina says of the best way to contact her:

You're more than welcome to come in and drop off a resume, but don't expect to actually see me up at the front desk. If you've sent in your resume and want to check if there are any openings, don't do it the next week. Do it once a month. If you're calling to check up on your resume, say that. Don't sneak past reception. That's a big turn-off. Anybody that turns in or mails in a resume, I keep for a year and then I purge. Don't bother sending me demos of your work just because you're going to be getting coffee for the next couple of years.

Always be mindful of who might be reading your inquiry. Radio and podcast producer Sarah Stacey says:

> I've seen people send emails into radio stations looking for work where they will basically beg for someone else's job. It happened to me once. It was like, "I've noticed that you do podcasts on your radio station, and I could be the person to do that for you." I'm literally reading it like, that's my job. So, don't do that.

When calling, Nathan Lively suggests doing some research to know who to contact:

> At Hotel AV companies, the person doing the hiring is called the labor booker or the scheduling manager and there might be a manager above [them] . . . The reason that's important is I'm reaching out to these people, to these companies, and saying, "I'm Nathan Lively, I'm a sound engineer, and I'm looking for work." That sounds a little bit like an amateur. I'm just identifying myself as an outsider. But, when I call and I say something a little bit different, and I say, "I'm looking for the labor booker," then I'm identifying myself as an insider. I know how the game is played. At the circus, it was the director of audio operations. At most theaters, it's the production manager. For a lot of tours, it's the tour manager or artists management.

Studio owner Catherine Vericolli shares how to communicate in a cover letter:

> Good way: clear, concise, professional email. Friendly, proper grammar and spelling, if you can . . . I don't need your entire story or your whole life in an email. I just need concise information about what you're interested in.
>
> Worst way: Asking me questions about recording and contacting me incessantly . . . Don't ask me questions about recording if you're interested in picking up a job . . . It's not an appropriate time to do that.
>
> ("What Makes")

Film sound designer Aline Bruijns suggests writing a personal message:

> If I get something that I can tell is a copy/paste mail, I throw it out. Make it personal. Let people know that you really did your homework that you really like what they do and that you want to learn from them . . . You're halfway in the door ready because people will feel it, people will know it.

The Relationship Approach to Finding Opportunities

If you have ever seen a popular club with a line of people waiting and someone cuts to the front of the line of the line and walks in, how did they do it? Relationships.

"Going in, either you need to know someone, or you need to find someone to trust you. After that, the quality of your work is what keeps you there," says Mehrnaz Mohabati.

As an emerging professional, the goal is to *hear* about opportunities. As a professional, your goal is to be *recommended* for opportunities. As an established professional, you will be hearing about opportunities *and* recommending people for them. This is why it's important to network with people at all career stages. Recording engineer James Clemens-Seely says:

> "Being in front of me right now" [when a candidate is needed] is a big part of getting that job. So much freelancing is being at the top of that stack. The most recent contact a person had is the first person they're going to think to contact again.

Andrew King mixes sports for broadcast television, where a recommendation is key for landing work:

> If you try to look up the top A1 [mixers] in the industry, all you'll find [online] is maybe some interviews with them. They don't have websites or anything like that . . . I'll get random calls sometimes like, "Hey, this other person gave me your name. Are you available for this show?" That's really how this part of the industry works – word of mouth. Reputation is everything. Your name starts floating around the market. If your reputation is bad, you don't work.

A shared experience can open the door with a cold contact. Aline Bruijns shares how she landed one work opportunity in the Netherlands:

> I got into a big trailer company located in Amsterdam because I saw the founder was having a lecture . . . [After attending], I looked him up on LinkedIn and I sent him a message. I said, "I really enjoyed the talk. If you're ever in need of a sound designer, please let me know." [He replied], "Thanks for contacting me. Would you be willing to send me an example?" So, I [put] something together and sent it to him. He said, "Okay, this is cool. Would you like to come over sometime and have a talk?" I got accepted. It's good to be kind and respectful – instead of shouting from the room, "Take me! Take me!"

Chris Bushong has worked in live sound for over 20 years with almost no web presence:

> I don't do social media. How do I get gigs? I've got two email addresses and a phone number. There's probably stuff I've missed out on by not having it, but I'm also of that weird generation that when we started out, we didn't have that. Being in Nashville helps . . . It's really all about who you know.

There are a lot of people that have the connections that don't necessarily have the right skill set but their friends bring them along . . . It's that "scratch my back I'll scratch your back" thing. So, network as much as you humanly can.

Landing a Full-Time Job from a Car Ride: Sal Ojeda

Early in his career, Sal Ojeda worked at multiple recording studios in Los Angeles. Sal landed a full-time job in post-production through a chain of events that started with a car ride after a recording session.

> The guys from the band were just visiting [from Chile] to record and they were going back that week after, so they didn't have a car. They asked me if I could drive them to drop off [borrowed] amps to a friend. This friend was working at a post-production studio in Burbank. I drove the guys and their friend gave us a tour of the studio. He was really nice and joking and very welcoming. At the end, I said, "This is a pretty nice studio. Are you hiring?" I was just joking. He was like, "Yeah, we are. We're super busy. You seem like a cool guy. Just send me your resume, and I'll see what I can do."
>
> I applied, and they called me back the following week to do a test mix. It was Spanish and Portuguese dubbing. They offered me a full-time job. The day before they offered me the post gig, my next task at another studio was repainting the wood outside the studio. The other owner was just looking for stuff for me to do.
>
> Sometimes it doesn't work if you're straight up looking to get something out of people. That's the one time that it worked . . . Get to know people and talk to people just for the fun of talking because you never know where the next thing is going to come from.

One way to land opportunities is to partner (informally) with people who do work that compliments yours. Mastering engineer Piper Payne explains:

> Most people that are just starting or are having a burgeoning career in audio . . . what they can get is a production partner – someone who is in the same career level but maybe a different discipline. They can join forces. My buddy and I did this. He's a producer/mixer, and I'm a mastering engineer. If someone was like, "I need a mixing engineer before I send it to you for mastering," I would suggest him. I would never send him something that wasn't appropriate. We had this spoken pact to attempt to send each other as much work as possible.

Music producer John McLucas also uses production partners for his work:

> For an independent musician, it's a headache and a half to hire [individually]. The smarter producers, the more business-savvy ones, know that by bundling and building relationships you sell one thing to the client. So you're buying into the whole squad instead of shopping five times for five independent contractors. It's just about getting people from A to B. People don't care what the process is. They just want it to sound good.

Bringing work to studios and other professionals can also open up doors to future opportunities. James Clemens-Seely explains:

> Be a revenue source, not a resource drain. Call up the studios and be like, do you have an overnight when you're not using the studio, or a day that's not booked that I can come in and mix a project? I'll pay you. It's a little weird because it's pay to play, but honestly, it's super sensible. You start a relationship with "Here's what I can do for you," rather than "What can you do for me?" You're going to get way more bites from that. If you show up and you say, "Hey, I've got cake. Do you want cake?" They're going to be like, "Oh, awesome. Everyone usually shows up and asks me if I've got any flour to spare, and you show up with cake!"

Using a Referral

A **referral** is a recommendation provided by another professional. Referrals can be used as a way to get a foot in the door for a job opportunity. A referral can occur in several ways:

- A professional sharing your resume on your behalf ("I'll send your resume to my friend who works there.")
- Sharing contact information ("Get in touch with this person who I used to work with.")
- A direct introduction to someone else in-person or by message ("I'd like to introduce you to someone I think can help.")
- Giving approval to share a name or relationship ("Tell them I sent you," or "Mention you know me.")

The idea of sharing a referral could feel egotistical or like "name dropping" (sharing names of people you know for the sake of appearing important). However, a referral is intended to be shared, especially when cold contacting. Tina Morris explains, "When you do get a recommendation, you 100% go up to the top of the list. I make sure even if I don't have openings, I meet that person for future hiring, so I can keep their name around."

Devyn Nicholson, Manager of Audiovisual Services at McGill University, says of receiving resumes directly:

> It wouldn't necessarily put them ahead of the pack. It would get them a read, though, for sure. I would definitely be seeing that CV because it came to me directly. Our recruiters are doing a lot of the first pass filtering.

Referrals can seem friendly or casual but should not be taken lightly. Catherine Vericolli says:

> I lose sleep over it. I will straight up say [to someone I'm referring], "I am putting my name on this, so that's an extra big heavy baggage for you. If you're going to not be on time, or thorough, or respectful, or you're going to push your [demo] to a client or something weird, then that comes back to me because I am recommending you for this gig." So, I don't recommend people to my friends and colleagues lightly, and I assume that my friends and colleagues don't do the same in my direction.
>
> ("What Makes")

Emerging professionals can utilize their business relationships to find a referral. For example, asking, "Do you know anyone at this company? I'm interested in applying for a job there," may help find someone who can refer you. Professionals may refer you for opportunities without you knowing, as well.

Finding Work Through Referrals: Britany Felix

Britany Felix, owner of Podcasting for Coaches, is a self-taught podcast editor and consultant. Britany's business specializes in helping women entrepreneurs, coaches, and consultants create podcasts.

> When I first started [in 2016], it was super easy to get a client from a Facebook post. There would be maybe five people who would comment. If you had the better pitch and you had a website, you stood out and you were almost guaranteed to get that client. Now, there are between 30 and 50 comments. Just relying on those cold leads is not enough anymore. The vast majority of my business (and most other well-established full-time editors) comes from referrals. So, networking is absolutely crucial.
>
> Surprisingly, a lot of my referrals have come from other podcast editors. So, when I talk about networking and building relationships,

it's not just with your ideal clients. It's absolutely with other editors, because we all have our niches. When other editors have somebody who approaches them that's not a good fit, I'm the person they recommend because we've interacted and they know what my niche is. Now I have a warm lead because somebody has recommended me. So, having that kind of relationship with other podcast editors definitely helps the bottom line . . .

One thing I love about the podcast editing community is, for the most part, it is a very active, supportive community. People are constantly asking questions, and there is a ton of valuable information. More experienced professional editors are trying to help guide the newbies. So being extremely, extremely involved in those communities (not just audio engineers – but specifically podcast editors) will be invaluable . . .

If you're only doing one particular type of thing, like if you are only audio editing, you will absolutely get people who are like, "Do you know somebody who can help me create my artwork?" Or, "Do you know somebody who can write the show notes for me?" When you can refer them to the right people, you're now also seen as more of an expert. That can be another form of revenue because you can work out some referral arrangements with other service providers.

Email Etiquette: Nadia Wheaton

Nadia Wheaton is a composer and music production manager based in Seattle (USA).

There is no shame in sending emails asking for job opportunities. It's all about tact, but the fundamental idea of reaching out should not be embarrassing.

General Golden Rules on Emails

- Your email should always be straightforward, concise, to-the-point (especially in the subject line)
- The addressing line needs to be correct with the correct name (if available)
- Your sign-off signature should be appropriate
- No direct links to music samples or reels, unless directly asked
- Proofread for typos, spelling errors, clarity

If you receive a response:

- Your response time to emails should be within 1–3 business days, and *at max* one week
- Anything later than that means suspect/very low priority
- If you're applying for a job, you should be responding within one day

The General Email Structure

Greeting

One sentence summary of the point of the email

Further explanation, if needed

General thanks, follow up information (if applicable)

Sign off

Sample Cold Contact Email

Hi _____.

I hope you've been well! We met several times last year while I was interning for your client _____. I remember your great advice about networking at GDC and have kept it with me since.

My internship with _____ ended last month. If you have any clients that need help, full-time or part-time assistance, I've attached my resume for reference. I specialize in music-related work, but I am also extremely proficient with general admin/social media assistance.

Thank you so much for reading, and I really appreciate you taking the time to do that. Please let me know if you have any questions for me.

All my best,

Nadia

Notes

- Phrasing: A nice, gentle way to ask for an internship: "If you're open to having an intern," [or] "Would you be open to having an intern?" Even a simple "Do you offer mentee programs?"

- A previous boss gave this advice: "I read these emails and one thing I skim for is: What can they offer me? Are they offering me anything? Nothing? Moving on."
- Make sure your thanks is honest and thoughtful.
- Attach a resume right off the bat. Easy, accessible, and no one needs to respond and ask you for it.
- Praise or specific long-form facts about how said person's [work] changed your life is not needed in initial emails . . . Add it if you want, but I'd keep it to around one or two sentences (anything more seems awkward). Don't ramble about how much you love the person's whole discography – it often doesn't get read.
- Never waiver in your confidence, but the tone of your emails should always be clear, thoughtful, and appreciative.
- Keep your biography to the hard facts: location, what type of composer, genres, or projects you've worked on.

Sample Email: Following Up After Meeting

Hi ____!

My name is Nadia Wheaton, and it was great meeting you at the IASIG party at GDC. I'm a freelance composer's assistant, and my most recent video game project was _____.

If you're free to grab a coffee in the coming weeks, I'd love to learn a little more about your career path and would appreciate any insight you may have. I am a big fan of your video game scoring work and went through college watching a lot of your library tutorials! I'm also based in Santa Monica, but I can meet wherever is most convenient for you.

Hope your GDC was great and looking forward to hearing back!

Best,

Nadia

See nadia.audio/random/ for more sample emails.

Interviews and Meetings

If you get a response to a job application or cold contact, it may be to schedule an interview or meeting. A meeting can be a form of interview but possibly less formal. Before a meeting or interview, research the company and people who you will be meeting with (if known) for information about their work, artistic or creative style, culture, and also what their business needs are. Check social media for people you know in common.

Some jobs will have a formal interview, but you may not get the typical questions from job-seeking resources (like, "How would you describe yourself?" or "What is your greatest weakness?"). Instead, interviews tend to focus on work history, your technical skills, and evaluating social skills and temperament.

Matthew Florianz, Lead/Principal Audio Designer at Frontier Developments (Cambridge, UK), gives these interview tips:

> Try to enjoy it! It is a unique opportunity to talk to (fellow) industry professionals so feel free to be curious . . . Prepare questions. It's unsatisfying to talk to someone when they have nothing to add to the conversation – it might come across as "not really interested in the job." . . . If nerves are getting in the way, make this known as you might otherwise come across as uninterested or unengaged. Points aren't deducted for being nervous. We've all been there.

Jeff Dudzick, who worked with dozens of contractor audiobook editors at Audible (USA), suggests:

> The person should really analyze the job posting/description to suss out what the company's "problems" are and then communicate how they can help solve them. The best candidates I interviewed were the ones who could connect parts of the job with things they've done in the past, at least making those things relevant enough, to show their experience.
>
> Candidates (and this is hard but necessary) should plan and rehearse ahead of time answering lots of different questions of the "tell me about a time when _____ happened, and what did you do, and what was the result" variety, even if the result wasn't ideal because you can still say what you learned in the process. The best candidates were prepared and could tell those stories. Storytelling ability (natural or learned), an obvious good quality at Audible, would likely serve someone extremely well in almost any job pursuing situation.

Devyn Nicholson is Manager of Audiovisual Services at McGill University in Montreal (Canada), where he manages a team of ten. Devyn recommends answering questions directly during an interview:

> There are a lot of people that when you ask a question, you end up with several of your other questions being answered, but the one you asked about wasn't . . . It's not just what your answer is – It's *how* you answer. It's coherency in the ideas you're putting down in front of me and how you complete these ideas entirely, A to Z, and stay on track.

Don't [BS] me, because I can see that a mile away, too. So without even really trying, I can throw these questions out that seem nebulous, but they're nebulous on purpose because it's those questions where you see the true colors of an applicant.

A Path to the West End in Small Steps: Pete Reed

When Pete Reed finished university in London, he thought it would be easy to get a job in the West End, one of the premier locations for theater work in the world. Instead, he found himself working full-time at a coffee shop to cover his expenses. He eventually reached his goal of working on West End theater shows, but it took coming into it from a different path.

I would ring stage doors [at West End theaters] and ask for the name of the heads of departments. I would write a short letter to each show's sound HOD [head of department] and hand-deliver my CV (resume). I never had a response. It was really disheartening!

I came to the realization that I was aiming too high and decided to lower my sights. I knew I wanted out of serving mochas, so started looking for anything I could to do with audio. I was applying to all sorts of jobs, but where I started to get interviews and call-backs was from cruise ships and holiday entertainment venues. At an interview, I was asked where I saw myself in five years, and I answered with the truth: working in the West End. This is not the correct answer and needless to say, I did not get the job.

Then, a month later out of the blue I got a phone call offering me a job in Spain. That was the first step to getting work in the industry I wanted to work in. There were further steps, but that was the first. It wasn't a problem that I didn't live in Spain as the company I worked for provided all staff with accommodations. There are a lot of opportunities for that type of work throughout all of Europe, though it is seasonal. I ended up working in Lanzarote, then Turkey, and finally Tenerife while I worked for that company. Everything was provided for me – flights, accommodation, and food. It was a great gig!

My second step was getting a job at an "off the West End" theater. I was working in Kingston, Greater London . . . I was hired as a production technician with an audio focus. This meant I got to work with and operate for a lot of highly respected sound designers. For one show, the sound designer was a West End show's head of department. I ended up mixing two of their shows in that venue.

> Almost five years after I said in a job interview (that I didn't get), "In five years I will be in the West End," I had applied, interviewed, and got offered a job at *Matilda the Musical*. It was the same show that the sound designer I had worked with before was the HOD at.
>
> I must stress it was not easy. I had applied to almost 90 jobs just in the West End. I remember because I wrote them all down. In the end, getting my first West End gig was mainly because they took a chance on me . . . [Getting started], start at small shows, tours, cruises to get some experience under your belt. Get yourself out there and learn as much as you can.

Fake It 'til You Make It

"Fake it 'til you make it" is a saying that implies you should say "yes" to opportunities (even if you are not qualified) then figure out how to do the job. When successful, these opportunities can help your career, but it comes with some risk. Music studio owner Catherine Vericolli says of faking it 'til you make it:

> If you get really lucky, that's cool, but you're going to have to figure shit out really fast . . . It's like the one hit wonder recording engineers who get really lucky with one thing and then you never hear about them again. It's because they don't really know what they're doing. They also won't have the tools or the experience to do what's right to figure it out (even if it's minimal), or to admit that they don't know what they're doing.
>
> ("Interview")

Devyn Nicholson shares what can happen without proper troubleshooting skills:

> Usually what happens when someone fakes it until they make it is they end up standing in front of a 600-person class looking like a total idiot. Not only is it them looking like a total idiot, but it's also the professor who's waiting for it all to work again standing right beside them . . . and they're very angry.

Early in his career, broadcast mixer Andrew King was offered a mixing gig, but turned it down because he had never been on a professional television broadcast truck. Andrew says if he had tried to fake it on the job, "I would have lost my entire career because I didn't know nearly enough at that point . . . If you say you can do it and then they can't go on the air because of you, your career is over. They're not going to let you anywhere near that room again."

Instead, Andrew was upfront about his skill level and asked to shadow instead. The company agreed and later hired him as an A2 (assistant). Over time, he worked his way to the same mixing position he was offered originally.

"Fake it 'til you make it" is different from **imposter syndrome,** which is a self-limiting belief that you are not as skilled as others perceive you to be. Kevin McCoy, who mixes for touring Broadway shows, says of imposter syndrome:

> Don't tell yourself that you're not qualified for a job if you are just on the line of doing it. If you're obviously not qualified for a job, or you don't have a basic understanding of how sound works or signal flow and someone's asking you to be the head of audio on something, then there's a problem. But if you're just not sure that you're qualified for a job, maybe try faking it, but you have to back up the faking it with a desire to learn.

> I didn't think I was qualified to be A1 on *Motown*. I didn't think I was qualified to work on *Hamilton*. Those were definitely both stretches for me, but successful stretches. They worked out in the end. You have to balance between: Am I thinking that I'm faking it because I'm actually not qualified? Or am I thinking that I'm faking it because of imposter syndrome?

Live sound engineer Michael Lawrence suggests asking for help instead of faking it:

> We all want to learn, but we don't want to ruin a show. We don't want to create a potentially unsafe situation. You have to be really careful. So, just saying, "Hey, I'm really interested in learning how to do this. Can you show me once? Can I look over your shoulder when you do this?" . . . Show me how you want it done, and I'll go ahead and take care of the rest of it . . . The worst thing that happens is that you screw it up. Let me tell you something: I have screwed up so many gigs. No one who is successful in this field is going to become successful without screwing up gigs.

Fake It 'til You Make It – As an Apprentice: James Clemens-Seely

At the Banff Centre for Arts and Creativity in Alberta, Canada, James Clemens-Seely manages a team of apprentices (called "practicum") who are given on-the-job opportunities to learn and practice audio skills. James tells the practicum to try taking gigs even if they don't know how to do something – but to always consider the consequences.

> There are some jobs where the consequences, the likelihood of catastrophic failure, is too high. A live broadcast mix is something where you can really screw it up and no amount of burning midnight oil

afterward can save it. Most of the time, the likelihood of catastrophic failure may be fairly high but the negative consequences may not be that bad. In pure music recording, most people don't have a very refined sense of whether it's good or not, so you can get away with it not being super good but it could still be successful.

When you're in that [early] survival phase, it feels like if you say no, it's going to eventually cascade to cut off a whole quadrant of the universe to you. So, you feel like you have to say yes to everything. But, it takes a certain amount of maturity and wisdom (in self-assessment, understanding what the job is, and what the consequences could be) to know that performing poorly could also cut off a quarter of your [future work] universe.

It's hard to make that calculation of what gigs you should say, "I'm not ready for that, but I would love to assist on that." If it's taking you to the next tier of career or professionalism but you're not ready to make that jump, the job might boot you back out hard, which is a risk. So, it isn't always worth doing, but at the same time, it would have to be a special situation for me to recommend not to say yes.

Most of the time, I'm the person in the room who has the highest level of required excellence. Even when I was a student I had a pretty high level of acceptable excellence . . . If there was a gig where I went home afterward and thought, "I really blew it – the bass mic was completely unusable," Often, the [client] didn't really care because it's not about the bass mic. They have other priorities at that point, like, "We're finished here. Pizza? Let's do pizza."

If someone needed [their audio work] to not get screwed up, they should have allocated more money to the project so that they could guarantee a higher level of proficiency. But if you're getting paid low (but not unreasonably), take it. The reason it doesn't pay well is because it can afford to be a bit of a learning experience for everyone involved. If it couldn't afford to be a learning experience, it would pay top dollar. Maybe it would be inappropriate for you to take it because you'd be claiming to sell a service that you can't actually provide.

When you're early on in your career, saying no to an opportunity will be an exception. Generally speaking, you should probe why you would say no, rather than just think, "Oh, no, that's not for me."

Reasons to Turn Down a Gig

The search for work is not only about finding work, but also knowing how to respectfully turn down opportunities. "Say yes to everything" is common advice, but everyone will be approached with some opportunities that are out of line with their career goals and interests. As discussed in Chapter 6, the six reasons to consider taking an opportunity are money, learning or improving a skill, relationship/connection, creative/personal fulfillment, professional recognition/career advancement, and personal priorities. There are other additional reasons to say "no" to an opportunity.

Scheduling conflicts can occur for any freelancer who is working for more than one business or client. Generally, it is appropriate to turn down a work offer when you are already booked. However, there are also circumstances where it is appropriate to back out of work you have already committed to. Game audio lead Jack Menhorn says of backing out of work:

> It's difficult to say what is ok and what isn't. It comes down more to the reason why you did, and if it is for the good of both yourself and the client. I have done this once and it felt terrible. I had overcommitted myself and was worried I would drop the ball and underperform on both projects. After I backed out, I recommended someone else for the gig and then sent the client an apology card and bottle of scotch. It worked out for me in the end because the person I recommended for the gig got it and then reciprocated later on and got me a career-changing job.

Theatrical mixer Kevin McCoy shares a story:

> I had been waiting for a while without a sound job so I took a gig as a production manager. I was interested in doing management stuff and it was a way to try, and it was a favor for a friend . . . I took that gig at much lower pay than I normally get for running sound, but it was just for a few months. So, it felt like not a big deal. A week after I got that job, a Broadway tour called and said, "We need a last-minute sound person for this tour that's out. Are you available?" I said, "No, I literally just took this job." Sometimes that just happens – it's just how it is. It's sort of the nature of the beast . . . If you're working at a small professional theater in Minneapolis and you get called for a Broadway job, or if someone calls you and offers you a job that pays like three times what you're getting, employers just need to understand . . . You can't be known for jumping ship all the time from job to job, but you can't be afraid to do it when it's really demonstrably the right thing to do. There's definitely a balancing act there.

Film sound designer Shaun Farley says of backing out of commitments:

> If a project hasn't been able to hold its schedule and conflicts with other commitments, I think it's ok. If some emergency has come up, I think it's ok then as well. I try really hard to not do this, though. The few times I have, I've called around to some people I feel comfortable handing the project off to first. I want to have a name or two I can refer my client to. If I think I have to back out, I make sure I can give them a solution to the problem before I commit to it. If I can't do that, I don't back out. Happy clients may or may not refer other people to you, but an unhappy client will always tell people to avoid you.

Red flags are issues that make a job or gig more difficult than it needs to be. For example, an amateur client who has never worked with a sound person will likely need more education and time than a professional client. You may see signs someone is a challenging personality. A venue or studio may be unorganized, have technical issues, or not schedule enough setup time. However, what one person considers a red flag may not be a problem for another, and your tolerance may change based on your circumstances (such as needing income or how busy you are). Video game composer and sound designer Aaron Marks says:

> Companies that don't quite have their act together will end up driving you crazy, asking for the impossible, and cost you time and money. Initially, you'll want to take every job that comes your way, but sooner or later your gut will tell you when it's time to pass this headache off on someone else and let your competitor pull their hair out over it instead.
>
> (68)

With any red flag, consider the consequences of walking away from the gig, or if it would even be possible. Fei Yu, who works with composers and filmmakers, says of this evaluation:

> If I quit this job or project, then I don't get the money that I should get. Maybe I'm also destroying my personal relationship with a director as well. So, that's why I like to finish the project, and then next time, I have a choice if I still want to work with them or not. My mom always taught me that you have to finish this one. You will learn on this one, but you could choose not to work with them next time if you really feel like they're not respecting you.

When there are signs of questionable content, morals, or people, there are absolutely times when the right thing to do is to turn down an opportunity (or quit a job). Susan Rogers says of inappropriate work situations, "Normal doesn't feel terrible. If your gut instinct is saying, 'Get me out of here – This is wrong,' Your gut instinct is right. That's not normal."

Film sound designer Shaun Farley shares:

> I do remember one time when I honestly told a client that I didn't think I was
> the right person for the gig. I was uncomfortable with the subject material
> and that was going to affect my motivation to do the work. I explained why
> I was uncomfortable and told them that they needed someone who could
> be invested in their project and give it the effort it deserved. I'd rather they
> go somewhere else than risk putting in a half-hearted job that would reflect
> on me in the future. I don't know whatever happened with the project, and
> I've never heard from that filmmaker since. I don't feel bad about losing
> that client because I had no relationship with them to begin with. I might
> feel differently if it was an established relationship.

The best etiquette for turning down an opportunity is 1. always respond; and
2. do it in a timely manner. Turning down a gig can be as simple as: "I'm not
available for this project, but thank you for asking me." Podcast consultant/
editor Britany Felix says of turning down work:

> What I say at the end [of a consultation] is that I just don't think it's a good
> fit, which is true. Now, I will admit, sometimes I take the coward's way out
> and I do it via email, but that is all I have to say.

> I had a business coach early on who said you never have to apologize for
> any decision that you make if it is authentic to who you are (in terms of
> who you work with and what you charge). So, when I send that email, it's
> not like, "I'm so sorry. I didn't mean to waste your time." If there's some-
> body who wasn't the right fit for me but I don't think they're going to be
> a terrible client overall, I will offer to recommend or refer them to others.
> That way, I'm not completely leaving them in the lurch. But if I got some
> serious high maintenance/toxic person vibes during that call, I'm not going
> to put any of my other colleagues in that position to work with them. It's
> just going to flat be, "It doesn't appear that we're a good fit for each other
> and I wish you all the best" . . . I've had people that I turned down refer
> other people to me. If I tell them "I'm fully booked at this time" or "I can't
> take on any new clients," then they're not going to refer their friend who
> was also looking for an editor.

Handling Rejection

It is possible you will not get a response if you did not land an opportunity. Jeff
Dudzick, Content Creation Manager at Audible, shares:

> If I think back to some jobs that I applied to that I didn't even hear
> back, now I would understand why. Even if I would have been a great fit,

I wouldn't be surprised now that someone didn't see my email or just didn't see what they needed to see quickly enough. Some things are just a quick glance. It's so many factors . . . It's not a fair system, but I'm sorry, we're all too busy.

Honestly, a lot of times people are just not experienced enough. They're too fresh – or they're too fresh to us because we need some experience. We still don't have the time to train or onboard . . . We would love to have a lengthy training program and to really get everyone who comes in on the same page, but we're all too busy for that.

If you've been rejected by a potential client, it means you've been rejected by *one* client and there will be future opportunities. "I've lost work to low ballers. Should I say lost work? If they are going for someone that charges so little, they're probably not aiming for much. So, probably dodged a bullet there," says Mariana Hutten.

Balancing Multiple Opportunities

When you are working with multiple clients, it can be an ongoing process to fill a schedule with work. This process should begin before you find yourself out of work, and it is normal to ask your existing clients if they have any upcoming work. Kevin McCoy says of a production company he worked for, "Towards the end of the season (which was the spring) I would start talking to them. 'Do you have a place for me next year?'"

Podcast editor and consultant Britany Felix says of seeking out work from other professionals (in podcasting):

Most other editors are okay with you doing your own thing as long as there's no poaching of clients or conflict of interest in terms of going after the same type of people or those kinds of things. The podcast editor that I outsource to in Jamaica has her own separate business doing exactly what I do. Once COVID hit, her business took a dip, and so she was able to take on more work for me and make up that lack of income. I'm willing to do that because she's reliable.

While networking and seeking out work, it is still important to protect the time needed to do your existing work well. "If you have a main gig, you've got to be ready to put that first – not even in matters of money but matters of exhaustion and your own mental capacity," says Becca Stoll.

Many disciplines of the audio industry have an ebb and flow of work throughout the year, which may require seeking out more opportunities. These shifts could be related to anything from the weather (for outdoor concert and touring work) to holidays that affect consumers' buying or entertainment habits. Some trends are consistent year to year; however, the ebb and flow of your work may be different from someone else who does similar work. Knowing these trends can help you find complimenting work, such as a touring live sound engineer who works local corporate gigs during the off-months.

When work is slow, the best solution may *not* always be to branch out into other disciplines. James Clemens-Seely explains, "Just because you have a different experience or a different certification doesn't mean you're going to have access to a different client base."

Major Events That Affect Work: Patrushkha Mierzwa

Patrushkha Mierzwa has worked on over 80 movies and television projects and was one of the first woman boom operators in Hollywood.

There are industry and world events that we may not have anything to do with personally but that affect our livelihoods:

Labor strikes: Historically, labor strikes occur every 8–13 years . . . When a strike occurs, expect a pause in your monthly cash flow for at least six months.

Pandemics/regional conflict: The year 2020 is a great example of what can happen to all industries when there is a health crisis (a pandemic) or a regional conflict that is controlled by the government. The pandemic was affecting all production worldwide; however, local governments all had different strategies for handling the problem. For example, in Sweden, the government decided not to impose a quarantine. In the United States, the country was in various stages of quarantine depending on which state you were physically present in . . . As a result, you may have had a surprise in your cash flow due to a local problem that was controlled by the local government.

Seasonality: Entertainment projects have start dates and end dates that are different for each project. Some freelancers have the relationships to jump from one project to the next without much gap in time regarding cash flow. Others may have a much bigger gap in cash flow that is 3–4 months long in between projects. As a freelancer, you need to be prepared for a 4-month income drought if you do not find your next project after your current project ends.

> **World Events**: The Financial Crisis of 2008 impacted the film world for 18 months because the funding vehicles that financed film projects dried up. As a result, there were fewer projects shooting, and the budgets were smaller since the funders did not want to take additional risk in film projects.
>
> (268–269)
>
> *From "Behind the Sound Cart: A Veteran's Guide to Sound on the Set"*

References

Bruijns, Aline. Personal interview. 21 June 2021.

Bushong, Chris. Personal interview. 27 Sept. 2020.

Chirnside, Kyle. "Getting an Audio Education, Landing the First Gig, and More." *Signal to Noise Podcast*, ep. 12, ProSoundWeb, 6 Sept. 2019, www.prosoundweb.com/category/podcasts/signal-to-noise/.

Clemens-Seely, James. Personal interview. 20 Dec. 2020.

Dudzick, Jeff. Personal interview. 1 Oct. 2021.

Farley, Shaun. Email interview. Conducted by April Tucker, 15 July 2020.

Felix, Britany. Personal interview. 22 June 2020.

Ferguson, Eric. Personal interview. 2 Dec. 2020.

Finan, Kate, panelist. "SoundGirls Career Paths in Film & TV Panel at Sony Studios." *YouTube,* uploaded by SoundGirls, 26 Oct. 2018, youtu.be/Y_g3drC2yyA.

Florianz, Matthew. "How to Stand Out When Applying for a Job in Game Audio." *LinkedIn*, 25 March 2018. www.linkedin.com/pulse/applying-job-game-audio-matthew-florianz/. Accessed 14 Aug. 2021.

Gillmore, Kirsty. "Five Ways to Make Your Freelance Applications Work For You." *SoundGirls*, soundgirls.org/five-ways-to-make-your-freelance-applications-work-for-you/.

Hutten, Mariana. Personal interview. 19 Nov. 2020.

King, Andrew. Personal interview. 15 Dec. 2020.

Lawrence, Michael. "New Co-Host Chris Leonard & Early Career Advice." *Signal to Noise Podcast*, ep. 16, ProSoundWeb, 9 Dec. 2019, www.prosoundweb.com/category/podcasts/signal-to-noise/.

Lively, Nathan. "From the Sound Up Lesson 9: How do you get it?" *Sound Design Live,* 28 March 2017. *SoundCloud,* soundcloud.com/sounddesignlive/sets/from-the-sound-up-how-to.

Marks, Aaron. *Aaron Marks' Complete Guide to Game Audio: For Composers, Sound Designers, Musicians, and Game Developers.* 3rd ed., A K Peters/CRC Press, 2017, p. 68.

McCoy, Kevin. Personal interview. 22 Dec. 2020.

McLucas, John. Personal interview. 25 Jan. and 8 Feb. 2021.

Menhorn, Jack. Email interview. 30 June and 26 July 2020.

Mierzwa, Patrushkha. *Behind the Sound Cart: A Veteran's Guide to Sound on the Set,* Ulano Sound Services, Inc., 2021. pp. 268–269.

Mohabati, Mehrnaz. Personal interview. 29 Jan. 2021.

Morris, Tina. Personal interview. *What Makes You Stand Out,* SoundGirls, 20 May 2020, youtu.be/wYamiOK8Y6A. Accessed 21 May 2020.

Nicholson, Devyn. Personal interview. 23 Nov. 2020.

Ojeda, Sal. Personal interview. 27 Sept. 2020.

Payne, Piper. Personal interview. 29 May 2020.

Reed, Pete. Email interview. Conducted by April Tucker, 24 Nov. 2020.

Rogers, Susan. Personal interview. 12 Oct. 2020.

Stacey, Sarah. Personal interview. 19 June 2021.

Stoll, Becca. Personal interview. 27 April 2020.

Tipp, Nick. Personal interview. 22 Sept. 2020.

Trifiro, Jack. Personal interview. 4 June 2020.

Wheaton, Nadia. "(Actual) Helpful Words of Advice on Writing Emails." *Nadia.audio,* March 16, 2020. nadia.audio/random/2020/03/16/actual-helpful-words-on-emails/. Accessed 1 Aug. 2020.

Vericolli, Catherine. Personal interview. 29 May 2020.

Vericolli, Catherine. Personal interview. *What Makes You Stand Out,* SoundGirls, 20 May 2020, youtu.be/wYamiOK8Y6A. Accessed 21 May 2020.

Yu, Fei. Personal interview. 9 Nov. 2020.

Zúmer, Javier. Personal interview. 24 April 2021.

8
Marketing and Sales (How to Land the Gig)

When some people think of "sales" they picture a pushy person trying to get them to buy something they don't need or want. The phrase "foot in the door" comes from that image of a door-to-door salesperson literally trying to put their foot in a doorway so it can't be closed during their sales pitch. The audio industry uses the term "foot in the door" to mean gaining someone's attention to be considered for a work opportunity. This need to capture attention (and stand out) applies whether you are seeking out your own clients or collaborators; trying to work freelance (for a company, studio, or venue); or seeking out a full-time job. Activities that help get a foot in the door, such as sales and marketing, are especially important in competitive fields or when you plan to utilize social media to land opportunities.

It is not necessary to be aggressive in sales (in-person or online) to have career success. Game sound designer/composer Harry Mack explains:

> Cold-calling [a stranger] has never, ever worked out for me. I have never picked up the phone or an email and ever had someone say, "Why yes, I actually do have some work for you. I'm so glad that you emailed me." I hear some people that rant at conventions, "You audio designers are always emailing me. I don't want to listen to your sound reel. I've already listened to 30 today."

Sales and marketing are an extension of networking and relationship-building. Recording engineer James Clemens-Seely shares:

> The idea of having to sort of "salesperson" myself makes me kind of queasy. I've always tried to find myself in situations where I can overlap with people that maybe I would have a reason to work with in the future. It probably means I missed out on some opportunities where I wasn't a pushy sales-person, but it's never felt like something that I could do in any authentic way. I'd just come off as a jerk.

DOI: 10.4324/9781003050346-9

Video game composer/sound designer Ashton Morris explains how an existing relationship turned into a sale (landing an opportunity):

> Here's how I got a composing job: A year [prior] I saw an interesting [game developer] on Twitter and followed him. I occasionally saw him there and probably commented and retweeted a few of his tweets. One day I saw him say, "I should probably start thinking about adding music soon," so I PM'd him and said I'd be interested in collaborating with him. He then replied, "I had already listened to your whole Bandcamp when you originally followed me, and I love your work." That's how it works: long-term involvement within the community.

Sales and marketing can take place in-person or online, and the process can be similar regardless of which route you use. John McLucas, who uses content creation and social media to attract potential clients (in music production), explains:

> There's a misconception that audio people aren't marketing, but the elbow-rubbing, the events, going to shows to spot the artists you want to work with – it's all the same . . . The core principles haven't changed at all . . . The only thing that's changed is that the watering hole is the Facebook group, not at a concert. The principles of talking with the artist after the show do not change when you're in the DM [direct message], despite what really [sales-aggressive] engineers try to push on people . . .
>
> Some people look at it like, "I'm not a YouTuber. I'm an audio engineer." Yeah, but you just told me you go to shows every Friday for six hours to *maybe* find the right person to work with. So, you'll put in six hours there, but when I tell you to take TikTok seriously, you're laughing because you don't want to put time into TikTok . . . It just builds relationships at scale. It's a different way to do exactly the same thing. All opportunities come down to if people know, like, and trust you.
>
> ("Interview")

While it is possible to build a career without formal sales and marketing training, understanding the basics can help find and land more opportunities. Phebean AdeDamola Oluwagbemi explains:

> A lot of people just follow, "develop your skills, develop your skills, be good at what you do," and forget about the fact that you have to now be able to sell yourself to get the business that you want. You don't have to wait until somebody recognizes you – [like] waiting on an artist's success before you become successful.

Podcast editor/consultant Britany Felix adds:

> You have to have a desire to learn how to market your services, how to brand yourself, how to have a website, how to set up your rates, and how to collect payments – all of these things that come with the back-end business side of it. If you don't have a desire to learn those things, you could be the best audio editor in the world. But, if no one knows you exist because you don't want to have branding or marketing, it's going to serve you no good. It's going to be pointless.

Marketing Yourself

Marketing is the process of promoting your services and products. Marketing can be done purposefully, such as sharing a post on social media about something you worked on, writing a blog post, or creating a video or podcast. It also includes how you present yourself to the public, such as your company logo, your website design, the tone and vocabulary used on your website or social media, and any public photos of you.

Marketing can also be person-to-person. Networking with professionals (in-person or online), building relationships, volunteering, or participating in other industry-related activities let others know your professional interests and pursuits. These interactions can result in work opportunities. At times, real-life relationships and shared experiences (on and off the job) are still more critical than having a presence online or on social media. Langston Masingale explains:

> Building relationships with people is the one algorithm that is always going to be successful . . . being near people, being able to shake their hands, patting them on the back, tell them good job. Those sorts of things are far more powerful than an algorithm on Facebook or Instagram.

A **target audience** is the group of people who would be most interested in your expertise, experiences, and services, whether they are online or in person. Audio engineer Phebean AdeDamola Oluwagbemi explains:

> The first thing to determine is who your [potential] clients are and where they can be found. Not all clients can be found on social media even if *you* have a social media presence . . . There are a lot who are not on social media and are making money. It depends on who you're trying to target, really . . . Then, determine how you're going to reach those clients. Those are the key things that will determine if social media works for you. For music producers, you need the power of social media because that's where a lot of artists spend their time.

Marketing for Utility Sound Technicians: Patrushkha Mierzwa

Patrushkha Mierzwa is a utility sound technician and one of the first women boom operators in Hollywood (USA). She has worked in production sound for films and television shows for major directors including Robert Rodriguez, Quentin Tarantino, and James Gray.

How do [you market yourself] as a 2nd assistant sound? The same way as a boom operator: Let everyone you meet know that you are a utility sound technician/2nd assistant sound. Attend trade meetings, use social media to contribute, offer help, or ask questions. Find a mentor. Get to know people in the rental departments of equipment houses and film schools.

Most film schools don't teach production sound; however, there is often a directory or place to post your [business] card. Take those jobs, knowing that you can make a quick call to your mentor from the set when you get stuck. Let your instructors know you have chosen sound as a career. So few people want to work in sound that your name will get around.

Join a camera group. Watch for news of productions that will shoot in your area and contact the production company to send your résumé to the sound department. Many cities and countries have film commissions to promote their area and publish information on upcoming projects. Go to film festivals and approach people who you feel you connect with. If you post on Facebook, saying, "Hey, I'm looking for work. Thanks" . . . well, that's not enough for professional people to consider. Tell me your skills and talents. Why do we need *you?* . . .

The most common ways of working with established teams are by being recommended by other utility sound technicians or boom operators; they naturally work in more circles more often than mixers. Word of mouth can be helpful; not every referral comes from a sound person. Remember, we all have friendships with people in other departments; grips and electricians and dressers know sound people who are professional and pleasant to work with. You can make an introduction to the sound team directly, as there are always times when one of them may need to find someone to fill in.

(281–282)

From "Behind the Sound Cart: A Veteran's Guide to Sound on the Set"

Branding

A **brand** is how others perceive your business. **Branding** is how you shape the perception of your brand. For example, if you are known locally as a drummer but you want to build up clients as a podcast editor, your brand (the way others perceive you) is currently for your music skills. Branding is the efforts you would take to build awareness that you are also a podcast editor, such as making a website, telling your friends and family you are seeking podcasting work, and networking with other professional podcast editors. "The idea of building a brand isn't how your mix sounds, or the weird stuff you do to mic up a piano . . . You're cementing yourself in the community," says John McLucas ("Interview").

Branding shows how you are different from the competition, or **differentiation**. Branding can be focused on the business (business branding) or on the person (personal branding). Personal branding is when *you* are what people are interested in and involves showing your personality and interests. In the creative sectors (like music production) or jobs where you may spend a lot of time with a client, personal branding can help you stand out.

For example, an audio company that uses business branding is focused on the company's name and products (for example, Sennheiser, JBL, Neumann). Another company making similar products might also incorporate personal branding, especially of the company's owner (for example, Eveanna Manley of Manley Labs or Steven Slate of Slate Audio). Recording Studios like Capitol Records or Abbey Road have a reputation based on their name (business branding). Professionals like Leslie Ann Jones or Finneas are recognized for their personal brands.

Differentiating Your Business: Tom Kelly

Tom Kelly of Clean Cut Audio (Denver, Colorado, USA) uses his image to differentiate himself and his podcast company in a highly competitive space. In Tom's YouTube videos and at events, Tom wears a sequence jacket that used to be owned by Taylor Swift.

> My logo is just my glasses and my beard. Can't miss it . . . Everything is marketing in disguise. I went into [podcasting] with the thought in mind that I can't just fall into the obscurity of all these other people. What is my unique thing? For a while, the only unique thing I had was the sequined jacket, but it worked . . . You're going to know the guy in the jacket is the audio guy. So, I've taken every opportunity for

> some kind of brand authority/brand awareness. It's super important online, especially because there are so many editors. If you're in a group and [someone asks], "Hey, I need help editing my podcast," they get 300 messages from a bunch of random people. You have to stand out with a little thumbnail and it's tough. Online presence has got to be a huge priority.

Social Media as a Marketing Tool

Social media can help market yourself through:

- Exposure/visibility – people who you may not reach otherwise can learn about you or your work
- Onboarding into the industry – showing that you understand the norms and practices of your discipline
- Reputation-building – When done effectively, engaging in social media (and content creation) can show you are knowledgeable and serious about your work and career
- Staying in people's awareness – Your name may come up for a recommendation or referral if others remember you

Mariana Hutten shares about using social media for branding:

> I don't think it works if your [social] is just stuff that you worked on. I see a lot of engineers just putting album covers that they did, like, "I mixed this." I think it's important to have the album covers be a kind of portfolio, but also sell your lifestyle a bit. Sell the overall "this is what I do," and the entire package of your daily life in this industry. In this century, the lines between your work life and your life are so blurry . . . It's about balance. Don't have your entire Instagram be pictures of your dog, but also not have your entire Instagram be pictures of your gear only . . . I have friends who are like that. I haven't seen their face in like six months.

John McLucas adds:

> There's a difference between somebody who only posts about themselves versus somebody who's creating a community based on helping others or being valuable or entertaining . . . There's more to it than just "look at how [amazing] I am." . . . There's community value in being informative, educational, or thought-provoking.
>
> ("Drewsif")

> **Pro Tip**
>
> Good promotion is, "Here's what I've been working on," tag some
> people who were involved, promote other people's work that you
> like, or promote people you've worked with. Bad promotion . . .
> I have seen people promote things where they say something bad
> about a competitor or another company. Obviously, that's something
> you should never do.
>
> – Sarah Stacey

In some places around the world, there are limited professionals, venues, and/
or structure within the entertainment industry. This makes social media content and presence more crucial to building a career. Phebean AdeDamola
Oluwagbemi says of the music industry in Nigeria:

> The companies that are really notable in the country are few – less than 10.
> They can't employ more than 100 people. When you look at the number
> of people who are getting into the industry, the best thing will be to get
> yourself on social media. Then you get credibility, start finding your own
> clients that you can begin to work with, and start building your own circle
> of influence. That's what we do here majorly. Then, we'll try as much as
> possible to get international recognition because if you work with an inter-
> national brand, it gives you some sort of credibility. "I worked with BBC"
> or "I worked with Universal Music Group" – even if it's just a small job you
> did with them, it gives you some kind of credibility.

Phebean adds that the quality of your work contributes to your brand:

> Own your craft in such a way that when people hear what you do or see
> what you do, there's this excellence and quality that people can't second
> guess. That gives you credibility faster than just churning out any [social
> media content] . . . Show your skills in such a way that when you put out
> content about the things you do, or about a job, or when they hear some-
> thing that you've done, they can say, "this is good" and can recommend
> you to other people. That's the power of the audience, too – they refer you
> even if they don't have the power to help you. They can refer you because
> they know what you do.

Different platforms can also be utilized for different purposes. John McLucas
shares how social media platforms were being utilized in 2021:

> YouTube is more reputation awareness, the top of funnel marketing.
> Instagram and Facebook are a lot more like watering holes, like when

you want to actually have a conversation. On YouTube, you're probably not going to have a conversation in the comment section, and you can't message creators directly. That's why you do an Instagram or a Facebook . . . Maybe Discord is going to make more sense.

("Interview")

Professional Social Media: Alesia Hendley

Alesia Hendley has worked in live sound, AV, and IT for AV systems. She is also a content creator and multimedia journalist. Alesia discusses how social media platforms can be used for business.

Instagram

> When utilized correctly, Instagram can be a visual resume of what you can do and how you do it – with a hint of personality to give a little sense of who you are. Pictures and videos are one the best ways to show prospective clients and employers the gear you've worked with, the types of training you've participated in, or even the school you've attended. It's living proof of you working with a particular console or at a specific gig. Always be sure to have social media postings cleared in a contract or have final approval from the overall facilitator of the event. Do not overdo the picture-taking. Remember, you're working so you will have other important things taking place. Find the right time and prioritize accordingly.

Twitter

> A great way to take advantage of Twitter is to post about your work. Share tips on how you EQ a particular instrument or share a blog post about a console. Incorporate a known hashtag to correspond with the post . . .

> When you post pictures and videos on Twitter, it will create more engagement. You want to be active so that people have a reason to follow and connect with you. Of course, you must be professional on this platform as well. Don't tweet about how unorganized a client is or how difficult a band was during a show. Your public timeline is not the place. Your professionalism and personality have to equally shine and reflect you in the same light across every social media outlet . . .

The best thing you can do on Twitter is to engage in a Twitter chat. Twitter chats allow you to connect with like-minded individuals in the industry. It also allows you to become a part of a community where you can share your knowledge, collaborate, and learn from others . . . Ask questions to show you want to grow, have a strong passion, and are willing to learn.

LinkedIn

LinkedIn is the social platform created for professionals. This is the place where you always want to remain professional no matter what. This is not the place to post too many pictures or add videos. Make sure you keep your profile up to date and keep any gaps in your work history to a minimum. Fill out every section in its entirety, especially the skills section. This will allow people within your network to endorse your skillset.

This platform is professional, but you still need to find a way to incorporate a hint of personality. You can do this by publishing articles on LinkedIn. Here you can write about audio and anything else you have a passion for. This way, anyone looking at your profile for the first time gets your work history, schooling, and a little taste of who you are as a person. LinkedIn is not the place to down-talk your previous employer or talk about any other personal business.

From "Professional Social Media", soundgirls.org

Creating Content

Content for social media is the digital presentation of information (including text, photos, videos, and/or audio) to an audience. A **content creator** is someone who creates original content. Original content can also be as simple as sharing a photo or short video or briefly sharing an element of your work. John McLucas explains:

Content does outreach for you. You want your 12 touchpoints? Just post 12 ten-second clips of you working on something and what it sounds like from your phone. I can't tell you how many people make decisions because they watched my Instagram story highlight of pop music and then reached out to me. I ask, "How did you find me?" They'll say, "I found your own hashtag and I listened to stuff from your highlights, and then I hit you up." They literally only heard my work on their phone . . .

Content creation is just the other version of marketing and watering holes. It's still a big-time commitment, but the time you're putting in mopping and cleaning bathrooms is not necessarily honing your craft. But, you're certainly putting in the time because you know that it means that you will grow in the long term.

("Interview")

Phebean AdeDamola Oluwagbemi adds:

As an audio engineer, I can just shoot videos of me working in the studio, working with a client, and talking about things I do. Be yourself because at the end of the day, you don't want to build something that looks different from the person you are.

Content creators who build an audience (or build "authority" as being trust-worthy on a topic) are called **influencers**. For audio professionals looking to use social media to reach potential clients, content creation is primarily about personal marketing, not income. "It's possible to monetize, but I wouldn't count on a platform to bring you revenue as the way you're going to make your living . . . [Income] is not the focus," says John McLucas ("Interview").

Content Creation Tips: John McLucas

John McLucas is a pop music producer and content creator. John has over 200 videos on his YouTube channel, has hosted over 100 episodes of the Modern Music Creator Podcast, and has built a career as a music producer with a specialty in vocal production. John uses content creation for self-marketing, but also offers his expertise in content to others.

Focus on One Platform

It can be really overwhelming to go across multiple platforms. I would dedicate two hours a week to put together everything you would need so you don't get overwhelmed. You'll still start to build that authority. At the core of this is the routine-ness . . . it keeps it really simple and you don't have to think day to day.

("Lost")

Simplify the Process

If it works better for you, you can also consider taking two hours every Sunday morning and just prepping content for the week. That goes for every platform – Edit the photos, do all the filtering, all the captions, all the hashtags, tag people. It's literally like meal prep but for [content].

("Quickest")

Who Do You Want to Appeal to?

For me, that's singer/songwriters, solo artists . . . Where are they hanging out [online]? Then, what problems are they encountering that I can solve? A lot of these people are trying to start up their own freelance businesses based around their music . . . I created a podcast as something that can add value to their lives and their [challenges] . . .

I make content that shows people, "John knows what he's talking about when it comes to pop music and especially vocal production." Content, in that awareness, is the same thing as being in any other watering hole that's physical. It's just like overhearing my mix as you walk down the hall at a studio . . . I make content that demonstrates that skill set and demonstrates I know what I'm doing over and over and over. That builds a reputation as John, pop guy, big vocal, big drum guy, content, beard.

("Reasons")

Ideas on Content

People just want to see that you're doing [work]. It's the same as the restaurant that has nobody sitting in it next to the restaurant with a bunch of people in it. The food could be delicious, but everybody wants to go to the one with a line. So when you create the line yourself, people see you as somebody who's in demand.

If somebody says, "John, I need your help," and I don't have a full plate, then I go and help. If they want to pay me $50 to be on a Zoom call with me for an hour to learn about compressors, that's fine. Then, I'm taking a picture, putting it on social media, and saying, "Great session with Jimmy Bimi." I can leave out all the details I want, but I had a session with the guy. So, I can still create the perception and the truth that I am working with people, even if it's not my ideal capacity . . .

You could make content around why going to the studio is important . . . Instead of saying, "You need to go to the studio to get better drums," post a video of the [before and after] drum sound you got from when John Band came in the studio and they were unsure.

("Interview")

Get Permission

I would make a stipulation, especially working for free, like, "I need to be able to post that I'm doing this for you – a little snippet of it, even if it's just the guitars. I need to be able to show that I'm cooking for people." . . . Some people are cool with me posting everything. Some people don't want any vocal but they're cool with instruments. Some people don't want anything. If I can even post the drums just so people know that I'm working on stuff, that's super valuable to me and it does let everybody know what I've got going on in my world. That is how I pitch it to people. You can even ask, "I would need you to post this snippet as well, since I'm going to do this for free." That's fair. Then you can start getting like more shares and getting more social reach that way.

("Rebranded")

Business Niches

As an emerging professional, it can benefit to be open-minded to opportunities and learning as broadly as possible. However, when it comes to marketing – the messaging that will reach potential clients – being too broad can be problematic. John McLucas explains:

When I bring on a client, they're excited because I know exactly how to speak to them. They know that [my work] is made for them. A lot of people are very bad at that. They're the equivalent of like, "We do Italian food, burgers, sushi, and change tires, and also do your pharmacy refills." Well, sorry, I want the number one bolognese in Los Angeles tonight. If I need my tires changed, maybe I'll come to you. That's a really bad offer and really bad messaging.

("Interview")

A **niche** is an area of specialization. A niche can be broad, such as a discipline (such as live sound or video games), or specific within a discipline. For example, a niche in music production could be a genre such as hip hop, country, or classical music. In post-production sound, a niche is divided by job roles, such as dialog editor, sound designer, or re-recording mixer. A niche can be even more specific, such as an A2 sound operator for fringe musical theater shows.

Podcasting is a discipline where having a genre niche can help you stand out. Britany Felix explains:

> The cliche of "the riches are in the niches" is 100% true. When you try to serve everyone, nobody connects with you. You're just a voice shouting into the void, and nobody zeroes in on your voice. But when you are known as the go-to person for a particular topic or style of show or anything like that, then you build a reputation for that . . . I know one podcast editor who only works on personal finance podcasts and he is the go to guy . . . I work with female coaches and consultants. There's another editor I know who works only with people who live locally in Kentucky. It can really vary.

VR sound designer Megan Frazier, who used to work as a recruiter, says of niches:

> I had seen it before as a recruiter where there are these little clusters of really insular little communities . . . Injecting oneself and being visible in those communities is incredibly easy if you know how to do it . . . There is a segment of "everybody knows everybody", like, this is our art style. This is our thing. These are our technical constraints. It's just defining what your specialty is ahead of time. I took a gamble [on a niche], and I won. I'm sure there are people out there who have taken a gamble and lost.

In some circumstances, *not* having a niche can be an approach. For example, in Bogota, Colombia, an independent studio would need to offer services across disciplines to survive – anything from recording voiceover for television to recording and mixing music (Pinzon). Jeff Gross, who has been in the audio industry in Los Angeles for 30 years, says of not having a niche:

> I had this on a business card: "All Things Sound." I bounced around to so many different things. It's stupid to write producer, musician, writer, Foley mixer, audio mixer, whatever. It's a survival thing. I've literally done everything from putting in a sound system in a theater to court case audio restoration to ghost hunting (audio cleaning) with some friends. It's the same tools, just a different approach.

Picking a Niche: John McLucas

Picking a niche means "I want to reinforce the perception of certain things in my work as my main things." It's not getting married. It's not making a crazy decision. It's just slight tweaks to content we put out, how we talk to people, and how we describe ourselves. It's to highlight the main 80% of our work and not necessarily need to discuss all the other little things that do come up for us.

The fear is when you "pick a niche," it is sabotaging other opportunities on the table, and that can be scary. The truth is, one or two of those things are going to be what you're most known for and brings in the majority of your income. Not that you can't do the others, but there's the reality of your reputation and what you are known for.

It would be really difficult to understand who I'm serving if I didn't have my niche well defined. The more that you can appeal to a specific group of people, the more emphatic and excited they're going to be about inquiring to work with you and hire you. Here's an example:

We help modern pop artists 100% self-produce amazing music without years of traditional learning or online education.

Does that appeal to modern pop artists? . . . I concisely laid out the importance, the audience, and the goal they want to achieve in that sentence.

We help people make dope songs. I also help out other businesses with their marketing. I do some drawing too. I do some YouTube stuff, and I do gear reviews.

Do you notice how different those two sentences sound? If you were a modern pop artist, do you think you would hire that second person?

Niching down doesn't necessarily have to mean "I only do this and I don't do anything else." All it means is to say, "these are the main three things I want people to think of when they think of me." What are the three things you feel are your strengths, your superpowers? Or when you ask other people, what do they come back with? Your brand is the perception of people around you.

After that, lean into it with content to solidify yourself as an expert, as a specialist. Everything should be cohesive with these few things you want to be known for from your bio to your email signature. Then, at the same time, you will always get inquiries for other things outside your niche. It's inevitable.

Leaning into one specific area isn't a squashing of your diverse self. We are all diverse people. It's scattered when you feel obligated to put every little thing out online about yourself. If people don't know any-thing about what you do, then this is your time to hone it in a little bit.

("Why Picking")

From the Modern Music Creator Podcast

Sales

A **sale** is an exchange of money for a service (or landing a gig and getting paid for it). The **sales process** (or "**sales**") refers to any activities leading to a sale. Sales come naturally to some people and contributes to their career success. Music producer Nick Tipp, who has been business-minded since a young age, says:

My interest and long-term experience in business, making deals, and trying to make money wherever I can has led me to everything in audio. Who are these people? What are they doing? How are we going to interact? Can I provide services for you? What's going on in your life? Oh, you should get a better sound system. I'll sell it to you.

While it may seem the goal of the sales process is to convince others to hire you, fundamentally, it is about solving problems. A problem can be as simple as, "I need help from someone who knows audio." Megan Frazier, who used to work full-time in sales and recruiting says:

The purpose of sales isn't to actually sell yourself. It is to help the people through the decision-making process of finding the person that they need . . . It's about them and helping them through their process and through their pain (because it is such a pain in the rear to find people).

In the audio industry, the sales process is leading a relationship toward asking to work together. Finding clients and opportunities can be an ongoing process, and timing is crucial. John McLucas says of music production, "With this ser-vice, this specific industry, people don't need [your services] all the time. So, you can be nurturing but they just aren't ready to record" ("Interview").

There are times to ask directly about opportunities (or working together), but typically not early in a relationship. Violinist/arranger/live sound engineer Hiro Goto explains:

> You have to build a relationship first before asking someone if they have more work for you. With people I've known a long time, they know my work. We've exchanged a number of years of earning trust. Then, I would be comfortable asking. If I work for someone and they never called me again, it doesn't come to my mind to ask them for more work. The bottom line is trust. You have to build a relationship with them through working together.

The sales process can be online, in-person, or a combination. John McLucas often starts the process online (in social media groups or creating content and interacting with others), but moves to a video chat or phone call to build a personal connection. John explains:

> I talk with [someone] on the phone for probably 30 to 60 minutes before I even allow them to possibly work with me. That's how I want to run my business. But, if you're doing a US$200 mastering gig, you literally can't afford to be on the phone that long. I like a business where it's very much based on getting along and they like me.
>
> ("Interview")

Building a website, having a social media presence, or creating content can be helpful marketing tools to connect with potential clients, but these are not meant to do the entire sales process for you. Mariana Hutten says:

> I find on average once a month somebody reaches through the "book me" form on my website. The vast majority of my work comes from word of mouth, or from somebody that came to the place I work at and I recorded them, and then it becomes a relationship. It's pretty minor that somebody hires me through my website.

Sales At the Wrong Time: James Clemens-Seely

James Clemens-Seely has worked as a recording engineer with a wide variety of musicians from garage bands to electronic music artists, award-winning musicians and producers, and nearly every professional orchestra in Canada.

> In pure music recording situations, which is most of where I spend my professional life, the artists are also filled of anxieties. This is a stressful, financially expensive, and emotionally taxing time in the studio where

they are trying to do something that will outlive them. It costs more than making music on a normal day. It involves more strangers or people that they don't necessarily have in their inner circle.

The big people in the situation are also filled with anxiety. Turns out that they also worry every year that maybe they'll never work again, and maybe this will be their last project ever. Even though they've been working forever and making money and their walls are full of trophies, they're still filled with anxieties that are rational for any freelancer to have.

I've seen the [most junior] person in the room start handing out business cards. If you're bringing that energy into the space (where it's usually a pretty intimate, fragile, emotional space), something like that can really derail it . . . It gives a lot of people a bad taste. Alienating professionals by selling yourself as competition to them when you're there by their graces – that's a pretty boneheaded move.

Treating [a recording session] like there's something financial, salesy, or network-building – that scares me. I want it to be a situation where we build a shared experience. Then afterward, I try to embed that into my network, but I don't try to build the network while I'm actually doing the work.

I try to build a relationship in my interactions after the project and before the project, like being fast in replying to emails, writing clear emails where I answer every question they have, and raising questions . . . in a courteous way. I try to do a lot of before-and-after stuff that reeks of thoroughness and care and being considerate. But in the session . . . I don't want to bring any of the "what can they do for me" side of that discussion [there].

Sales Funnel

A **sales funnel** is a step-by-step process to turn a potential client into a paying client. Video game sound designer Akash Thakkar explains:

When we're talking about finding clients and finding great work, your goal is to move potential [clients] through this funnel by talking to them and following up. This is done in a friendly, sleaze-free way. You're not con-stantly elbowing them in the ribs saying, "please hire me" because that's way too much. I hear too many horror stories of other composers and sound designers who are just begging for work. That's not the way to do this.

Your goal is to actually form some sort of business (or friendly) relationship with these people. You want to talk to them, and you actually want to get to know them. You want to know their needs. That's really important. You want to know how you can actually solve their problem. This isn't just a gig. You are solving their problem because they don't have music, they don't have sound, and they don't have voiceover or implementation for their game. It's really important for them to actually have that off of their plate so they don't need to worry about it.

Your messaging and communication with a potential client can vary depending on how close you are to a sale (where you are in the funnel). For example, John McLucas starts many relationships online but transitions to a more personal conversation (through video messages, phone calls, or meetings) once he sees the potential for working together:

That person who says [on social media], "DM me for rates" – they're doing sales badly. There actually is a good way to do it, but it involves social media first, and then human-to-human connection off of that . . . The human connection is probably the most important part of my business.

("Interview")

The four elements of the sales funnel are leads, prospects, opportunities, and clients.

Figure 8.1 Sales funnel.

Leads

A **lead** is any potential client. Leads are the broadest of the categories and could include anyone you could see potentially working with (who does not know you). It is possible a lead knows *of* you (through reputation, website, social media, etc.) but does not know you personally. For example:

Emerging professional job role	Seeking out	Leads
Production sound mixer	On set work (student films, professional film and TV projects, commercials)	Filmmakers (professionals and students), local media production companies, other sound mixers
Classical music recording engineer	Paid recording opportunities (live performances, recording sessions for public release or auditions)	Venues and concert halls, classical music ensembles, musicians, students, other engineers
Podcast editor	Paid work on podcasts	Individuals or businesses who may be interested in making a podcast, podcast production companies, other podcast editors

When Your Leads Are Top of the Industry

If you are targeting work at the top of the industry, it can take years to build the relationships needed *and* for an opportunity to open up. These opportunities include highly competitive jobs at well-known companies or with well-known professionals (such as worldwide-known recording studios, sound facilities, live sound companies, television networks, etc.)

Pete Reed has worked at the top of the industry in both theater and video games. As an emerging professional, it took Pete five years and applying to 90 jobs in the West End (London) before landing a position on *Matilda the Musical*. When Pete later decided to pivot into video games, it took three years to land work on a AAA game title. Pete says:

> After two years, I started to hear back from applications and got phone/Skype interviews . . . In my third year of applying, I was getting to the test stage of the interview process and then face-to-face interviews. I received three AAA offers in practically the same week.

Shaun Farley started working for Skywalker Sound, a premier facility for post-production sound, ten years into his professional career. Shaun's volunteer work for the website *Designing Sound* (designingsound.org) helped him build professional relationships and to become more recognizable within the film sound industry. Shaun says of getting a foot in the door at a major facility:

> You can't get hired if you can't even get an interview. If you're relying solely on your resume and reel, you've got a major uphill battle ahead of you, especially in the early stages of your career. If you're recognizable in ways beyond that, you've got a better shot of making it to phase one of the hiring process. Relationships will help you be more recognizable.

Seeking Out Leads in Emerging Technologies: Alesia Hendley

Alesia Hendley is an audio professional and content creator with experience in live sound and AV. Alesia discusses areas of the audio and AV industries showing signs of growth and opportunities, and how to find leads.

> I would pay very close attention to these three things that I feel are going to be huge in the next few years:
>
> 1. Immersive events. Everybody's creating these immersive events, and you can't live without sound in an immersive environment.
>
> 2. Gaming. eSports is so huge . . . You've got 16-year-olds making millions of dollars in tournaments . . . Tournaments can't live without audio, either. Even if they're all virtual, there's still an audio person behind the whole video team running it.
>
> 3. AR and VR. There are sound operators that create and design and orchestrate the sound that goes into those VR headsets. Artists are even doing virtual shows in those spaces.

Seeking Information on Leads

> Who's the integrator for those spaces? Let's Google it . . . [Let's say] I never heard of that sound company. Let me go see their Facebook and their Twitter. Let me see who on LinkedIn works for that company that I can connect with. It's all about connecting the dots.

> There's a PR company or an agency that the companies work with
> that creates these environments. Look for the design agencies.
> There's an integrator in there somewhere. There's a third party in
> there somewhere that's designing it. Get in with an agency. You
> help design in some form or fashion. You meet somebody at that
> event that works at [a big company], and you're in the door. Stay
> connected, find opportunities some way, somehow, and three years
> later, you work there when VR is huge. It's all about the environ-
> ment, and being in the right place at the right time. Gradually grow
> with the industry.
>
> ("Interview")

Prospects

A **prospect** is someone you have identified as a lead and have begun to
interact with (this interaction is what makes it different from a lead). This
part of the sales process is essentially the same as networking where the goal
is to meet people and start building a relationship. A prospect could turn
into a work opportunity immediately, months or years later, or possibly not
at all.

The difficult part about this step of the sales process is recognizing when it is
appropriate to talk about sales. Jeff Gross says of people who turn to sales talk
too early:

> It's off-putting, but no one would tell you that's bad etiquette. It's kind
> of like the guy who doesn't wear deodorant. He doesn't know it, but the
> people around him know it, and nobody really wants to say anything. It's
> like, "Whoa, dude," but you don't want to make him feel bad. But he still
> stinks. People who act like that – they're the deodorant guy.

John McLucas suggests:

> Be a good friend and a good listener. Don't be a salesperson until it's the
> appropriate moment. The appropriate moment is completely different in
> every single situation. When somebody has an active interest in learning
> about you, that is the right moment to have a pitch ready to say, "I'm a
> music producer, and I work primarily in pop rock and metal music." But,
> I don't ever, ever say that unless somebody is interested in hearing that. If
> I take an interest in them first and want to get to know them, then they'll
> typically reciprocate.

Then, don't push it . . . When they need the services you provide (or get asked for a recommendation), it's going to go so much further that you are just a cool person that also happens to do what they need – more than the person who was really adamant about selling themselves onto you. That is so off-putting and it'll ruin a relationship so quickly.

("Building")

Opportunities

An **opportunity** is when a potential work arrangement is in consideration. A prospect turns to an opportunity when there is a discussion about possibly working together. In some cases, an opportunity exists already and you are trying to present yourself as the person to do it (such as Ashton Morris' earlier example of reaching out to a game developer who was seeking a composer). In other cases, a prospect will come to you with an opportunity. An opportunity may be discussed through a formal meeting or an informal conversation. Akash Thakkar explains:

> An opportunity is when clients (or potential clients) come to you and say, "Send us a demo," or, "We're interested in working with you," "What's your rate," that sort of thing . . . When you have an opportunity, basically people are asking you to pitch to them in some way, shape, or form. So, you're being asked to show your stuff . . . What would it be like working with you? Anything like that is an opportunity.

This is the time to make a sales pitch, or sharing the reasons why you might be the right person for the work. Britany Felix schedules a consultation where she discusses her services and also what a client will gain by hiring her. Britany might say to a potential client:

> When you're thinking about hiring an editor, don't think about what you're getting in terms of what [an editor] is going to give you. Think about what you can do in your business with the extra time you're going to gain. Yes, an editor may cost you US$500–1,000 a month to do your podcast episode, but it's also going to free up hours of your time every single week. So, how much revenue can you earn with that additional time?

John McLucas also does a consultation where he discusses the process of working together, but is selective about what he shares with potential clients at this point:

> One of my mentors said, "You have to sell them on what they want and give them what they need." That's my approach to it: You're going to need

to know some of these things, but I'm not going to tell you about it yet . . . Once you see that it's worth it and you have belief in the process, then I give you the un-fun stuff. For example, grammar in music. Like, you've got to know your grammar [when recording], but nobody likes thinking about that [when closing a sale].

("Interview")

This meeting can also be a pitch to show a potential client the quality of your work and start building a relationship. Mastering engineer Marc Thériault explains:

> If you come to me, "Marc, give me your best sound," I'm going to lose my time. If somebody asks for a pitch, I will say, "Fine, you want me to spend an hour on your song? You need to spend an hour with me in the studio. I won't do a pitch blind. That means you're using my time, and I might be wrong because I don't know what you're looking for" . . . Then we talk, and then we target exactly what you're looking for . . . The thing is that mastering – you need communication with the artists, and you need to know what they're looking for.

Even if you do not land the opportunity, these meetings can be useful for learning what your potential clients are looking for. Britany Felix explains:

> If you're listening to the people you're talking to, they're going to tell you exactly what you need to do. Because once you get on your fifth client who has come to you for a consult, asked you about your services, and that fifth person has said, "Do you also do [podcast] show notes? Do you also do this?" If every single time you say no and then they don't hire you, there's your reason why. Just listen to what they're asking you for, because that's what they want and need, and that's what they want to pay for. Your potential clients are going to tell you everything you need to know. Then just regurgitate that language back to them and offer exactly what they're asking.

What to do after the meeting – from setting rates, sending an offer (such as a written proposal), negotiating, and more are covered in Chapters 10 and 11.

Ways to Create Opportunities

Sales are not a matter of letting people know you want them to hire you. In the modern audio industry, education and sharing your expertise can lead to opportunities. "If you offer to give [people] the knowledge not to need you, often it gives them the knowledge to know how much they need you," says James Clemens-Seely.

Mixing/mastering engineer Mariana Hutten adds:

> I actually do get a lot of work because I taught people something. It's kind of weird. It feels like you'll lose work if you're giving out information, but for me, it's been the opposite. I give out information, and I do workshops. It sells my expertise instead of losing work . . . Some people are actually trying to become better in engineering but still would send me their stuff for me to do. Of course, there are a few detailed secrets that I have that I don't give up.

John McLucas finds teaching his clients also improves the quality of what he receives later:

> I've helped many people with their vocal chain or direct monitoring problems (for them to record their vocals). Is that what I want to be sitting on my day doing – talking about signal flow and how everything's distorting on every track? No. But, I know when I do that, when they're feeling grateful, then it helps them finish the project. I get better tracks back [for mixing], they get a better recording, and they get something that they can take away with them for the rest of the songs they record for the rest of their life. There's tons of reciprocity effect.
>
> ("Interview")

Podcast editor Tom Kelly, who has shared his expertise through content creation (YouTube videos and a podcast) and social media groups, suggests:

> I encourage people to at least try to offer as much free help as you can. Get your name recognized in these groups . . . It got me a lot of clients very quickly and clients gave me multiple clients then. That's been super helpful . . . People would reach out from my YouTube channel [asking], "Can I pay you for an hour of your time?"

John McLucas has also found creating original content has led to work opportunities in music production:

> I've gotten more opportunities with bigger people [through content]. I'll pop up on somebody's "for you" page or in their recommended, and then they messaged me, and they're a fan of me and my content . . . It has taken the hierarchy away in so many great ways . . .
>
> All my content works for me now. They're just continually bringing in business over and over. "I found you from this video." So, I have this recurring churn of leads that I don't have to touch, and I [create content] now because I like it . . . That's the means to having the great clients that I like, and having the business that I want to have is the same way someone else will spend three years [working in fast food] waiting for their chance. I spent three years building a business [on content].
>
> ("Interview")

Participating with an industry community can lead to opportunities, as well. Shaun Farley co-ran the popular industry website Designing Sound from 2011 to 2019, and says of the experience:

> It basically plastered my name all over the web and associated it with sound design . . . I was even being interviewed by other sites simply because of how visible I already was. People in the sound design community (film, TV, and games) knew who I was because of it. That visibility goes a long way. I was completely search engine optimized before the term actually existed.

Phebean AdeDamola Oluwagbemi says of creating Audio Girl Africa (to support African women trying to enter the audio industry):

> Audio Girl Africa has given me networking opportunities. I can't really say it has given me work, but it has . . . really established me in the eyes of a lot of people as a professional audio engineer who you can reach out to for professional help and jobs.

Helping others solve problems can lead to future opportunities. "This is a people industry. This is about getting to know people and them getting to know you, and you solving their problem," says Megan Frazier.

Sometimes this may mean recommending someone else for an opportunity. Music producer and Foley mixer Jeff Gross explains:

> I look at people for what they need. If I have a chance to put something in their path for them to be able to use, maybe they will do the same [for me] down the road. That's not the reason I'm initially doing it, though . . . It's utilizing the people around you to move something forward. Like, if someone needs tech work, I'll say, "Do you have a tech? I know somebody who can help." I look at everybody in that sense.

Doing what's in the best interest of others helps build trust, which can lead to opportunities over time. Music producer Nick Tipp explains:

> Engendering trust is probably the most important thing you could do – more important than doing your job. Every time I've ever done lectures or presented that to college students, they laugh. But, it's actually true. You're not going to have an opportunity to do your job if people don't trust you . . . My goal in each interaction is to get somebody to trust me so that I can do a really great thing for them and they have a great experience. That's Entrepreneurship 101 – get the client to like you.

Turning Leads Into Sales on Social Media: John McLucas

Music producer and content creator John McLucas suggests a plan for someone getting started in music production, seeking clients through social media, and open to doing some free work to build a portfolio.

First, look for Facebook groups. Write down on an Excel sheet 10 to 20 group names that have musicians in it in your genre. Regional groups are great, too. We need to budget out probably one to two hours a day. All you're going to do is go into the group and scroll through them. This is what I call "embarrassing value." When there's a post that you could respond to and help somebody out, make sure that your response makes everybody else look like a chump. For example, I responded to someone who was [asking], "Should I hire a producer?" I took four minutes and typed out a two-paragraph response. It was well organized and articulate. Professionals don't type improperly – we speak with full sentences with proper punctuation. It was so well done I made all the other comments look like chumps. They were just like, "Just go for it," or "Bro, you've got this," or like, "I'd hire a producer. DM [direct message] me, I am a producer." It was a lot of surface-level responses. I had 10 full sentences organized and I didn't sell anything. I just left value. Because when you're good and when you're an authority, people read that comment (whether it's that person or people in the group) and they start to build an opinion of you. They click on you [to learn more].

Where I see a lot of [new people] mess up is they think they have to sell themselves right away. A lot of producers try to sell themselves at the end of their four-sentence comment. They would present the solution and then sell them that solution, like, "You should try a subscription-based format with a producer. I actually do that for my clients. You can DM me about it." I don't even ask for [work] and I get reached out to. So, that's the difference.

Go through as many of those comments as you can, just 10 to 20 minutes of leaving embarrassing value. Then, every single person who has relevant music to you (even if it sounds good but you can help with), start a conversation with them in the DM. Show them that you actually care about their music. So, you say, "Hey, Michelle, I absolutely love your last single, The Jangly Guitars." You just talk about something so they know that you care, and then you leave it at that. A lot of people just say, "Hey, thanks," but a small percentage of those will turn into relaxed conversations. Now we've gone from Facebook groups to conversations, and all the people that will reciprocate are [prospects].

Then, you're going to track all the people that are kind of interested in working together in an Excel sheet. You're not going to put every conversation in there, but you know who could potentially be a good fit because you're building rapport and you just have a normal conversation. If it goes nowhere, that's fine.

A small percentage of people will then ask you, "Hey, what do you do? What's your thing?" I like doing all my main outreach on video. You'll see in a video clip, "I just got out of audio school and I'm just trying to be involved in as many projects as I can. I'm not interested in getting paid. I just want to be involved with great quality creatives and I see you as somebody like that. So if you're open to it, I would love to do something together just to have it be part of my world, if you're open to it. If not, that's great, but I thought I would put it on the table. I really loved your last single, Papaya, and I would love to do Mango." Then sign off.

That's a pretty disarming ask. Whereas a lot of people come in and they just cold DM. Cold DMs work, but I hate that life. It's so gross. Instead, we're only presenting this as a free offer for people who care and are interested speculatively. Then, you are able to close some free projects. If you do that 30 times and nobody says yes, then your portfolio is really bad, or you need to [check your profile] . . . It should be professional. From there, you should be able to close a small number of free projects.

("Interview")

References

AdeDamola Oluwagbemi, Phebean. Personal interview. 10 Jan. 2021.

Clemens-Seely, James. Personal interview. 20 Dec. 2020.

Farley, Shaun. Email interview. Conducted by April Tucker, 15 July 2020.

Felix, Britany. Personal interview. 22 June 2020.

Frazier, Megan. Personal interview. 11 and 21 Aug. 2020.

Goto, Hiro. Personal interview. 13 Feb. 2021.

Gross, Jeff. Interview with April Tucker and Ryan Tucker. 14 Jan. 2021.

Hendley, Alesia. Personal interview. 17 Dec. 2020.

Hendley, Alesia. "Professional Social Media." *SoundGirls*, soundgirls.org/professional-social-media/.

Hutten, Mariana. Personal interview. 19 Nov. 2020.

Kelly, Tom. Personal interview. 22 June 2020.

Masingale, Langston. Personal interview. 14 Sept. 2021.

Mack, Harry, guest. "Career Advice For Freelance Designers." *Sound Design Live*, 16 July 2013. *SoundCloud*, soundcloud.com/sounddesignlive/career-advice-for-freelance.

McLucas, John. Personal interview. 25 Jan. and 8 Feb. 2021.

McLucas, John. "5 Reasons You're Failing & How To Win." *The Modern Music Creator Podcast*, ep. 3, 30 July 2018, pod.co/the-modern-music-creator.

McLucas, John. "Building a Career in LOS ANGELES in the Modern Market – My Story." *The Modern Music Creator Podcast*, ep. 2, 23 July 2018, pod.co/the-modern-music-creator.

McLucas, John. "Drewsif on Common Band Mistakes, DIY Business, Touring Misconceptions." *The Modern Music Creator Podcast*, ep. 46, 27 May 2019, pod.co/the-modern-music-creator.

McLucas, John. "Lost With Making Content For Music? Focus Up NOW With This." *The Modern Music Creator Podcast*, ep. 89, 21 Dec. 2020, pod.co/the-modern-music-creator.

McLucas, John. "The Quickest Way To NOT Grow Your Network." *The Modern Music Creator Podcast*, ep. 9, 10 Sept. 2018, pod.co/the-modern-music-creator.

McLucas, John. "Why Picking a 'Niche' Is Stupid & Brilliant At The Same Time." *The Modern Music Creator Podcast*, ep. 91, 1 Feb. 2021, pod.co/the-modern-music-creator.

McLucas, John. "How I Rebranded My Friends Audio Business By Accident – Value-First Thinking." *The Modern Music Creator Podcast*, ep. 53, 15 July 2019, pod.co/the-modern-music-creator.

Mierzwa, Patrushkha. *Behind the Sound Cart: A Veteran's Guide to Sound on the Set*, Ulano Sound Services, Inc., 2021. pp. 281–282.

Morris, Ashton. "Freelance Game Audio: Getting Started and Finding Work." *Ashton Morris*, www.ashtonmorris.com/freelance-game-audio-finding-work/. Accessed 1 Aug. 2021.

Pinzon, Luisa. Personal interview. 10 Nov. 2020.

Reed, Pete. Email interview. Conducted by April Tucker, 24 Nov. 2020.

Stacey, Sarah. Personal interview. 19 June 2021.

Thakkar, Akash. "Successful Freelancing in Game Audio." *YouTube*, uploaded by GDC. 9 Jan. 2019. youtu.be/93ggs7hwJeU.

Thériault, Marc. Personal interview. 4 Dec. 2020.

Tipp, Nick. Personal interview. 22 Sept. 2020.

9
What Makes You Stand Out?

Talent, as it relates to faders and knobs, only gets you to the conversation. There's a whole bunch of intangibles that keep you there. So, you might get noticed because you can mix really well, but you won't stay at the highest level if all you can do is mix really well.

– Jason Reynolds

The audio industry can be highly competitive, especially in the creative disciplines. When a company or employer receives hundreds of resumes for a gig, or job inquiries daily, how can you increase your odds of being considered? If you are trying to build a business when there are over 400 million active websites, over a billion users on social media platforms, and five billion videos viewed on YouTube *every day*, how do you stand out?

Soft skills are personal traits that are desirable on the job, including social and communication skills, personality, and character traits. **Hard skills** are job-specific skills that are an asset, such as technical expertise, language skills, and more. Many of the same soft skills and hard skills are brought up by professionals as vital traits – regardless of their job or discipline.

Good Attitude and Doing What Is Needed for the Job

Canadian boom operator/sound utility Camille Kennedy built her experience by saying yes to opportunities others turned down:

I took every job. It didn't matter what it was. In the middle of the winter outside at night? Sure. Nobody else wanted to work on a Friday night standing in a field at minus 20°C [–4°F]. I'll do it.

Claudia Engelhart, who has worked with jazz guitarist Bill Frisell for over 30 years, says:

DOI: 10.4324/9781003050346-10

I'm happy to bring you bottles of water if that makes it easier for your day. I'll go get coffee if you want me to get coffee. You really have to be willing to do that. They like you, also, because you went out of your way.

Meegan Holmes, Global Manager of Eighth Day Sound (one of the largest touring sound companies in the USA), says of her early career:

I just did what people didn't want to do. Dude didn't want to mix monitors? I'll mix monitors. Didn't want to deal with RF? I'll do the RF. I'll do the intercom. You don't want to mix front of house? I'll do it today.

As a hiring manager, Meegan sees a willingness to work on anything as an asset:

The most useful people we have on our roster do not care what they are doing. I call them and say, "Are you available September 16 through 19?" "Yeah, I am." There's no, "What am I doing? What is it? What's the gig?" They might ask eventually. But, there's a group of people who work for us that are like, "I'm there. I'll do whatever. I just want work."

Game Audio Lead Jack Menhorn shares:

It can be off-putting to work with someone who doesn't seem all that interested in their job and is looking for a way to step up to a "better" role or job. I have worked with some people who were in a role that they weren't very interested in because they were trying to move up, and it was unpleasant for all involved.

When music producer/music editor Fei Yu works at recording studios around the world, she remembers the assistants who maintained a good attitude:

After a week of working with you, I want to feel like you really love your job. Next time when I come back, I'll want to work with you again. I think that's the most important thing. Most people are super enthusiastic at the very beginning but after a week, I feel maybe they don't want to do this. They don't want to be an assistant at all. They just wanted to show they were trying at first. They will not be very willing to help you.

Jack Trifiro (live sound engineer and tour manager) adds:

Even if you think you've made it, you're going to work just as hard as everyone else. As a tour manager, I expect anybody to do any job that I ask them to do without turning it down. I've had crew members that I've had to let go because of that. If I'm working hard, I expect you to be there right beside me working hard as well.

Good Attitudes: James Clemens-Seely

Recording engineer James Clemens-Seely has overseen teams of audio practicum (apprentices) at the Banff Centre for Arts and Creativity in Canada, where they record music in the studio and at live events. James emphasizes having a good attitude regardless of your personal interest in a project.

> At one time of year, there's a lot of very contemporary music that may not appeal to the people in the engineering program that they have to work on. One of the skills that I demanded from people was the ability to turn on your love and compassion for whatever project you find yourself in right now. Your responsibility . . . is to support these humans. If you can't bring yourself to behave like that just because it's not the kind of music you like, or you can't see how it relates to your career goals . . . I think you're bad at being a person [in this industry] if that's your attitude at this moment . . .

> Sometimes they'll say, "I don't care that I'm bad at this job because it's not the job I want." I want the best of people to shine through at all times, and so here's a moment where the best of that person is not shining . . . They have the ability to pay attention to something. That's what you have to do right now and not for any other reason than that. Not for any outcome or expected reward, but just because the assignment right now is to pay attention.

Work Ethic and Drive

Kate Finan, supervising sound editor and co-owner of Boom Box Post, found her work ethic helped her stand out while working at a post-production sound studio:

> I was a secretary at my first job, and I got promoted within one month because I used to come in early and I would stay late every day. I would ask every single day, "Who can I observe today? While I'm eating my lunch, can I sit in on a mix?" When I got told [at the end of a shift], "Go home," I would say "No, I think I'm going to stay and I'm going to check all of your cables for you." Those are the things that I did. That does not go unnoticed because you're not just saying "what am I doing for my career?" every day. Instead, you're saying, "what can I do to make your job easier for you?"

Michael Lawrence says of standing out in live sound:

> Even when the gig doesn't go well, people can really spot a person who is trying hard to make it work. "Hey, we had this guy come in and he showed up, he was on the ball, and we didn't have to ask him for anything. He was polite and he was on time." Just that feedback will put you in a very high percentile in this field.

Your attitude about the job can affect your work ethic and drive. Karol Urban (re-recording mixer and sound editor) says, "When I am working on whatever job that is given to me, regardless of what it is, I make it as important as the biggest job that I could ever work on."

Jason Reynolds, live sound engineer (Shaggy, Bob Marley) adds:

> Treat every gig as if it's your dream gig. Would your stage look like that on your dream gig? Would your cables be run like that? . . . If my production manager or if my tour manager is okay with it, what about your standard? Work hard and be the best that you can be, but have that personal standard of excellence. Those things are important and often understated.

Work Ethic in Live Sound: Karrie Keyes

Karrie Keyes has been a touring monitor mixer for over 30 years (including for Pearl Jam, Red Hot Chili Peppers, and Neil Young). She is co-founder and Executive Director of SoundGirls, an organization that supports women working in professional audio. Karrie gives this advice about how to stand out through your work ethic.

- **Be on time.**
- **Show up for work.** "You would be amazed at how many people actually don't show up for a show they said that they were going to do."
- **Don't cancel a gig because you get offered a better gig.** "The only time that this doesn't apply is if you get a tour. Everybody's going to go, yes, you just got picked up three months of work. Go!"
- **Do the work without complaining.** "Some gigs suck. Sometimes you don't get time to go eat dinner, or you didn't get time to go do whatever, but no one wants to hear about it because everybody's been in that position."
- **Load and unload trucks.** "That is the number one thing of the job is to get the gear out of the truck and set up. There are A-listers that load trucks. So, you'd better be willing to load the truck and get dirty."

- **Be prepared.** "Make sure you have some snacks, a water bottle, sunscreen, and a jacket in your backpack so that you can survive a 16-hour day having no idea what type of elements you're going to be dealing with."
- **Ask questions** including about your responsibilities. "Don't ask a million questions, but ask questions . . . Ask, what is my role here today? What do you expect of me? People are going to tell you."
- **Don't overstep your role.** "Don't get off on the wrong foot with the front of the house engineer. Your opinion is not warranted . . . It's not your place and no one asked you, and now it's just irritated people."
- **Make friends.** "It's better to make friends with whoever because most people are willing to show you and teach you. Maybe not that day because they don't have time. If you're generally respectful and say, 'I'd really like to sit down with you,' most people are open to that."

Curiosity

In the audio industry, there will always be something new to learn. Jason Reynolds explains:

> It's important to always have this attitude of learning because you don't know it all. You'll never know it all. I look for [people] who have this attitude of "Yo, I want to learn. I'm always trying to get better, always trying to get to that next level."

Catherine Vericolli (music studio owner/manager and engineer) adds:

> I'm interested in how much a person wants to soak up and suck up knowledge. What are you reading? Are you studying stuff? Are you getting into things that are new for you? Are you building gear? There are lots of different weird things that somebody could be doing when they can't actually go to work. In what ways are you educating yourself?

Recording engineer James Clemens-Seely has found curiosity to be an asset on the job:

> The notion of having as many skills as possible and as much variety in your experience as possible – It has served me well. I love knowing enough about how to play the bass to talk to a bass player, and I love knowing how to program a lighting board enough to talk to a lighting designer. I didn't have the energy to get good enough to do any of those things at the high

level that I would have liked to, but I could talk to high-level people, identify with them, and help them be comfortable and feel understood. I like
doing that.

Soft Skills and Introverts: Devyn Nicholson

*Devyn Nicholson is Manager of Audiovisual Services at McGill University in
Montreal (Canada). Devyn oversees the AV needs of construction projects on
campus and is involved with hiring crew for his department.*

People who are introverted – it may zap their energy more than
someone who's extroverted, but the communication skills needed
are not being boisterous. You don't need to be charismatic. You don't
need to be a spokesperson. I'm looking for: you were asked a question,
and you responded in a friendly and kind manner with information
that was accurate and of value (given the situation).

A message can be delivered in many different ways. The message
being delivered in an extroverted way versus an introverted way is
really not an issue as far as I see it, unless it is someone who is so
introverted that they literally can't even answer. That might be an
issue. But, the same is true for an extrovert on the opposite end of the
spectrum. If they literally cannot stop themselves from gushing all
over the person with an answer that just never ends, we might have
an issue with that, too . . .

A lot of the time, it's really the calm people that we're looking for.
It's people who do not get flustered in the middle of the walls burning
down around them – especially for our techs who get deployed into
the field to have to stand in front of 600 people [in a class] who don't
want to be there in the first place. If you lose your cool, it's exponentially harder to solve your problem.

Being Respectful and Working Well with Others

Treating others with respect – both clients and your colleagues – is an important
soft skill. Chris Leonard, Director of Audio at IMS Technology Services (a
concerts and live events company) explains:

No amount of modeling numbers that you can rattle off to me or console that you've mixed on – I don't care about any of that. If you can't speak to me or other people like decent human beings or with respect (regardless of whatever is going on) you're not going to last.

Recording engineer James Clemens-Seely adds:

Most of the people who are looking to hire someone like me don't really care about the difference between excellent sound and okay sound if I'm nice and efficient enough. But, they do care if someone wasted time or was a [jerk].

Live sound engineer Chris Bushong says of being mindful of others while on tour:

I know a lot of people on the road that have lost gigs. They were very competent at their job, but afterward, they got on the bus and drank too much, or wouldn't keep their bunk clean, or something like that. When you're 10–12 people living together in 80 square feet? No. Part of it is just being an adult.

As an emerging professional, one skill to be developed is the ability to communicate about audio (such as your preferences and what you hear). Scott Adamson (touring front of house mixer) explains:

If someone has an undue amount of confidence (if someone's too arrogant) that can be very off-putting. When people think, "Oh, I really should be mixing this," or "I should be doing this" because of me, me, me, me – That doesn't work because we work in a service industry. Ultimately, we're serving artists and trying to help them realize their vision of their show. So, when someone makes it more about them than about being involved in music and being involved in what we're all trying to create collectively, that's a bit off-putting. I wouldn't put them forward as much as someone who had a better attitude.

Karrie Keyes adds:

It's super irritating to have been doing this for 30 years and I hear [someone] say, "I could mix this. If I was mixing this, I would do it this way." The house engineer does not want to hear that. My attitude is, that's fantastic, but I mix for the bands, and this is how they want it . . . It's the same as assisting in the studio or interning at a studio. Know who's in the room and don't overstep your boundaries.

Getting on Broadway by Being Nice: Kevin McCoy

Kevin McCoy is Head Audio for Hamilton ("And Peggy" Company). Kevin landed his first Broadway tour earlier in his career thanks to a relationship he made during technical rehearsals for a show.

> I was teching a show in Duluth, Minnesota. One of the local crew was normally a touring person on big, big shows, and he just happened to be home on a hiatus and worked as a local crew for our tech production period. That was definitely a case of "fake it 'til you make it," because I had never worked on a Broadway-style show. I even went to him and asked, "Hey, what's an A2 [assistant] supposed to do? What do I do during the show? Do I just like, listen to these microphones?" He said, "Yeah, you just listen to those microphones." On that show, I didn't know what [line-by-line] mixing was. That was the first show where I had to learn to do that.
>
> So, that was definitely faking it. We had a good enough time, but then we went on tour, so I didn't see him again. Six months later, I got a call from him looking for an A2. It was a big step up because it was a First National (the first touring production of a show), which is a prestigious thing. When I asked him later why he hired me, he said, "We did that show in Duluth and you were nice." I said, "Even though I asked you how to do the job? I had no clue what I was doing." He's like, "Yeah, I didn't care about that. It was just that you were nice."

Teamwork is expected in many environments. Gabrielle Fisher, Audio Designer at Disneyland (USA), explains:

> What makes you stand out is a willingness to collaborate. You feed off each other's energy and knowledge and you're going to be working with other audio people. [At a theme park], you're also going to be working in full-scale entertainment production. You're going to be working with broadcast people . . . Lighting people . . . So, you've got to be able to work well with others.

Kristen Quinn (Audio Director, Polyarc Games) adds:

> Being able to take feedback is so critical. That is a skill unto itself. It was hard when I started – I used to get a little defeated or beat down or think they didn't like my stuff. All I want is their approval, but it's not personal.

I might make the best sound in the world and it doesn't work for the thing I'm trying to build this out for. So, it's not a testament to whether your stuff is good or bad. Everybody is trying to collaborate to make the experience better, impactful, and deliver the information they're trying to convey.

While confidence can be a helpful quality, there is a threshold where it may work against you. Sal Ojeda (re-recording mixer and sound editor) shares:

Everyone thinks that right after graduation you're going to work with the next big artist and that studios are eager to get your talent. Of course, you have to be confident in what you know and your talent. But at the same time, you have to humble yourself and remember that you're nobody. Even if you know how to do whatever gig you're getting, there's so much stuff that you don't know.

Having a good attitude and being respectful are especially important in creative environments. Chris Bushong explains:

You can learn the skills and the technical stuff from the people above you. But if people don't want to work with you, then it doesn't matter. You can be the most brilliant engineer in the world, but if you're a [jerk], nobody's going to hire you. It has to do a lot with the way creative people interact. They all have a passion for it, and if you're the guy just screaming in the corner for no reason, they're going to be like, "Yo, this dude's got a bad vibe. I don't want to be around him." They want somebody that's going to be into their project or into whatever is going on.

Even those who are freelance or work alone most of the time need these soft skills. Radio and podcast producer Sarah Stacey (who works with independent clients and for the BBC) explains:

[People who wouldn't be good at this work are] people who are not great listeners, like, they're not great at taking on feedback or working with other people. Because even if you do work alone a lot of the time, you still need to have sounding boards, and you still need people to tell you where you're going wrong, or where you could improve. So, if you're not open to that feedback, then it's not a great fit.

Being Trustworthy: James Clemens-Seely

James Clemens-Seely has worked with artists from garage bands to major orchestras and recorded everywhere from storage lockers to concert halls and churches.

Anytime I'm working with new musicians and I'm placing microphones one on one, I'll say, "Is this okay here?" You can make a really great recording of an uncomfortable musician and it'll sound like an uncomfortable musician. I try to hammer home the idea that I'm here for musical reasons, and [tech] should not be getting in the way of the music . . . I think it sets them at ease in a way that is really important to do. It's very easy to overlook.

This applies with literally any musician. Rock and roll drummer with a synth: "Is there a cable that you're tripping over?" It sets that tone for a supportive presence rather than another source of stress . . . For a front of house mixer, if there isn't feedback in the first five seconds and when somebody goes up to talk to a mic and says, "can you hear me?" and that mic is live – that stuff builds trust right away . . .

It's doable in subtle ways. You can just be clear and thoughtful in your communications before and after with a client, in your delivery, punctuality, in a whole bunch of non-musical/non-artistic ways like that. Most people, even subconsciously, they'll see that you were kind in your communication with them, and that you were offering support and that you were understanding, and all of that helps build trust. I think it's through a million tiny things that you can build a trusting relationship almost immediately.

Navigating Difficult Situations

Working in the audio industry can require troubleshooting skills – but not just related to equipment. Difficult situations will arise where you will be expected to interact with other people (from clients to co-workers) while solving problems. One question asked in interviews by AV manager Devyn Nicholson is "You're an installer, and someone comes up to you in the room. They're mad, and they're complaining. How do you handle that?"

Music producer and Foley mixer Jeff Gross adds, "You have to know when to take a step back and realize that you're dealing with somebody who thinks that they're a great communicator, but they probably are not."

Problem-solving can be a natural part of some work. Kristen Quinn, who works in sound for video games, explains:

Conflict can be healthy. There is a very healthy conflict that exists innately in a lot of game development, like someone trying to stay on schedule versus someone wanting to build a really amazing, emotional design experience. Or an audio person saying, "I want to spend this much longer on this because we can make it better." All those things are innately in conflict. So, you're constantly having to work with people and be in conflict. But, it doesn't have to be adversarial. It's not an "us versus them" thing. We're just a team. We're all a group of people working together.

People Skills in Jazz: Claudia Engelhart

Claudia Engelhart is a touring sound engineer and tour manager for jazz artists.

> There's so much psychology that goes on. If somebody is having a bad day and they're moody or they're being grumpy, you've got to know when to just let them go and chill out, or bring them whatever it is they need to make them feel better. It's a lot of just accommodating people and trying to keep them comfortable throughout the day, really. That's how I feel about mixing sound – it's making it comfortable for them. They don't have to think at all about sound. You want to just give them the space to be able to just be creative people out there because it's a pretty intense process to play music live every day, especially improvising . . . Jazz is heady music and a lot of jazz musicians are in another world, so you really have to be sensitive to their space. You have to be ready and willing to be there, but at the same time, know when not to be in their face all the time and just let them be, and be creative, and do what they do that's so great.

Survival Mindset

The audio industry can require some creativity to learn skills, gain experience, and find ways to sustain financially. As an emerging professional, Alesia Hendley recognized a need to learn equipment that wasn't available to her but had to think outside the box to get hands-on time:

> When I was in school, I got to work with SSL consoles and it was amazing, but I knew those boards weren't at venues. I couldn't walk into these places saying, "I'm an audio engineer but I have no experience with the consoles you have" . . . One of my classmates actually had a job at a Guitar Center. I'd go in and get some work in. We were bouncing ideas off of each other and improving together.

Mariana Hutten found diversifying, or working on a variety of projects, has helped her sustain herself through difficult times:

> I'm the type of person that I like to have several things going on. I'm so thankful that I'm this kind of person because that's what saved me from COVID. I was always working [through COVID] because I'm doing all these things. I just don't feel comfortable doing only one thing. I have that added constant fear that I'm going to lose work. I need to be doing more than one thing to feel secure, to feel safe.

John McLucas adds:

> [You're] jack of every single trade you can possibly think of because you don't have a full calendar. For myself, not only was I recording artists and bands, but I was recording voiceover work for voice reels, I was doing overdubs for web series, mixing web series, doing drum charts, mixing, recording, session work, going on the road, helping out at a merch booth – whatever I could get my hands on.
>
> ("Should You Be")

Being open-minded can pay off later in a career. "Others go hard for a smaller group of opportunities, where I said 'yes' 800 times – and eight of them were so pivotally important that they blossomed out and created the kind of life that I wanted," says John McLucas ("interview").

Sometimes there is a need for a skillset and no one to fill the role. Fei Yu was building a career as a music editor (sound for picture) in China when she noticed the Chinese market was missing one important role: music supervisors. Fei started to learn music supervising and offering that to her clients in addition to music editing and music producing:

> Everybody started to become aware of how important publishing is, how important it was to clear the license for the copyright, so that's why I started. There were no professional music industry people doing this kind of thing, so that's why I started to see myself as a music supervisor.

Ways to Differentiate: Rachel Cruz

Rachel Cruz is Program Manager (Electrics) at Fender Musical Instruments Corporation. Rachel applies her expertise in learning what customers want to how an emerging professional can stand out.

> When you're on the outside and you're trying to get in, ways to differentiate could mean how you interact with people. The basic

stuff – being known as reliable and easy to get along with. They shouldn't be differentiators, but they are. Then, the differentiation is constantly showing how you do the work . . . I know that these things don't seem very interesting and they're pretty obvious. But, I found even when I was young, a lot of people were not good at the basic stuff. It's not rocket science: do the basics really well, do them repeatedly, and you will differentiate yourself.

Let's say you get an entry-level job at a game developer or studio. Never make it about you. Look and see what the company needs. See where the gaps are and then figure out how you can serve up value to the company by plugging that gap. People aren't going to fall over you for being an awesome audio engineer. All anybody cares about is how you perform in service to the company. That's actually a big way to differentiate because most young people don't think about it themselves that way. They think, "I've got to be awesome, and I've got to be the best. They'll understand that I'm the best." But you've got to think about yourself in relationship to the company . . .

A lot of audio engineering is very subjective, so your success is really dependent on your reputation and your interpersonal skills. Reputation is not necessarily as important at the front end as people think it is. It's always important, but it's more important as you're moving out of that starter phase of your career and going into the middle of it. Building your reputation is about positioning yourself for the next thing. But when you're starting, you don't have much of a reputation. All you have are your differentiators – your basic stuff . . .

I've found that if I can just get myself into the environment, like into the room or into the studio, I can figure out how I can add value into the environment. There are a million ways to differentiate once you're on the inside.

Hard Skills

Hard skills are job-specific skills (including technical skills). Any hard skill – from a driver's license to knowing CPR – could help you stand out as a job candidate or on the job.

Broadcast engineer Ann Charles says:

> You can become the world's expert at something really quickly because it's such a small industry. For example, if you like accessibility [tools for sight- or hearing-impaired] and radio systems, there aren't that many people doing it, so very quickly you can become the one who knows more about it than most people. It's a really friendly and supportive community.
>
> (qtd. in Gaston-Bird, 96–97)

Music producer, music editor, and music supervisor Fei Yu has built a business pairing Chinese filmmakers and English-speaking composers. Fei's business requires her to be bilingual, and she has found it an asset to have a bilingual assistant. Fei shares:

> Since they understand both languages, they can help me more. For example, If I am dealing with jetlag, a time difference, or if I need to deal with a composer, I may need an assistant to get the music to the director.

Language Skills: Claudia Engelhart

Front of house mixer Claudia Engelhart had a once-in-a-lifetime opportunity in Brazil with jazz great Miles Davis – thanks to her language skills.

> Miles Davis was starting his tour in the same venue that we had just finished doing a weekend in Rio. I speak Portuguese and the crew in Rio asked me, "Could you come down the first day and help us translate? None of us speak English well." I volunteered, and I ended up being on tour with them for two weeks in Brazil just as a translator [for] the crew . . .
>
> I established this relationship with this whole crew of guys that was probably one of the best things I've ever done in my career – as a translator! Because I knew sound, I could translate all the technical stuff . . . I had the skills of being able to speak the language and I had the skills of knowing what they were talking about. I got more gigs off of that – just volunteering for two weeks. They all liked me and trusted me, and I got back to the States and boom, my phone was ringing . . . Those guys were super great to me.

Aline Bruijns found her training as a jazz singer helpful as she pursued a career in post-production:

Because of that study . . . it's so easy to talk to composers. I could talk to (or direct) voice actors because you understand how the voice works, and also feel what is logical when you need to insert a breath . . . You understand your instrument, you understand the vocals, you understand where you need to have a breath or when it's really hard within your vocal range.

For live sound engineers interested in going on tour, tour or production management could be a hard skill that helps land work. For example, Jason Reynolds learned how to write riders for his own band. When Jason was touring (as monitor mixer) and saw their rider could be improved, he offered to help – and it later grew into a job. "If you're prepared and an opportunity comes, you get the gig. The problem is a lot of us are not prepared, and then opportunities pass us by because we're not prepared," says Jason.

Unique Skills on Set: Patrushkha Mierzwa

Patrushkha Mierzwa has worked in production sound for over 80 movies, including Quentin Tarantino's The Hateful Eight *and* Ad Astra *(directed by James Gray).*

> When *The Hateful Eight* production went to Telluride, Colorado [USA], to shoot, we started a search for an intern . . . I think there were over 600 applications for 30 jobs. I called the one person I knew who lived there but hadn't worked with him in 34 years; not only was he still there . . . but [he remembered] me! (Be kind to everyone.) He gave me the name of someone to interview, and I called her. She was working for a film festival and stressed her student directing. I said, "You have no idea what you're in for, so don't even tell me about your directing." We talked more. I needed to know what kind of person she was, and could she handle the stresses of major moviemaking?
>
> Why was she the one hired over everyone else? She had been on the [high school] ski team. Think about it. Does a sound team need a student director on a Quentin Tarantino movie? Or do they need someone who could run to the truck 1/4 mile away through snow to get a windshield? I knew the trucks would be parked down the hill and imagined that someone acclimated to high altitudes and who owned snowshoes would be invaluable. She did a great job and was hired again for *Ad Astra.*
>
> (274)

From "Behind the Sound Cart: A Veteran's Guide to Sound on the Set"

Unique Musical Skills: Hiro Goto

Hiro Goto has found having unique musical skills and an audio engineering background has been a way to stand out. Hiro earned a degree in music performance (violin) and music production and engineering (from Berklee College of Music, USA), has jazz and improvisation skills, and plays electric violin with a rig of effects processing and a guitar amp. His unique skillset attracts attention anytime he plays a live show (with bands ranging from pop to metal, jazz, or country).

"It's big that I can play and record because a lot of violinists or musicians don't know anything about audio engineering," says Hiro. "They know their instruments very well, but it doesn't mean you know how to mic yourself or understand the sonic aspect of it . . . A lot of people know me as a violinist who can improvise and also engineer (so I have the proper gear to record myself)."

Being conversational at shows (especially with musicians and live sound engineers) has led to gigs in live performance, as a studio musician, live sound engineer, and music engineer. Hiro explains, "When I play on stage, I get hired by other groups. That happens quite often. A lot of times I perform and then they find out I'm also an engineer. It's like, 'Oh, can you do this too?'"

Hiro also acts as a resource for other musicians interested in learning similar skills, and by offering help, he has received recommendations for work. "Violinists who want to do electric violin will contact me or contact someone that knows me. They'll say, 'Ask Hiro.' When people ask me questions, I'll just answer them. There's nothing to lose – just give them information. It's not a secret. You don't want them to have crappy sound and make the show's engineer struggle. You can help make it better," Hiro says.

Technical Ability

Troubleshooting skills (and being able to solve a technical problem) can help you stand out in a memorable way. Camille Kennedy, who works in production sound on-set, shares:

> In one situation, I was told, "We need to do playback, and we can't hire someone. Everyone's booked." . . . I figured it out in ten minutes, went to

the director, and said, "We got it working. Check this out." They said, "You're on for the rest of the show." It's how you get jobs in studios. You always hear that – how did you get the job in this mixing studio? Oh, I was walking down the hallway, and something was broken and they asked me if I could fix it, so I fixed it.

Eric Ferguson adds that attention to technical detail is important for building trust with clients (and potential clients):

I do a lot of music mixing for mostly Thai clients, and it's all about the deliverables. It's all about when it gets there, how it looks, and is it in the right format? Is it downloadable really quickly? When people go to your website and click something and it doesn't work, how can they think that you're going to deliver the right sample rate and bit depth? Can you name it correctly, encode it correctly, have it on a secure server, all of that kind of stuff? File names, where files are saved – your own organization is crucial. That's critical in our business now because we all work remotely.

When working as part of a professional audio team, you may be required to use the technical tools dictated by others. Being proficient with industry-standard programs, plugins, and gear (even if they are not your personal preferences) can be the difference between landing some gigs or not.

Using Industry Standard Tools: Sal Ojeda

Sal Ojeda is a re-recording mixer and sound editor for film and television. Sal also has a music background and has worked on the sound crew for major US live broadcast productions such as the Emmys, Oscar Awards, and television musicals.

I'm the new guy coming into their world, so I adapt. I work with whatever they use. That's a reason why you should know what's going on in terms of technology. If you're really serious about coming to LA or New York and getting into the industry, then you're going to have to learn what people use. When I got to LA, I never used a trackball before. As soon as I noticed that's used in all the studios, I bought one and I got used to it. At that point, I didn't care if I like trackballs or not. I saw that's what the studios used, and I bought the same model. That's something that can make your day go twice as fast or slow you down. That's the mentality that anyone should have.

In television, no one is going to pay you to learn on the job. It's really hard to get a gig like that, but it happens every now and then . . . I knew that I had to somehow get my hands on that piece of gear or that plugin to learn it to be able to do the gig when the call came. You can't call yourself an engineer if you don't know how your system works. Could you break everything down and put it back together again? I really enjoy finding out how something works, and having that confidence that if something happens, I can troubleshoot it.

Standing Out in Audio School: Leslie Gaston-Bird

Leslie Gaston-Bird taught for nearly 15 years at the University of Colorado – Denver (USA). Leslie could identify the students that stood out early in the school year by their attitude and self-initiative.

We drilled over and over in our students that you are paying for this space, so use it. What would happen is the first 10 weeks of the semester, nobody was in the lab. But that 15% [who stand out] were in there, like, "I saw the lab was open, so I just came in here and fiddled with the buttons for four hours" . . .

The students who get the high grade are the ones who said, "I believe in this song, and I want it to sound right. I'm going to go that extra mile to make a project I'm proud of. I'm going to do some research on how to put the mics correctly. I'm going to use my ears and I'm going to build my ears." When they showed up and played their project in class, I was like, that's it.

If we do our jobs as educators and we show students what the prize is, and if a student can keep their eye on the prize (this lovely thing that you can make with these amazing tools) – if they get it, then the world's their oyster. The guest lecturer comes in and the student is talking to them and getting their contact information. A student is collaborating interdisciplinary with film, games, theater, live sound. They're offering to do PA for a guitar-playing friend.

As a professor, you're more inclined to put students in touch with resources if they show that they know what to do with the gifts that you're giving them. If it's sound design, then I might introduce them to some people in the film program and they work on their first film. If they're into games, same deal.

Let's be real. We're talking about the top motivated students. Most students don't even get that far. They show up, they get their grade, and they go home. I can bring in guest speakers, I can bring in guest lectures, and I can talk about what's going on [in the industry]. But most students have an idea of what they want to do, get their grade, and I don't hear from them again. It's that top 15% who I can put resources into – and those resources are networking resources.

("Interview")

References

Adamson, Scott. Personal interview. 14 May 2020.

Bruijns, Aline. Personal interview. 21 June 2021.

Bushong, Chris. Personal interview. 27 Sept. 2020.

Clemens-Seely, James. Personal interview. 20 Dec. 2020.

Cruz, Rachel. Personal interview. 9 Oct. 2020.

Engelhart, Claudia. Personal interview. 8 June 2020.

Ferguson, Eric. Personal interview. 2 Dec. 2020.

Finan, Kate, panelist. "SoundGirls Career Paths in Film & TV Panel at Sony Studios." *YouTube*, uploaded by SoundGirls, 26 Oct. 2018, youtu.be/Y_g3drC2yyA.

Fisher, Gabrielle. Personal interview. 18 Aug. 2020.

Gaston-Bird, Leslie. Personal interview. 2 Oct. 2020.

Gaston-Bird, Leslie. *Women in Audio*. Routledge, 2019, pp. 96–97. doi.org/10.4324/9780429455940.

Goto, Hiro. Personal interview. 13 Feb. 2021.

Gross, Jeff. Interview with April Tucker and Ryan Tucker. 14 Jan. 2021.

Hendley, Alesia. Personal interview. 17 Dec. 2020.

Holmes, Meegan. "Career Paths in Live Sound." *YouTube*, uploaded by SoundGirls, 1 Oct. 2019, youtu.be/dbBxX9PddhI.

Hutten, Mariana. Personal interview. 19 Nov. 2020.

Kennedy, Camille. Personal interview. 21 Feb. 2021.

Keyes, Karrie. Personal interview. 19 June 2020.

Lawrence, Michael. "Getting an Audio Education, Landing the First Gig, and More." *Signal to Noise Podcast*, ep. 12, ProSoundWeb, 6 Sept. 2019, www.prosoundweb.com/podcasts/signal-to-noise/.

Leonard, Chris, guest. "Chris Leonard, Play Well with Others." *Church Sound Podcast*, ep. 11, ProSoundWeb, 16 July 2020. www.prosoundweb.com/podcasts/church-sound-podcast/.

McCoy, Kevin. Personal interview. 22 Dec. 2020.

McLucas, John. Personal interview. 25 Jan. and 8 Feb. 2021.

McLucas, John. "Should You Be a Jack of All Trades or Master of One? The Answer is More Complex Than You Think," *The Modern Music Creator Podcast*, ep. 13, 8 Oct. 2018, pod.co/the-modern-music-creator.

Menhorn, Jack. Email interview. 30 June and 26 July 2020.

Mierzwa, Patrushkha. *Behind the Sound Cart: A Veteran's Guide to Sound on the Set*, Ulano Sound Services, Inc., 2021. p. 274.

Nicholson, Devyn. Personal interview. 23 Nov. 2020.

Ojeda, Sal. Personal interview. 27 Sept. 2020.

Quinn, Kristen. Personal interview. 30 Sept. 2020.

Reynolds, Jason. Personal interview. 27 July 2020.

Stacey, Sarah. Personal interview. 19 June 2021.

Trifiro, Jack. Personal interview. 4 June 2020.

Urban, Karol. Personal interview. 17 July 2020.

Vericolli, Catherine. Personal interview. *What Makes You Stand Out*, SoundGirls, 20 May 2020, youtu.be/wYamiOK8Y6A. Accessed 21 May 2020.

Yu, Fei. Personal interview. 9 Nov. 2020.

Part 2
Operations

10
Rates and Knowing Your Worth

"How much should I charge?" is a very common question asked by emerging professionals. Discussing money with clients (and colleagues) can be uncomfortable, but is necessary to build a professional career. Catherine Vericolli explains:

> Knowing your worth is the foundation of how you will survive in this industry. It's also the most difficult thing to pinpoint. It's hard to talk to other people about money. It sucks and a lot of people don't like talking about it. With that being said, the more to the point and transparent you can be about it, the better. Do your research. Ask people that you know, ask friends of yours, and be open and honest about that stuff . . . But, you have to know what's appropriate and also what you're okay with for your own worth. It's tough to navigate through.
>
> (Holmes et al.)

James Clemens-Seely adds:

> I hate the money mystery in the arts. You go to a lawyer's office and you know what the money deal is before you talk to them. I want it to be like that in music as much as possible, because we're doing this because we love it . . . I always want to be very clear if I'm [hiring] someone what they will be paid and what they will be expected to do. If I'm contracting for someone, it's the same thing: sharing what my rates are and what the costs are.

Rate is the amount charged for a service or product. It is possible to set your own rates or have a rate set by a client or employer. Your rate can vary based on your personal preferences, relationships, and much more. A **market rate** is the typical rate (or range of rates) charged for a service. Rates, like personal priorities, will differ for everyone, which can make it difficult to find a market rate. Looking to public job listings, advertisements, or rates posted by other

DOI: 10.4324/9781003050346-12

professionals will not give the full story of a market rate. Recording studio owner Catherine Vericolli explains:

> In a studio situation, we have a multitude of different freelancers at different levels. If I have a client that can afford to pay [higher], then great. If it's a good match, then cool, but if I know their budget is low and I have somebody willing to work [lower], then cool.
>
> (Holmes et al.)

Michelle Sabolchick Pettinato says of rates in live sound touring:

> There is a very wide-ranging pay scale in this industry and there is always someone who will work for less than you will. Every tour is different and has its own budget. Working for a relatively unknown band on their first national tour is unlikely to pay the same as a very fast-moving, up-and-coming band selling out every show. That being said, I have done tours for very well-known artists with multiple platinum albums who have paid less than other artists who were far less successful.
>
> (55–56)

Market rates can vary based on where you live. Video game sound designer Javier Zúmer (who has worked in Spain, Ireland, and England) shares:

> I remember looking at rates from the US and thinking, this is too much. I get the idea that in the US, salaries are higher, but expenditures are higher . . . those numbers don't make much sense in Europe sometimes.

Rates can reflect the quality of your sound equipment or the work involved. Kelly Kramarik says of her rates for production sound:

> I'm priced a little bit lower than the other people [in the area]. But I don't feel comfortable charging as much as the other people that have Sound Device recorders and Electrosonic wireless systems. I have a Zoom recorder and I have Sennheiser wireless. There is a US$10,000 difference in our kits for those two things alone.

Even a genre of music could influence rates. Mariana Hutten says of her rates in Toronto (Canada), "In the hip hop scene there's more low balling for sure, but there's quite a jazz scene here in Toronto and it's kind of the opposite. They get suspicious if you're charging too little."

For touring professionals, being away from home is a consideration. Front of house mixer Scott Adamson shares:

> If I'm going to spend an entire week away from home, I'm not going to be able to see my friends or my family or my pets or whatever, what is that worth to me? . . . It's extremely difficult to renegotiate salary once you've

started. If it's someone you expect to work with for a long time, figure out what makes it worth it for you to go.

Rates can vary based on experience. Music producer John McLucas explains:

> With very high-end people, you're paying [for] their expertise and their ability to make intuitive decisions right off the bat that are so perfect and on-point every single time . . . That's the premium you're paying for. Let's say a photographer can give you 40 great [or] perfect photos in an hour because they're so good at taking photos. They're not going to be the same as an amateur who's probably going to need hours and hours to get 15 usable ones that you're probably going to have to clean up a lot.
>
> ("Day Rates")

Camille Kennedy says of experienced boom operators (for production sound):

> It takes 10 years to learn it before you can walk on set and say, "This light is going to be an issue," or "I'm going to boom from this side." You could walk on and boom right away, and you wouldn't need a rehearsal. You can definitely negotiate your rate [at that level].

The six reasons to consider an opportunity (from Chapter 6) may influence your rate. Tommy Edwards, Vice President of Product Development at Warm Audio, explains:

> Is this a financial project where I'm going to make money? Is it a project where it's resume building? Is it a project where if I do this person a favor and do this project, it could lead to something else? There are a lot of different things in that calculus for doing that deal. You have to balance those sometimes.

Michelle Sabolchick Pettinato adds:

> For me, money is not always the deciding factor. I have taken smaller tours for less money because I really wanted to work for that artist. I have turned down larger tours for more money because I knew they were going to be unpleasant, and taken smaller tours that I knew I would enjoy even though I was making less money. Everyone has different priorities and mine are more about mixing the kind of music I want to mix than getting the biggest paycheck.
>
> (55–56)

It will take some experimenting and practice to find what you can charge for your work (or even *when* you can start charging for your work). Whether you compromise on your rates is a personal choice.

Game composer/sound designer Ashton Morris shares:

> One of the lowest paying gigs I got when I first started out has, over the years, paid as much as my highest paying gig, mostly through the attention it received and the ease it made to meet new developers. This helped to create new clients, fans, and also sell many soundtracks.

Catherine Vericolli says of the risk of charging too low for your work to land gigs:

> Don't undercut people's rates. This is an industry where it's difficult to make this a full-time career. Budgets are not what they used to be, especially for in-studio stuff. The best thing that you can do for yourself and for the industry is to take a look at the rates that are around you and try to be competitive within those rates. [Charging too low] doesn't help the industry and it doesn't help the client. So, support your fellow engineer. The more support that you give to the people in this business who do what you do, the more successful you're going to be.
>
> (Holmes et al.)

The Risk of Underpricing: Tommy Edwards

Tommy Edwards is Vice President of Product Development at Warm Audio and has worked for Logitech, Blue Microphones, Line6, and Alesis.

> The mistake a lot of people early in their career make is they underprice themselves. I've seen this numerous times. They underprice themselves, they get into it, they realize how much actual work it is, and now they're not necessarily performing well . . . You made a mistake, you underpriced yourself, but at least the experience of this will be a big resume builder. It's easier to think, it's not about the money.
>
> From my side of the fence, because I hire consultants and engineers, I don't want them underpaid, because I realize on my end, I'm not going to get the results I want . . . I don't want to hire somebody for a job I know [they underbid] only to have them discover they underbid and now they're miserable. Now it's a drag and I can't get deliverables from them. That's not good for me, either. So, it is important to understand that that's why you have to do a lot of work [to know and learn] the fine art of asking somebody what their budget is.

Podcast editor Britany Felix says of compromising on her rates:

> I don't offer discounts. My rates are what they are. I'm worth it, and I still get business. I'm coming into it with that attitude. I am not going to

compete with another editor in a race to the bottom, as far as rates go. My clients, being in the industry they're in, they understand that, and they appreciate that. The person who is looking for the lowest price editor will never be my ideal client, and I'm not even going to go after them.

At the same time, there is a risk of losing opportunities if your rates are too high compared to others at your skill level. Front of house mixer Claudia Engelhart explains:

If you're going to play hardball financially, you'd better have some skills. You've got to have something to back yourself up with. If you're just starting out, you've got to be willing to do anything – whatever it takes. Then it'll start generating and the word will get out, "that person is really great."

Pro Tip: Setting Rates

What's your rent? What are your insurance costs (business, renters, auto, etc.)? What's your monthly cell phone bill? What do you spend on food each week? What software renewal/upgrade costs will you have each year? What does all of this add up to for a year? I set my rates based on that. I figure that needs to be covered even if I can only get six months' worth of work – and even then, that year of expenses should probably only be 50–60% of what I make in that six months of work. Because, don't forget, you've got to pay taxes on your freelance income, and you need to pay about a third of your income to the state and federal tax agencies [in the United States] . . . Once you've got those numbers, start doing the math to figure out what your weekly or hourly rate needs to be to make those numbers. That's your starting point. Your rates can't be any lower than that.

– Shaun Farley

Framework to Set Rates

A **set rate** (or fixed-fee rate) is a pre-established rate provided by a client. Live sound engineer Jack Trifiro explains, "House gigs have set pay. You can't argue it – That's the rate. If you don't want to do it, don't [take] the work." A **sliding scale rate** is when you offer a different rate for the same service to different clients (for example, mixing a song with five tracks versus 500).

Podcast and radio producer Sarah Stacey says of the difference:

> Some people dictate [rates]. This is how much I have, so if you can work within this range, that's ideal . . . Some people will ask, "What's your day rate?" Figuring out what to say is quite difficult. It's worth having a figure that you won't go below that.

A client may have different set rates for different projects. Recording studio manager Tina Morris explains:

> We have an agreement with the [record] labels . . . If I hire you as a free-lancer, I'm very upfront about what we pay, and you know how much we're charging the label . . . If we're hiring you, we also have a [profit] margin that we have to make, as well. I'm not going to give you all that money, unfortunately.
>
> (Holmes et al.)

Every opportunity requires some investigation to find what is a fair rate. For example:

- A house of worship may only have volunteers running sound while a megachurch could have the budget for a full-time sound person
- A songwriter who is self-financing an album (for creative fulfillment) may have a small budget compared to when they are producing similar music for a jingle house (who is paying the bill)
- A filmmaker or game developer's budget is influenced by whether they are self-funding, crowdfunding, or if they have investors
- A full-time job may have a lower rate (or salary) than a freelancer because of other benefits provided (the stability of a full-time job, retirement contributions, health care, etc.)
- There may be other comparable benefits provided (free meals, free or reduced-cost housing, per diem, training, access to equipment)

The **5W's and H** are Who, What, When, Where, Why, and How. These questions help to gather information used to set rates for both small and large projects. For example:

- Who is the client? Who will I be working with?
- What is the project? What services are needed? What equipment is required?
- When does the gig take place? Or, when is the deadline?
- Where will the work occur?
- Why is this project happening? (This may give clues to their budget.)
- How much will this pay? (Or, what is their pay range?)

The 5 W's and H in Live Sound: Samantha Potter

Audio engineer Samantha Potter suggests these questions to gather information for a live sound gig:

- What is the minimum wage in your area?
- How much do meals cost on average in your area?
- What does the gig entail?
- What is your position?
- Are you expected to load in and load out [equipment]?
- Are you being hired only as an engineer?
- Show size?
- Hours – Is it a 16-hour [commitment] or a 3-hour show?
- Corporate? Concert? Theater?
- What do freelancers with similar experience make in your area?
- Will you need to travel?
- Will they feed you?
- Is there a per diem?
- What is the size of your client? (How many people does the company employ? Or, what is the venue's occupancy cap?)
- What is the absolute lowest you're willing to make per hour?
- How much experience do you have?
- Does the gig pay hourly or a day rate?

Hour Rate, Asset Rate, or Flat Rate?

While it is impossible to gauge the exact time needed for every scenario, it is helpful to have a general idea of how quickly you can do the work required. If you are pursuing work as a podcast editor, how many hours does it take you to edit a 30-minute podcast? For a mastering engineer, how many hours does it take for you to master a basic song? The time required will change from project to project and will likely change as your skills improve. Rates can be given in several different ways:

- **Time needed.** This could be an hourly rate (where you are paid for actual hours worked) or a day rate (one price for up to x hours and overtime after that point).
- **Quantity of work required.** The workload can affect rate. For example, a video game sound designer may charge a rate per asset (sound created). A podcast editor may charge a different rate depending on the length of the episode.

- **Flat rate.** A flat rate is a single rate for an entire job or task regardless of how much time it takes you.

Rates can be calculated differently based on discipline, as well (see Part 3, Career Paths, for specifics). When multiple services are required (at different rates), they can be presented to a client as a package. For example, a song may require recording, editing, mixing, and mastering (each with a separate rate). This is presented to a client through a bid, quote, or proposal (see Chapter 11 for more).

Setting Rates in Music Production: John McLucas

John McLucas discusses the differences between using day rates and flat rates in music production work.

If your service is based around being in-person, like recording or production where they're sitting in-person with you . . . your direct time is being used up by them based on how prepared they are. For a lot of people starting out, the issue is you've charged a flat rate for which you thought was a 90-minute session, then [it turns into] a four-hour session. Your time just got cut into 30% of its value from what you expected it to be.

On an hourly rate, if they lag on being prepared, it's on them. If they're very prepared and it's shorter, then they're rewarded, and I'll tell them that directly . . . "This could cost you $400, or it could cost you $1,000. It's truly up to you."

Say we're recording a ton of vocals for a bunch of songs. I'll say, "If you come in prepared, you've got your lyrics ready, you've rehearsed, you know how you want to warm up, you can execute really, really well, and you know the song so well that we're just focusing on emotion, it'll go quicker. You're going to save money."

Once you have a main rhythm guitar tone set for 10 songs, it is 100% up to how prepared that person is. [I would say], "That first 30 to 60 minutes you're paying for me to dial something in, but the rest of it is literally based on how good you are. So, if you come in crazy prepared, you crush it, all 10 songs of rhythm guitar will be done in like two days versus eight days . . . You save a lot of money and then it makes the process a lot more fun."

Anything that's not in-person with them sitting next to me, that's all on me, as far as time. That's why I'll charge a flat rate . . . When a track needs extra time and attention, I just work till the job is done regardless of how long it takes. I don't feel pressured by saying, "I need four extra hours to tinker with this and I'm going to have to charge you more. Is that okay?" When charging a flat rate, you have to know how long it's going to take and feel like it's a fair price for that time . . .

There are projects where I will incorporate both. For example, it might be an all-inclusive package with two and a half days of in-person time, the mix, the master, the editing – It's all-inclusive from sending me a demo to the songs ready to be released (submitted to libraries and to labels and playlists). But, it includes a set amount of my time, which in essence is a day rate. I'll just say, "Any extra hours we need will be billed at this amount."

Some projects take longer, some take shorter, but it all works out . . . I like giving them that flat-rate package saying exactly how much time it contains with the caveat: if you go shorter, I can just discount the days by a half-day increment.

("Day Rates")

From the Modern Music Creator Podcast

Some preliminary information can be gathered on your own by searching online for the company or people involved. Then, the best way to gather information is by having a conversation with them. This will help determine if you are the right person for the opportunity and to communicate what you do (and don't do).

For example, some people do not know audio well enough to know terminology or job titles. In sound for picture, the term "sound mixer" is sometimes misused (a sound mixer works on production sound; a "re-recording mixer" works on post-production sound). A sound designer for video games has a completely different skillset required than a sound designer for sound for picture or theater. Tommy Edwards adds:

Anything that looks like the person is unaware of the actual time that it would take the average professional to do that job is a huge red flag . . . A good example: we want you to [edit and mix] a movie in two days. You see it in music technology, also: A mobile app takes a couple days, right? No, that's actually a couple months.

A phone call or meeting can also help you gauge how professional the project is and what the client would be like to work with. Sometimes there will be signs that a client is going to be more detail-oriented, unorganized, or need a lot of personal attention, all of which can result in extra hours on your end. Those concerns can be factored into your rate.

It is appropriate to ask directly about rates ("Do you have a rate or budget in mind?") if it has not come up already. Meegan Holmes, who manages employees and contractors, shares:

> If it's your first time working with somebody, it's totally appropriate to say to them, "How much money do you have for this position?" If it is a low number or a number you're not willing to take, it's perfectly acceptable to say, "Is there a chance that you could come up to X amount?"
>
> (Holmes et al.)

Rates in Practice: Samantha Potter

Samantha Potter, who works in live sound, recommends aiming for a starting wage of at least double the minimum wage. She offers three scenarios (in US Dollars – example minimum wage of $7.50/hour and starting wage $15/hour).

Scenario 1: A local 4-piece rock band wants to have a dedicated sound human with them at all their gigs to keep consistency high between clubs and bars. They don't ask you to help unload any of their stuff, but you are responsible for unloading and setting up all audio reinforcement gear. You deserve 1/5 of the full gig pay minimum. If the entire band makes $50 for the night, you're doing it for the love of it so make sure you love that band.

Scenario 2: A local production company wants you to come sub in on a theater production because their FOH [front of house] person caught the flu. You've got the script, gone over it, and feel prepared enough to [do the gig]. There's no load in, no load out – you come in, run FOH, and leave. In this scenario, preparation for the gig should count towards your work hours. The show is two hours long, and you spent two hours preparing in advance. $30/hour for a two-hour show would not be absurd here since you are not being compensated for your two hours of prep time. Other factors, such as venue size, estimated attendance, the size of production, and others may cause you to adjust your rate in either direction.

Scenario 3: A medium-sized corporation in the area has an end-of-quarter staff party. They need audio support for a jazz trio and MCs with presentations. They want you to load in all the equipment, run the show, and load out: a 10-hour day. The organization is well established, so I would revise my rule of thumb and make it triple the minimum wage. $21/hour times 10 hours is $210. *Minimum.* Another thing to consider in this scenario is meals. Will they feed you? If so, then don't worry about it. If they aren't feeding you, add the price of two meals into your quote (usually one meal for the first eight hours, plus another meal every four hours). In Kansas City [Missouri, USA], $10–$15 will feed you in the entertainment districts, so if meals are not included, I'd ask for $30 to cover meals. My personal day rate at the beginning of my career would have been around $225. Five years into my career, I would bump that to at least $30/hour plus food, or around $325.

Sharing and Negotiating Rates

Rates can be given verbally or in writing (see Chapter 11 for more detail). Recording engineer James Clemens-Seely focuses on clear communication with clients when discussing rates:

> I try to be as clear and upfront and encumbered about the financial side of things beforehand. I try to make that part of "it's nice dealing with me as a human." Here's what the cost will be, here's where any uncertainty in the cost lies, and what my prediction is (based on reasonable evidence) so that I can say it'll be between these two numbers. The session days cost you this much each. They're locked in a rate, and then the post-production will be by the hour. I'll say, "This record is probably going to be 10 to 30 hours of post, but I'm assuming it'll be 15."

> They know what the cost is, and they know what the possible overages are so I can discuss with them. Like, "Just so you know, it's going to cost more because the post is complicated in ways I hadn't anticipated." I try to do those with as much transparency as possible.

Music studio owner Catherine Vericolli adds:

> The more that you can passively educate your client about what you do, the easier it is to get paid what your actual worth is . . . If [your] clients understand the cost of the console, the room, and why the rates are what they are, it can be used to create value . . . I am super honest with them, and I've heard Piper [Payne] on the phone with her [mastering] clients

sometimes, too, explaining why shit is so expensive. It's like, look, the gear that we work with is very expensive. It is tailored to be extremely precise. To make a piece of gear do the thing that makes your record awesome, [it needs to be] extremely precise.

[My client says], "I want to press my record on vinyl." Let me educate the client on how a piece of vinyl is actually made and why it's so expensive. I'll also give them other options that they can take that might not be so expensive . . . Do not [BS] your clients about how much things cost. The clients that actually give a shit about that stuff really want to know where their money's going. They're going to Google that and they're going [to see] that's expensive.

("Interview")

Negotiating is discussing the terms of your agreement to work together. In most cases, you will be advocating and negotiating for yourself even if you are hired to work as part of a team. For example, a production sound mixer, boom operator, and utility sound technician (UST) may book projects as a team, but each person is hired separately. Patrushkha Mierzwa explains:

Boom operators and USTs think that mixers are very powerful and could get everything that they request approved in negotiations, including boom rate increases, generous kit rentals, and more . . . [However,] mixers are not your agents, and negotiating is yet another skill you will need to learn . . . Mixers do sincerely try to do what they can to better your deal, as it helps their work situation, too.

(254)

Negotiations tend to go in the right direction if you show you are trying to collaborate, solve problems, and have a genuine interest in an opportunity. Tommy Edwards says:

You want to be a problem solver, not a person who has problems. So, if you're able to come to that negotiation, to that offer, and put forth you're not complaining and you're going to be a problem solver, that's highly valuable.

Michelle Sabolchick Pettinato says of negotiating:

Always try and get the other party to give you a number first. When they ask you what your rate is, ask them what the job is paying or how much they are looking to pay. It takes some skill to negotiate well. They don't want to pay you any more than they absolutely have to, so they will not want to give you a figure if there is a chance of getting you for less. When pressed for a rate, you can offer them your salary range, which helps prevent you from pricing yourself out of a gig. However, sometimes you'll need to agree to a figure that is lower than your range just to get your foot in the door. There is always room to renegotiate once you prove yourself.

(55–56)

Negotiating Tips to Land a Tour: Claire Murphy

Claire Murphy is a touring guitar tech and author of the book, "Girl On The Road."

When negotiating your fee or joining a tour you will want to make sure you are speaking with the decision-maker. Who is it that makes the final call? Is it the artist? Is it the manager? There is no point putting your case forward to someone who can't actually hire you. You will make a much better impression when you speak to the person doing the hiring. If it is the artist, you can find out if they've had any problems on previous tours and how you can help resolve them. Are you able to help with the driving? Do they hate having to help with load in? Do they want their water bottle in the same place every day, yet the last tech could never do this? If it is the manager who makes the final call, what is their pain point? Do they hate it when crew don't answer emails in a timely manner? . . . If you can reassure them you can do this, you will have a much higher chance of not only getting the job but also getting paid what you desire . . .

If you come recommended, you need to remember that this is a big negotiating point. There are plenty of less expensive people out there that could probably do the job, but are they reliable? Recommendations from other people will give them peace of mind . . .

If you do want to take the tour but they really can't meet your pay expectations, ask if they can offer you something of equal or higher value, like your own room or slightly higher PDs [per diems], or maybe a better seat on a plane every time you fly. They might be able to hide this on a different line in the budget and you may end up getting a better quality tour [experience] in lieu of higher pay . . .

Don't be fooled into thinking that by not taking the job you are losing out, because you don't know what other job or tour might come up afterward that you wouldn't have been available for if you'd accepted the lesser paying job. This is a hard mentality to get your head around. You can say "no." Of course, in the beginning, you might want to say "yes" to anything to get your foot on the ladder.

(ch.4)

From "Girl on the Road: How to Break into Touring from a Female Perspective"

Landing an opportunity is not only about your client's interests. Tommy Edwards explains:

> Invariably, if you're giving somebody a deal on price, they need to recip-rocate with flexibility. You don't get to have both – you don't get to have access to me 24/7. If I work my butt off to hit a deadline and you're going to undercut the bid . . . you can't have both . . . Don't be afraid to say "no." Not every opportunity is a good one.

Working for Free

Working for free, or offering your services for nothing in return, is a topic where professionals have strong opinions on both sides (for and against). In some cultures and disciplines, it is the norm, while other cultures see it as offensive. There are no set rules or etiquette about working for free, and it is completely a matter of personal choice. Live sound engineer Aleš Štefančič says of offering his time for free:

> I made a conscious decision to offer my first gig for free – that would give the band a chance to meet me and evaluate my work with no cost to them. That one gig would then provide me with the opportunity to showcase my approach and my skills, but also see if I would enjoy working with the act. Some people are more compatible than others, so offering that opportunity gives everyone a chance to have a "casual first date" before committing to a long-term relationship.
>
> (18)

Music producer Jeff Gross says of working for low or free:

> I know that people don't have budgets anymore. Once it gets below a cer-tain point, then I have to weigh, is the material good? What's the possi-bility they might bring more work down the road? How's their personality? If they're an annoyance – or it sounds like ego over reality, then I don't want to touch their project. But if they're cool, then let's try to figure out a way maybe to make this work. Maybe we work out points on the contract . . . I always include: If you sell it, cut me in or pay me something (contrac-tually we work out the details). Or, I want to work with you on your next project. So, I always caveat the freebie aspect of it with "I'll scratch your back, you scratch mine" concept . . . [but] I wouldn't assume you're going to get work out of a freebie.

Sarah Stacey's approach to free work and favors depends on the client:

> [I'll do] a favor for a friend or for someone who's done a favor for me – I've done a lot of that. Usually, those are established relationships anyway, so I know there'll be more paid work to come from that. But if it's a stranger that I've never met, I'm definitely not [doing free favors].

John McLucas explains:

> Getting free work is still an accomplishment. Getting a random person on the internet to say they want to work with you and give you their time and energy, when you're unproven, is still hard . . . But, if you can't find paid work, you get free work. If you can't find free work, you need to work on your skills.
>
> ("Interview")

Holding the Line: Mehrnaz Mohabati

Mehrnaz Mohabati worked as a music engineer in Tehran, Iran, for seven years before moving to Los Angeles. Mehrnaz pivoted into post-production and has built a successful career as a dialog editor, ADR mixer, sound supervisor, and re-recording mixer for television and films – all without working for free.

> I haven't done a free job for nine years now [in LA]. People don't know the value when you do that. I think nobody should do it for free. Even if you're starting, you need to charge US$500 [for a short film].
>
> No matter how much you need, if you need to deliver a pizza it's better than doing a free job because you have bills. It's better to work for Uber than doing a feature for US$1,500 and teaching the producers that they can have that. If you are a pro, it's better to do the side job and mention, "quality work costs lots of time and money."
>
> I remember at the beginning of my career somebody [asked] me to do a short for US$200 [a very low amount for the work needed]. I said, "I prefer not to charge you, and I will ask a favor from you." I think sometimes you need to put those limitations.
>
> I am working on a docuseries where I really like the subject and the story. The director is a friend, and I know she has purpose in her life. I gave them a good discount, but it's acceptable. I still charge them. I think it's important to keep that.

Product designer and music studio owner Langston Masingale adds:

> If you don't have any examples of previous work (because you've never
> worked before) you shouldn't be chasing clients. You should be studying . . .
> Now, you should be out there learning the business, and you should be out
> there learning the practice of audio engineering. [Some] people don't even
> know how to wind cables . . .
>
> People of color – we have this phrase: "sis knew the assignment." The
> assignment as an engineer: the first thing you do is you listen. You learn how
> to understand what's being asked of you before you act. If you're working
> for free, you're learning from somebody . . . that knows what they're doing.
> That's when being free is good because normally we have to pay experts to
> teach us. It's called college. We pay a lot of money [for that].

In some cases, it can be difficult to gain the experience needed without doing
unpaid work. Guitar tech Claire Murphy says of touring for a week (unpaid)
shadowing a colleague:

> Whilst I didn't get paid, all my food and travel were covered and the experi-
> ence I was getting was invaluable. This is why I am not against working for
> free in the beginning. You need to have some experience, and unless you
> get very lucky, it's hard to get paid to learn on the job.
>
> (Ch.1)

An internship is a professional learning experience. While an internship may
be unpaid, it can be an opportunity to learn skills, build relationships with
professionals, and access equipment. Mehrnaz Mohabati, who works in post-
production sound, explains:

> I like the interns who are getting projects of their own because they can
> always ask their boss or their coworkers questions, and we always are open
> to answering. If somebody [asks], "Can you check [something I'm working
> on] for like five minutes and tell me what's going on?" I would definitely do
> that for them . . . A good intern can go forward very fast. If I see somebody
> good, I introduce them to others. We all are tied together. We say "Hey, do
> you know somebody that is charging this rate but can do this [skill] well?"
> You can introduce them.

However, some unpaid opportunities take advantage of those seeking experience. Radio and podcast producer Sarah Stacey explains:

> I have seen people offering full-time podcast production jobs where there's
> no actual money involved. I saw one in London recently, and it said, "You'll
> be working five days a week from our studio, eight hour days, but there's
> no budget. We have no pay." I just see it as exploitation, and I just don't

think it's good for anybody to take that kind of stuff. But if it's short-term opportunities, maybe a couple of weeks here and there, and there really is a great learning experience involved, then I would say go for it. But you don't want to make it a long-term thing because you do have to get paid.

Passion projects are projects where creative or personal fulfillment are a primary reason to take a gig. Offers to work on passion projects come about at all stages of a career. Sometimes these can be the most fun, creatively fulfilling, or have the most social impact of any work that you do. At the same time, just because someone else is sacrificing for their creation does not mean you have to – unless you want to. Live sound engineer Willa Snow says of passion projects:

> If I hear an artist, I am in love with their sound, I really want to help them, and they're just starting out, my day rate is not something that they can afford. But, I still want to work with them and help them, and I know that my services will help them grow. What is going to make it beneficial for both parties? It comes down to an exchange or a trade. So maybe I get 10% of their sales from their merch that night. Maybe it's an average of US$50 per show because that's what they can afford.

Game Audio Lead Jack Menhorn has a rule:

> If the project is going to be sold or others are getting paid, I won't work for free. Any free project I have worked on no one was getting paid, as it was a passion project for all involved. While experience is worth working for free, a credit (or "exposure") is not worth it. There are many ways to get experience in the game industry and lowering your own value (as well as others of similar skillset levels) is not a good solution.

Basic Guidelines for Free Work

Don't offer to work for free before you know their budget. "Never, ever say you're going to do something for free until someone says 'I don't have any money,'" says Piper Payne.

When appropriate, send an invoice with your normal rate and a discount. Catherine Vericolli explains:

> Figure out what your hourly rate would be and send them an invoice discounted to zero. At the very least, when they come back again (to make another song or another EP) after you've wowed them with your free record, now they understand how much it's going to cost this time.
>
> ("Interview")

Try to get something of value in exchange for your time. Video game sound designer/composer Akash Thakkar says:

> You can totally work for free. You're not devaluing the industry. But make sure you're not working for nothing . . . Most people work for nothing [at first]. They think, "Oh, credit, sweet." No, you have to get credit. In some places, it's the law. Get something else out of it. It could be tickets to an event like GDC. Maybe the developer can get you there. It may be a business card design. It could be web design [or] teaching you how to code – whatever it may be. Get something out of it other than credit . . . whether it be skill trade or something like that. That's really important.

John McLucas says free projects can be a great source of marketing content:

> When you have free clients, you can take screenshots and post little snippets. You're becoming somebody who's working on projects – even though they're free. It starts to build that perception slowly over time . . . the right people come through who will appreciate you and your time even if you're giving it for free. You'll have a good starting point. A lot of my free projects turned into paid ones.
>
> ("Interview")

Exchanging Value vs. Free Work: John McLucas

Music producer and content creator John McLucas suggests offering free work (in exchange for something of value, such as a testimonial, content, and demo material) when paid work is not available.

> Free work (when done effectively) is not giving away your time for nothing because you're new. It's actually quite the opposite – it's still an exchange of value, but it's not purely based on money. You're testing the market. It's saying, "I have a new business that I'm trying to launch. Because it's new, I am willing to offer a service for free. In exchange for this free service, I need X, Y, and Z." This is a one-time offer for very specific situations. For example, instead of going to open a restaurant, you're going to go set up your pie stand at the farmers market and sell pies, see which ones people like . . . build a reputation now that you know what people are interested in and what has the most impact on people . . . You will be asking for a testimonial . . .

John's Tips

1. Don't overstretch for a free project. Don't commit to a project that takes weeks of your time or even days and days. It should be concise, done, and move on.

2. Choose wisely who you offer this to . . . You are offering a tremendous experience for somebody for free or very little money.

3. Make it clear: this is not free work. This is an even exchange of value. It's important that it's discussed and framed properly [up front] . . . "in order for me to actually do this for free, these are things I need [a testimonial, exchange of service, etc.]."

4. Invoice accordingly. Make it clear that this is a one-time discount and show them [the cost] if they rehired you. If you crush it, they will come back again.

5. Ask for their value back immediately. As soon as the project is done, in that email or that message when you send the last stuff, send them your available times to book a call . . . Ask for a written two- to five-sentence testimonial in direct message or text (because that screenshot sometimes works better on a website).

[Organic content] can build authority, build reputation, build trust – there are so many amazing benefits of doing free work. From there, [you can] begin charging for your expertise, for your abilities, for your insight, and information.

("How Transitioned")

From the Modern Music Creator Podcast

Raising Rates and Trading Up Opportunities

Emerging professionals commonly ask how to transition from free work to paid work. Luckily, potential clients don't ask, "What did you get paid on your last project?" The question is typically, "What would you charge for this?"

As a freelancer, raising rates is simple: tell your clients you need to raise them. Meegan Holmes explains, "Don't go ahead and just automatically give yourself a pay increase . . . That needs to be negotiated and discussed. Hit it head-on, be straight up, and be very forthright with what you want and what you feel you deserve" (Holmes et al.).

Shaun Farley adds, "I never change rates while in a project, and none of my clients come frequently enough that they're totally shocked. I also don't raise my rates too often."

Promo mixer/podcast editor Kelly Kramarik adds:

> It's always an awkward conversation when you have to tell a client. My normal clients – we're all really good friends. So, once a year, I'm just like, "Hey, my rates going up to this. Hopefully that's okay with you." They've never said "no," so I'm clearly not as high as someone else.

There may be opportunities to renegotiate your rate once you are established with a client. Michelle Sabolchick Pettinato shares:

> When I got my first tour I was told "the job pays this much," and I took it because I desperately wanted to be on the road. After I finished the first six-week tour and knew the band was really happy with me, I renegotiated for a raise.
>
> (56)

When raising rates, be prepared for some clients to say "no." Podcast consultant/editor Britany Felix explains:

> If everyone tells me "yes," my rates are too low. If everyone tells me "no," my rates are too high. So, I need those "no's." If I'm not getting "no's," then I'm not charging enough. I used to charge way less. I was in the same boat and went through the same evolution. I got better rates literally by saying I'm going to increase it. I have never gotten to a point where I have gotten all "no's" and so I still continue to increase my prices . . . When I get a "no," it's not like, am I going to pay my bills? It's, "Okay, move on to the next one."

A free (or low-paying) client may work with you at your new rates, but it's worth re-evaluating the six reasons to take an opportunity. If you originally took a gig for credit/building experience but now have enough credits and experience to land paid work, what is the benefit for you? Are you willing to let the client go if they say "no"?

One reason to drop a free (or low-paying) client is to trade up for a better opportunity. **Trading up** is giving up a client (or type of work) for another that can offer you something different. James Clemens-Seely says of some of his past clients, "They needed to downgrade because they needed me when I was 20, not me when I'm 35. They're just hiring a 20-year-old now, and it's a perfect fit. I'm 15 years fancier."

Generally, clients and professionals are understanding about moving on to other opportunities that advance your career. Akash Thakkar says:

> You can give [clients] a bonus/old client rate (a cheaper rate), but if you don't want to work with that company anymore, it's a good idea if there is a hard cutoff and recommend someone else. If you want to transition them into a higher tier with you (say your rates are increasing), it's actually not a bad idea to say something like . . . "My rates are increasing to this number for clients now, but I'm going to give you this much because you're such a trusted client. Let me know if that works for you. If it doesn't, I'd be happy to recommend someone else who can do this work for you at a different rate." That can give you good karma and they'll love you for it even if you can't work with them anymore.
>
> [When] you recommend someone, you basically become Google to them. They'll keep coming to you for references. They will not cut you off if you give some sort of reference. You are kind of the person machine at that point . . . It works really well.

Music producer Nick Tipp says of letting clients go:

> Certainly, you don't want to treat this person in a bad way where they may say something about you. But even if you hate the music, don't want the work, and aren't being paid enough, you still want to be professional but indicate you're not interested in working with them anymore. That's how I let people go, but I have a real thing about comfort and ethics and wanting to feel good about who I'm working with.

Piper Payne adds:

> One thing I think that has really helped me and kept a good solid foundation of clients is making sure that even after Client X is out of my world (because maybe they can't afford me), Client X is still tied to other potential clients Y and Z. Don't ever, ever let a bridge burn. Never. If Client X sends me client Y or Z, I will take care of Client X's project at a deeply discounted or free rate. Expanding your network and expanding to potential clients is so much more important than pretty much anything else . . . I can't go to sleep at night unless I know that I did everything I could to keep my clients (and keep my clients happy) because happy clients mean more happy clients.

References

Adamson, Scott. Personal interview. 14 May 2020.

Clemens-Seely, James. Personal interview. 20 Dec. 2020.

Edwards, Tommy. Personal interview. 29 Oct. 2021.

Engelhart, Claudia. Personal interview. 8 June 2020.

Farley, Shaun. Email interview. Conducted by April Tucker, 15 July 2020.

Felix, Britany. Personal interview. 22 June 2020.

Gross, Jeff. Interview with April Tucker and Ryan Tucker. 14 Jan. 2021.

Holmes, Meegan, et al. Personal interview. *Once You Have the Gig – What Makes You Stand Out,* SoundGirls, 20 July 2020, youtu.be/QJhdm86kIlM. Accessed 21 July 2020.

Hutten, Mariana. Personal interview. 19 Nov. 2020.

Kennedy, Camille. Personal interview. 21 Feb. 2021.

Kramarik, Kelly. Personal interview. 18 June 2020.

Masingale, Langston. Personal interview. 14 Sept. 2021.

McLucas, John. Personal interview. 25 Jan. and 8 Feb. 2021.

McLucas, John. "Day Rates vs. Flat Rates – Marrying Both Pricing Methods While Not Dying, Fam." *The Modern Music Creator Podcast,* ep. 59, 21 Oct. 2019, pod.co/the-modern-music-creator.

McLucas, John. "How I Transitioned From Free Work To Full Time In The Music Industry." *The Modern Music Creator Podcast,* ep. 85, 23 Nov. 2020, pod.co/the-modern-music-creator.

Menhorn, Jack. Email interview. 30 June and 26 July 2020.

Mierzwa, Patrushkha. *Behind the Sound Cart: A Veteran's Guide to Sound on the Set,* Ulano Sound Services, Inc., 2021. p. 254.

Mohabati, Mehrnaz. Personal interview. 29 Jan. 2021.

Morris, Ashton. "Freelance Game Audio: Getting Started and Finding Work." *Ashton Morris,* www.ashtonmorris.com/freelance-game-audio-finding-work/. Accessed 1 Aug. 2021.

Murphy, Claire. *Girl On The Road.* Kindle ed., Self-published, 2020.

Payne, Piper. Personal interview. 29 May 2020.

Potter, Samantha. "Determining Your Day Rate." *SoundGirls,* soundgirls.org/determining-your-day-rate/. Accessed 12 July 2021.

Sabolchick Pettinato, Michelle. "Michelle Sabolchick Pettinato." *Get On Tour: A Sound Engineer's Guide*, edited by Nathan Lively, self-published, 2018. pp. 55–56.

Snow, Willa, guest. "Money Money Money." *Signal to Noise Podcast*, ep. 69, ProSoundWeb, 7 Oct. 2020, www.prosoundweb.com/podcasts/signal-to-noise/.

Stacey, Sarah. Personal interview. 19 June 2021.

Štefančič, Aleš. "Aleš Štefančič." *Get On Tour: A Sound Engineer's Guide*, edited by Nathan Lively, self-published, 2018. p. 18.

Thakkar, Akash. "Successful Freelancing in Game Audio." *YouTube*, uploaded by GDC. 9 Jan. 2019. youtu.be/93ggs7hwJeU.

Tipp, Nick. Personal interview. 22 Sept. 2020.

Trifiro, Jack. Personal interview. 4 June 2020.

Vericolli, Catherine. Personal interview. 29 May 2020.

Zúmer, Javier. Personal interview. 24 April 2021.

11

Offers, Bids, and Contracts

The first time you get burned and not get paid on a $10,000 invoice, you start to appreciate how important business acumen really is. I got burned spectacularly by a record company that shall remain unnamed. Thirty years later, I'm still pissed about it. If I had known about my rights, contracts, really understood the fine print and how these things worked earlier, and understood how the business really works to get paid, I would be richer than I am today. These are essential skills.

– Rob Jaczko

It is beneficial to start working like a business professional – from how you talk about money to how you document your communication with clients. By learning these basic concepts early in your career, you will look more professional, knowledgeable, and be better prepared as bigger opportunities come your way.

Written communication helps formalize what you are doing (or not doing) before you begin working. It also serves as a reference if there are any disagreements or confusion while working. Mastering engineer Piper Payne explains:

> I try and have everything in writing as much as possible so that I can have realistic expectations with my clients, partially because now I have a lot of clients. Even before . . . I would get things mixed up myself. Or, someone would not pay me on a certain schedule, and I could at least point to it in writing and say, "Hey, you said you were going to pay me on the first of every month and now it's the 15th. What's going on?"

AV Manager Devyn Nicholson adds:

> One thing that I've learned that is hugely important to me is that a paper trail means everything. If you do your business verbally, you've got nothing to stand on if anything happens. There are some individuals that I refuse to speak to verbally because there's just been too many incidents with them where they've said one thing and then the other happened. So, I'll respond

DOI: 10.4324/9781003050346-13

to them in email if they catch me on the phone: "As discussed in our phone conversation this afternoon, the following things were agreed upon by both of us."

An opportunity exists when you discuss with a potential client working together (see Chapter 8). For example, podcast editor/consultant Britany Felix schedules a consultation with potential clients where she describes her services and why she would be a good fit to work together. Britany says:

> After I wrap up that initial consultation with them, I have a [written] proposal that I send to them that outlines everything that's included and the rate for that. It outlines the billing terms of how long they have to pay, how the relationship can be ended – all of those terms of services, basically.

Britany has given her potential client an **offer**, or a promise to do something in exchange for something else (from another party). When a potential client asks you, "What would you charge for this work?" and you provide a specific rate, you have given them an offer. An offer can be verbal or in writing and can be as simple as a sentence or a detailed document. Even if an offer is made verbally, it is in everyone's interest to follow up with more specifics (such as deadline and cost) in writing.

Three different types of offers are a bid, quote, and proposal. Tommy Edwards says of the three types of offers:

> They're kind of similar . . . What they're putting forth is: we're going to do this work and it's going to cost this amount of money. It's a statement of work. Sometimes it's called a master service agreement. It's basically what the work is that's going to be done, the length of time we estimate that the work is going to take, and the amount we're expecting to be paid. Those are the three big items.

The three terms may be used interchangeably (by those who are not familiar with them), but each has a different intended use. When used appropriately, each has a different level of detail and formality. Which type of offer you use (and how you present it) will depend on the project.

A **quote** is an offer best used for simple gigs or projects, such as ones that only require giving an hourly rate, day rate or flat rate. For example: a mastering engineer quoting a price per song; tech work with an hourly rate; a live sound gig with a day rate.

A **bid** is used for projects that are larger in scale than a quote. A bid is ideal when multiple services are offered (such as editing and mixing), and/or a combination of costs (such as equipment rental, labor, or purchases required). For example, when producing a music album, there may be different rates for engineering, producing, mixing; expenses for hiring others (to do work such as editing/tuning or mastering); or other costs such as analog tape or studio rental time.

A **proposal** is essentially a bid with an additional sales element. A proposal includes additional information about yourself, your company, and why you would be a good fit. This can be helpful when there is competition, or to provide an offer when the highest quality presentation is expected.

An **estimate** is a best guess approximation of price but is *not* an offer. For example, if a piece of gear needs repaired and the cause of the problem is unknown, an estimate would give a general idea of the cost, but may not be the final price paid for the work. An estimate can be given verbally since it is not intended to be a formal offer. An example of the difference between an estimate and quote:

- Estimate for a podcast edit (verbal): "My rate is [amount] per hour, and on average an episode takes me between three to five hours. I can get back to you with a quote if you can send me a sample episode to look at."

- Quote for a remote recording (sent by email): "I can record the concert on [date] for [amount]. That would include a 4-mic setup and basic editing of the show (removing pauses, fading out the audience, etc.). I'll have the final version to you by [date]. If you would like any additional work after that, the cost will be [amount] per hour."

All offers require some investigation, similar to "The 5W's and H" process in Chapter 10. It is necessary to understand as much as possible about what is needed before giving an offer. This is especially important for projects with multiple services or if you may need additional help ("outsourcing" part of the work). What you find during the investigation will influence your offer. For example, your rates could be higher because the client expects you to work on an accelerated schedule. The quantity of work can affect an offer, such as doing original sound design for a video game that needs 50 assets versus 250. This investigation is also important to assess quality when you will be receiving audio that has been recorded or created by others.

Sample: Questions for a Post-Production Quote, Bid, or Proposal

For a post-production sound package (such as a television show or film), these questions will help determine the work required, time needed, and other information that will help give a competitive offer.

- What is the schedule/timeline?
- What is the total run time?
- What is the project being used for? Is it going to social media, a streaming service, broadcast television, or movie theater? Is it the final version or preliminary (such as a pitch for funding, or a film festival submission)?
- Is it scripted or documentary? What is the topic or genre?
- Who is the filmmaker and what is their background/experience level with sound?
- Does the mix need to be 5.1, stereo, or an immersive format?
- Production sound: Were lavalier mics and booms used? Was there lav coverage for everyone? Any known sound issues (like distortion, dropouts, or background noise)?
- Is there a budget for ADR? If so, how much does it need?
- Is there a budget for Foley? If so, how much? (full coverage or only where needed?)
- Is there a composer or music editor handling any of the music work?
- How much sound design is needed? Is there any creative/original/artistic sound design required?
- How are we going to review work? (In-person or remotely?)
- Is a home studio sufficient for mixing or is there a budget for studio time? How many days?
- Is there anything unusual in the content or about the film that might take more work for sound?
- What's the budget? Is there a set budget (or range) in mind?

From "Estimates for Post-Production Sound," apriltucker.com

How to Create a Bid or Proposal

To create a basic bid (or proposal), the steps are:

1. **Investigation:** Cover the 5W's and H; look online for information about the people behind the project; talk to colleagues who may know them; have a meeting or call to discuss the specific details of the project (that you would need to know to give an accurate bid to the best of your ability).
2. **Assessment:** List each element of the project required and the time or resources you believe are necessary to do each. This can include wishlish items (things you would like to buy to make the job go faster or better quality).

3. **Calculating costs:** Create a spreadsheet that includes the cost you would like to be paid for each service, any purchases or rentals required, other labor (hiring or outsourcing parts of the job, hiring an assistant), buffer (for overages/unexpected time needed), and wishlist items.
4. **Cost analysis:** Determine the total amount you will be charging the client, if it is reasonable, and making revisions (as necessary).
5. **When creating a proposal:** Write any relevant information about you (or your company or the crew you will be using) or why you believe they should choose you for the job.

For example, an independent ("indie") filmmaker is requesting a bid for their short film.

Investigation: The filmmaker's website shows they have done many short films before. A call with the filmmaker reveals the film is a 19-minute comedy and has been accepted into a well-known film festival. You were recommended by a mutual friend. The filmmaker is paying out of pocket so the budget is "modest," but they want it to sound as good as possible because of the exposure. They already got one quote which they said was too high. The festival is in six weeks, so they'd love to get started as soon as possible.

You will have to do all the sound work (editorial and mixing) and they might be interested in Foley. The production audio is good (per the filmmaker), but they were unhappy with one scene and would consider ADR, if it is not too expensive. The filmmaker would like to hear the mix on a dub stage (to hear how it would sound in a theater), but cost is also a concern. You ask to watch the film (for content) and for raw materials (to evaluate sound quality and check for possible tech problems) before giving a bid. When you receive it, the sound quality is decent, there are no obvious tech issues, and you like the film. So, you tell the filmmaker you will follow up with a bid.

Assessment: A quote would not give enough detail and the filmmaker seems serious about working with you, so a bid is the best option. Because the film-maker seems very concerned about cost, the bid will start with a basic package with some add-on options. Based on time tracking from school projects and short films you have done in the past, you have a general idea how long it will take you to do each task. You can work from your home studio setup for most of the project. You talk to a friend who works at a local studio (with ADR and Foley stages) for a quote, and plan to outsource the work to them (if it happens). You contact the same studio to ask the cost to forewall (rent) a dub stage per day or per hour. (Bonus: this is good for relationship-building with your friend and the studio.) There are a couple plugins you would like to pur-chase (to improve sound quality), if the budget allows.

Breakdown of tasks and time required:

- 1 day prep/organizing materials, editing music, and sound effects from the picture editor
- 1 day dialog edit
- 2 days sound design
- 2 days mix (home studio)
- 1 day mix review with client (home studio)
- 1/2 day fixes and output files for delivery

Add-on option available:

- Foley package – prep work, one day recording (with a Foley artist and mixer) and editing
- ADR package – mixer and stage for two hours plus editing
- Stage rental – one day mix review with client

Wishlist:

- Two plugins to help with the project

Calculating costs: This breakdown will be for your personal reference (not for client sharing). First, create a spreadsheet or table. It will have a row for each task with a rate applied to each. Then, tally up the total. (Note: rates/costs below are arbitrary numbers.)

Base Package

Item	Days	Rate per day	Cost
Prep work/editing	1	160	160
Dialog edit	1	160	160
Sound design	2	160	320
Mixing (unsupervised)	2	300	600
Mix review with client	1	300	300
Fixes and delivery	0.5	300	150
		Subtotal	**1,690**
Buffer (10%)			169
Wishlist (plugins)			300
		Base Package Total	**2,159**

Add-on Options

Item	Unit	Rate	Cost
Foley (package)	1	1,000	1,000
ADR (package)	1	450	450
Stage forewall fee	1	800	800
		Subtotal	2,250
Buffer (10%)			225
		Add-on Total	2,475
		Bid Total	4,634

Cost Analysis: The other quote the filmmaker received was over 5,000, which means you can beat the offer even with your wishlist items and a buffer. The cost of Foley, ADR, and stage combined may appear more than you will earn, but you have a sense the filmmaker will not pick them all. Outsourcing the work also has the purpose of building a relationship with the studio.

Calculating and Creating a Bid: Kate Finan

Kate Finan is co-owner and Supervising Sound Editor of Boom Box Post (Burbank, California, USA), a post-production sound company with a specialty in animated television.

[A spreadsheet] is for you to keep track of the number of services, time allotted for each, the subsequent cost to you, and percent profit for the project . . . Why do you need to make a profit on top of your rate? You need it to cover electricity used, high-speed internet costs, gear, Pro Tools updates, that extra bedroom or office space you rent. All of these costs of running a business (and, as a freelancer, you are a business!) should be factored into the cost of each project. Once you've come to a package price in your spreadsheet, ask yourself: does this number make sense for the level/length of project? If not, chip away at services.

Once you have a package price which includes the necessary services and gives you a decent profit margin, you will want to create a professional-looking bid to send to your clients as a .pdf. You can make this in QuickBooks, or there are plenty of templates for invoices in Pages. Just be sure that it is clearly marked as a [bid] rather than a bill.

You should include a line item for each service so that they know everything you are proposing to provide. Obviously, you will not list your profit as a line item. So, remember to increase the price of each line item by the percent profit on this document. Think about shopping at a [retail] store – the price tag doesn't tell you the factory direct price at which the store purchased the item. It tells you the price that they need you to pay in order to turn a profit.

Round each number to the closest $50 to keep it looking professional. Don't worry about getting the total price exactly [equal] to the package price. Go ahead and let it come out a little more expensive, then add in a discount. People love discounts! Save both the spreadsheet [and] this .pdf bid for your records.

At this point, you are ready to send your bid to the clients with a professionally worded email. They will either accept, decline, or begin to negotiate.

From A Sound Effect

What to Charge: Harry Mack

Harry Mack is a video game sound designer and composer based in Peterborough, Ontario, Canada.

You'll never know what [clients] are thinking for their budget unless they tell you. There's been so precious few wonderful occasions when they say, "Hey, we have a couple thousand dollars to do this, and here's the work. Is that something that you can do?" I love those types of emails because it's just so easy to say, "Yes, I can do it for that amount," or, "I can do it for that amount, but maybe we'd cut some of this because that would take too much." That's a great way to start a conversation, but it rarely happens that way. Mostly it's, "what do you charge?"

For the bidding process, I will never answer [with a number] on the first email unless it's a really obvious easy question. I will usually just say, "I would love to drop a quote for you. Can you tell me a little bit more about the project, such as your timelines and how big it is?" There are a million questions, and you just need to get these figures in order to even begin to think of a quote.

One of the tricks that I found is, usually they'll have a signature (or in their email they have their company name) and that can give you a lot of information. You can Google the studio and you can see that they've made a bunch of AAA titles, or you can't find the website and you don't know if they even are a studio. That will give you a lot of information whether to proceed, or how to proceed, and how to draw up your quote or your bid . . . If it's a new studio, I like to cut them a break. So, I like to say, "I give first client discounts, and I like to work with indie teams" . . .

If they're presenting to you a question of how much you charge but they don't know how much they need, it's impossible to draw up a quote for that . . . try to find out what you're going to need in terms of assets. If they still don't know, then you can just say, "Here's a ballpark figure I can charge per project" . . . But if they really don't know and they want per asset, like per minute of music, or per sound effects, then you give a range in the sense of like, "the low end, you'll probably need about 50 sound effects the high and 150" . . . There's going to be two figures, one of them is going to be the low end of the amount and the other is going to be the high end. Then they can kind of choose . . .

It's really tricky. Even after 10 years of this, I'm still like, oh my gosh, how much should I charge?

From Sound Design Live Podcast

Sample Bid Document for a Client

There are many ways to layout a bid document (or proposal) and templates available. This example bid is created from an invoice template but relabeled "bid" (as Kate Finan discussed earlier).

The numbers in the bid document will need some adjusting to equal the "Bid Total" in the spreadsheet. In this example, the "buffer" amounts (from both the package and add-ons) are included in the bid document on the "sound package" line item. The reasoning is that the filmmaker may pay the studio directly for Foley, ADR, or the dub stage, so those numbers need to be accurate. This approach – including buffers and increasing costs – also leaves some room for negotiation (offering a discount).

BID

Bid Date: March 24, 2022

Good Until: April 7, 2022

FROM	TO
Your company name	Client's company name
Address	Address
Phone	Phone
Email	Email

Description	Amount	Quantity	Total
Sound package for "The Best Short Film Ever"			
Includes: - Sound editorial (dialog and music editing, sound design) - Mixing (unsupervised) from home studio - 1 day mix review (home studio) - Stereo delivery (mix and stems) due May 1, 2022	2,450.00	1	2,450.00
Add on options:			
Foley package (recording and editing) at XYZ studio	1,000.00	1	1,000.00
ADR at XYZ Studio (mixer included) - 2 hours	450.00	1	450.00
Dub stage at XYZ Studio for 1 day review (replaces home studio review)	800.00	1	800.00
		Subtotal	4,700.00
		Discount	0.00
		Amount Due	**4,700.00**

Terms:

50% deposit is due prior to beginning services.

Once signed, please mail or email to the address above.

X _____

Print name:

Figure 11.1 Example of a Bid Document.

Putting Your Client's Financial Interest First: John McLucas

John McLucas is a music producer and content creator.

I had a prospect come in (now a client) . . . He wanted to do seven songs. I told him, I'm going to give you a quote for this, but I don't think doing seven songs right now is going to give you the best return for that investment of your money, your time, your effort, the stress, the anxiety, the loss of sleep that tends to go with a large project like that. I would recommend you do three songs, go out and gig on that for six months, then let's do the other four. You can space out the cost and you're going to get more bang for your buck per song. It's just a better return on the money you're putting in.

At the end of the day, I did talk myself out of a large chunk of money to do the other four songs right now, but I'd rather look after the interest of my client than try to do a bigger cash grab. As much as I'm a fan of business and the bottom line, I'm also so into the impression that I leave in somebody's mind before the project, during it, when they are done, a year after a project, and 10 years after a project. I would rather *that* [impression] be better and have them be just as happy as can be than getting that extra chunk of money right now.

What I do is more of taking care of people than taking care of songs. That's just how I've always viewed it, and I think that's part of what's gone so well for me. If you're worried about just getting the biggest possible project, they might be able to just barely afford it. You're probably not going to have that person come back at the end of the day. I want them to come back in eight months for those four songs. I want them to come back a year later for another EP, and then two years later for another album. You have a lifelong relationship with these people that is super fun. I've got their back, and they've got mine. That's a good business relationship.

From the Modern Music Creator Podcast

Tips for Offers

Ask for a client's budget or budget range. Budgets don't have be a game of poker. If you know their budget (or range), you can help a client get the most value for their money. It can be presented as, "I can work in a wide range of budgets, and 'low' or 'high' mean different things to different people." Tommy Edwards adds, "When you're a freelancer and you're dealing with [indie clients], I would probably start with, "Tell me what your budget is. I can tell you what I can do within that budget."

Put together options at different price points, if needed. If you're unsure of a potential client's budget (or know they are flexible), multiple price points or add-on items may help. This helps give options, especially for a client who may not know their budget (or wants you to share your numbers first).

Include a buffer, or plan for more hours than you need. A buffer helps cover unforeseen issues/changes without having to go back to discuss budget issues. Some professionals bid using a day rate but do not share the fact that a day is calculated at nine or 10 hours while they intend to work eight hours. Studio owner Catherine Vericolli schedules more recording time than she thinks a client will need but tells the client they will only get charged for the time used. "Guess what? They always need it," says Catherine.

Ask for deposits, when possible. This ensures you will be paid for some of your time, especially if the project cancels last minute.

Quoting low to land the job may not work. Tommy Edwards, who receives offers as part of his work in audio technology, explains:

> If I get a bid or a quote that's too low, I won't take it. It's like a bell curve. I won't take one that's way too high and I also will be hesitant to take one that's too low. It's a clue that the person may not have read the statement of work, might not understand the depth of work that I'm asking for. That's actually a red flag.

Offers take practice, and some will be turned down. Nathan Lively shares:

> I used to get upset about this a lot: I would get an idea for what I wanted to specialize in, and then I would immediately go out and try to find clients, and then no one would hire me. I would think . . . "the audio industry has failed me." Most of the time it just wasn't time yet. I was making the wrong offer to the wrong people, or it wasn't the right time.

Be prepared to justify your cost. Sound designer Shaun Farley shares:

> The few times someone has pushed back on an [offer], I explain my position. This is generally the explanation: I take pride in my work, and I take pride in giving my clients a result that they will be happy with. If I can't deliver that, then I don't want to put either of us in that position. My [offer] is based on my experience, and here's what I expect will have to

happen on this project (insert general breakdown of the amount of work needed). If you don't want to work with me based on this, I respect that and hope you're able to find someone who can give you something that will make you happy within your budget.

I did have one person continue to push back once. "This other place is offering to do the whole thing for less than what you're [quoting] just for the dialog edit." I responded with, "Did this place ask to see your rough cut, like I did? Have they had any exposure to your project beyond a five-minute phone call for an estimate?" They had not.

Negotiating on a Package: Kate Finan

Kate Finan of Boom Box Post shares ideas how to compromise with a potential client on an offer.

If you receive an email response asking for a lower package price, it is best not to simply lower it by giving a discount . . . You already established that in order to provide the services listed and make a profit, this is what the package price must be. Instead, this is the time to engage in a negotiation. You may negotiate based on several tactics.

First: you can alter one or more service. If they would like a lower price, then offer to mix it at your home studio rather than at a rented facility . . . You can also offer to do less of a certain service, like less custom design work or recording.

Second: you can add extra services for the same price . . .

Third: think outside the box and negotiate with things other than money or services . . . A good friend of mine often throws plug-ins and sound effects libraries into the negotiation . . . tell them it will save them on the design and recording side of things. You would likely buy it anyway, and this takes it out of your list of costs . . . Is the documentary you're bidding on riddled with bad dialog recordings? . . . Ask for them to splurge on Izotope RX for you, and tell them that it will save you tons of editorial time. Just be sure that all licenses are in your name. Items like this may not exactly equal the amount of money in question, but you will get to keep them forever. Often productions are more willing to spend on software and other tools than wages.

> Fourth: you can bundle services. This is an advanced negotiation tactic, but it really works . . . A great example of bundling . . . I was asked to work on a short film that was going to be used to garner investment toward a feature of the same name. In a scenario like this, you can hold firm on your short film bid but say that if they bring the feature to you, you will deduct the price of the short from that. This makes them more likely to bring you the feature, and also saves them a considerable amount of money on the feature.
>
> *From A Sound Effect*

Contracts

If a client accepts your offer (or you have accepted an offer from a client), you have entered into a contract with them. A **contract** is an agreement that creates a legal obligation. A contract can occur verbally (such as a "handshake deal") or in writing. The terms "contractor" and "contracted work" imply you are working under a contract to provide a service. In some instances, the company you are contracting for will provide you with a written contract to sign. For example, Avril Martinez says of some video game and VR work, "When you're looking at [contractor] jobs, a lot of times it's with a recruiter, so the recruiter actually has the contract already done for you." Sound for picture and music production work may use a contract called a "deal memo." A contract could also be a **letter of agreement**, which is a single page document with terms (details specific to the project). Some professionals have their clients sign the quote, bid, or proposal as a means of accepting the offer.

A written contract can be a formal legal document written or revised by a lawyer. Or, a contract could be informally written (and agreed on by both parties) over email or on a pizza box. In some instances, an initial agreement is informal (possibly verbal), but followed up later with a detailed written document, which is signed before starting the work. How formal a contract is (whether formed through a handshake deal, an email conversation, or a signed legal document) is a personal choice. Shaun Farley explains:

> If I've never worked with someone before, I get it in writing. If I've worked with someone a few times and have a good relationship with them, I probably won't bother. E-mail confirmation of estimate is enough. If I know I'm going to have to hire additional editors and supervise them, I get it in writing.

A written contract helps clarify important details before doing any work. Tommy Edwards explains:

> The big things are: When do you get paid? How do you get paid? How long does this last? How do I get out of it if this goes south? Those are really important things that you're not thinking about when you're excited about getting a recording contract or getting a contract to do a movie – what if this goes south?

Working without a written contract can have consequences. A contract can establish who owns materials, which can affect how you use your work later. Game composer and sound designer Akash Thakkar explains:

> I've seen a lot of friends who didn't have a contract. They say, "It's just a handshake deal. It'll be fine." [But what] if the game tanks, it doesn't come out – who owns the music? Who owns the sound? If they had a contract, it would go right back to them. It'd be no problem. But unfortunately, it was completely unclear. Someone I know actually had to go through some legal issues and it turned out the client owned it [because] there was no contract . . .

> If you're working on a free project and you're not getting anything out of it, it is really important to actually make sure you own those rights to that music or that sound. Or, if you're just doing a license, or you want to be clear that they own it, [include it in the contract] so there's no kerfuffle in the future . . . Use contracts even for free projects.

A written contract can also ensure payment when a client adds additional work. Mastering engineer Mariana Hutten shares, "My mentors told me, 'At some point, there will be a client that will remix their album five times, and you'll have to master five times. You should charge five times.'"

When working with independent artists, musicians, filmmakers, game developers, etc., it is typically up to you to provide a contract. Megan Frazier says of video games and VR, "Most indies, unless they are just really, really savvy, aren't going to know [about contracts]. They're going to want you to drive."

While it is possible to find sample contracts online or write/adapt your own, it is ideal to use a lawyer, when possible. Not all lawyers will be unaffordable, and a lawyer can help create a contract specific to your work that can be reused (like a template). Tommy Edwards says:

> If you don't have the budget for an attorney, there's a couple secret backdoors. A lot of attorneys that are experienced but still on the junior side (so they haven't made partner yet) are actually doing side work outside their firms. You might be able to get them for a more affordable price.

That's where networking really becomes important is trying to find those people . . . Attorneys can have that sliding scale, too. A lot of people don't realize that. They just see US$300 an hour and go, "I can't afford it" . . .

If you're staring at a contract, get a legal opinion on it. A lot of people don't, and that's risky. In my experience, what you'll see (in the music business or in business in general), is the contracts can be antiquated. They're rooted in a different period of time . . . I see lots of mistakes in contracts. There's not necessarily anything nefarious. Frankly, it's just cut and paste errors . . . So, if you don't have somebody look at it, there could be mistakes in it that then have legal consequences if everything goes south.

Piper Payne says of contracts, "It doesn't have to be this really big emotionally-charged thing . . . It's a proposal [to work together], and once it's accepted, then you just accept the money and everybody moves on with the project."

Terms are specific details included in a contract related to a project. Possible terms to include in a contract:

- Scope of the work – what services are included (or not included), what would be an additional cost or fee, working hours, the maximum hours included in a day rate
- Schedule, due date, and what happens if the schedule changes
- How both parties will know a project is completed, and what happens if any work is needed after that
- What materials are delivered and to whom
- Who is responsible for fees and/or scheduling (equipment, facilities, additional crew or talent, etc.)
- How many revisions are included (and the period of time where you will address notes), and any costs/fees for revisions after that time
- Payment schedule and dates (deposit and final payment)
- Cancellation fees and how to be notified
- How you would like to be listed in the credits (if applicable)
- Who owns the rights to the work you provided
- Archiving – what materials are you keeping and for how long
- Permission to use the material publicly for a demo or for promotional purposes
- What happens if a common problem occurs (materials with technical issues, materials delivered late). Who can give permission if additional work needs to be done and the cost of overages

Mixer/mastering engineer Mariana Hutten says of contract terms:

> Always put in a contract the things that bother you and make you waste a lot of time. Find the one thing that you hate dealing with and put that in the contract. For example, my contract has a deadline. From the moment I give you a mix and you approve it, you have one month to request a change. After that, you have to pay me to make it again as if it was never done.
>
> That's my approach because I'm so disconnected from the project. I've had so many artists where I was convinced the album was done and then six months later, they hit me up saying, "I added an electric guitar part to this song." To them, it's just a guitar. If I stop to reopen a project, I'm wasting time on the projects that I am working on right now, or wasting sleep hours, family time or friend's hangout time . . . I state in the contract you have up to two revisions. After two, you have to pay a fee for revisions because otherwise people will never finish the song.

Deposits and Payments

A deposit is a prepayment to reserve your time before beginning work. Catherine Vericolli says of deposits:

> An electrician does not fix your [outlets] and then say, "I'll send you an invoice and *maybe* you'll pay me." That's not how works. Most the time you pay that electrician up front to even come to your house . . . There's no difference for a recording engineer, or recording studio, to make a record for a client than there is for an electrician to come to your house and fix your kitchen outlets.

Whether you ask for deposits (and what amount) is a personal choice. "Deposits are a really fluid thing that you have to learn. In Los Angeles, a lot of people require payment up front – all of it," says Catherine Vericolli.

Mariana Hutten says of deposits:

> In Canada, it's really hard to ask somebody to pay 100% [upfront]. I have to do 50% . . . The reason why I do it is people cancel all the time. When I used to have my own studio, there were months that I was running on the negative [cash flow] from people just canceling last minute. Even doing deposits, I can't pay my rent and expenses of a commercial space with 50% of everything all the time. If my 50% was equivalent to covering all my costs, then it would be too expensive for people to want to hire me.

Re-recording mixer Sal Ojeda only asks for deposits from unfamiliar clients:

> One project was a friend of a friend, it was an independent project, and the client didn't live here (I didn't meet him until I was working on the project). At that point, I wasn't really sure how serious he was. We put together paperwork, and he was cool with it . . . If I don't know anything about the client, then I'm always cautious. I don't send out the final files, or if I send something, I send them low quality files. But 95% of the time it's been good. They end up paying.

Example Deposit and Cancellation Policy: Fivethirteen

Music recording studio Fivethirteen, owned by Catherine Vericolli, operated out of Tempe, Arizona (USA) for 15 years before relocating to Nashville, Tennessee (USA). Their policies are found on the studio's website.

> **Deposits:** We require a 50% deposit for a project at least 10 days in advance for new clients. Checks are accepted for deposits, but must clear at least 10 days prior to the session date.
>
> **Cancellations:** In the event that you need to cancel a session, we ask that you give us 48 hours notice so that we can accommodate and reschedule. In the case that this doesn't happen, there will be a fee enforced and collected before rescheduling.
>
> Any fees paid due to a cancelled session will be credited toward your rescheduled session. If you are a new client, and were required to make a deposit for your canceled session, your deposit will be credited towards your rescheduled session.

Podcaster Britany Felix requires payment up front for some services and uses an automated system:

> I never, ever let anyone just schedule a consultation call or sign up as an editing client or new launch client. I have to have a conversation with them first . . . As soon as they sign that proposal saying "Yes, I want to become a client," they're automatically redirected to a payment page, and they have to submit their payment there before they can get on my calendar to start their services. For my scheduling system, I use Acuity Scheduling. Then for my website where I collect payments, I have Squarespace connected to Stripe.

It may be important not to disrupt the creative process, which some professionals account for by not receiving payment until after the work is complete. Mastering Engineer Piper Payne explains:

> I never ever, ever want to hand somebody a bill after just taking care of like, birthing their baby out into the world after they've literally spent their life savings and multiple years of their life. We've just finished the culmination of all this work, the time, the artistry and they've sat behind me watching me do this thing. I can't turn around to them and say, "Okay, you owe me money."

Often times, an invoice needs to be sent to receive payment. An **invoice** is a document that shows the services that were provided and cost owed. After sending an invoice, the amount of time it takes to get paid depends on the terms set and the client. Some will pay an invoice as soon as it's received, others pay invoices on pre-set dates, and some wait until the final days before it is due. (See Chapter 12 for more on invoices.)

From Introduction to Finished Project: Catherine Vericolli

Catherine Vericolli started her recording studio, Fivethirteen Studios, in her home in Tempe, Arizona (USA) when she was 23 years old. Catherine managed Fivethirteen, which specialized in analog recording, for 15 years.

> When we book a session, we always sit down with the client. We never cold call – we always meet them [in person]. The reason why we do that is because it's in my house. You're going to be in my home making a record. Even if I live in a recording studio, I still want to know who you are, plus I want to make sure that this is the right place for you, because maybe it's not. We have walkthroughs, and they're free.
>
> We bring in the client and we talk to them. We say this is how much it's going to cost. Guaranteed they're going to say, "We don't have that much money." So we're like, "Cool, here's what we can do. You can either go and get more money and then come back, or instead of making a 12-song LP, we can do like a four-song EP, and have those songs be better." We explain to them how that process works, because we don't want to do our clients a disservice. We want to have the amount of time that we need to make a good record . . . 99.9% of the time the client will take some time to think about it. Maybe we won't hear from them for another year, but they always come back. They say, "now we're ready," because we were honest with them.

Then, we settle on the amount of time. That's the first question any client has: How long is this going to take? Then we say, "Well, how good are you?" We give them some very simple talking points to answer that question ("Can your drummer play to a click track?") We already have an idea in our heads because we've asked for a demo, which is like an iPhone recording in your house of a rehearsal, or a video on YouTube. Because we've done it for so long, we're able to go and figure out how much of a pain in the ass it is going to be.

That is just on-the-job learning, and you are going to [be wrong] a lot of times. You're going to be sitting in your studio editing drums for 12 hours straight and going, "I never want to do this again." That is how you make positive change. But you never put that on the client . . . The next time you [give an offer], you learn from it and change it. "It may be this much more money if you want to print to tape, or if we need an extra day of editing, or maybe we need to mix, so let's budget [for it]."

Asking for Money

People – and especially musicians – don't like surprises. They especially don't like surprises when it comes to money because they don't have any. Keep it very, very clear: this is how much it's going to cost, and this is the deposit [amount]. At the end of the session, you pay us and that's when you get all your [final materials] . . .

I want clients to think about money twice: When they book the session, then when they pay me at the end of all of it, and never again through the entire process. If I can achieve that, I'm doing my client a huge, huge service.

Anything that's money-related, paperwork-related, or legal-related: make it as simplistic as you can make it . . . You do not want the client to think about [that] when they're making a record. I'm talking about tracking when you have to create an environment for an artist to be comfortable enough to create at their best. If they're thinking about money, contracts, paperwork, or about how much time they've been in there, you're done. It's over.

If you're on your own and you don't work for a big conglomerate studio, even if you're just getting started, make it really, really easy to get paid. You can get paid in a [lot of] different ways – you can get paid in cash; you can get paid with a card . . . We have a Square system where it's really not that expensive. Also, Square business financing is awesome. We've had a couple bands take advantage of that . . .

I want to take enough money up front to not scare my client away. That has to do with being transparent, honest, and educating them. Then, I want to give them an experience of my services, my studio, and my staff that justifies the price so that at the end, it is much easier for me to take the money. If they try to pay me early, I won't take it and I will tell them I am superstitious . . . If I'm doing my job right and my staff are doing their job right, the client is going to be elated . . . They're so stoked that the money thing is like, "Whatever. Here, take the credit card." It's easier to take it after the experience.

The more you do it, the easier it is. The more that you hate it – actually, it's better. If you're coming across to a client that you're in this for the money, that's bad . . . People want to know that you're in it because you like their music, and because you're going to serve them in the way that they want. It's a service industry.

The studio gets paid up front because I have to pay for all of the things that are being used to make that record. So, I take the studio's chunk, which is a little bit more than half. Now the client's invested so if they [don't show up], I've been paid, and I'm not out [any money for the project]. My engineer might be out, which is very rare, but that does happen. I will pay him because that's the kind of manager I am . . . I think in 15 years it's only happened like two or three times where we didn't get paid [by a client], so we're doing something right. I think it's because people are afraid of me, which is fine.

References

"Rates." *513 Recording.* 513recording.com/rates. Accessed 2 June 2020.

Edwards, Tommy. Personal interview. 29 Oct. 2021.

Farley, Shaun. Email interview. Conducted by April Tucker, 15 July 2020.

Felix, Britany. Personal interview. 22 June 2020.

Finan, Kate. "How to Set (and Get) the Right Price for Your Audio Work." *A Sound Effect*, 10 Nov 2015. www.asoundeffect.com/sound-pricing/.

Frazier, Megan. Personal interview. 11 and 21 Aug. 2020.

Hutten, Mariana. Personal interview. 19 Nov. 2020.

Jaczko, Rob. Personal interview. 15 Oct. 2020.

Lively, Nathan. "Sound Engineer's Path Webinar." *Sound Design Live*, 16 Feb. 2015. www.sounddesignlive.com/sound-engineers-path-webinar/.

Mack, Harry, guest. "Career Advice for Freelance Designers." *Sound Design Live*, 16 July 2013. *SoundCloud*, soundcloud.com/sounddesignlive/career-advice-for-freelance.

Martinez, Avril. Personal interview. 28 Aug. 2020.

McLucas, John. "The Quickest Way to NOT Grow Your Network." *The Modern Music Creator Podcast*, ep. 9, 10 Sept. 2018, pod.co/the-modern-music-creator.

Nicholson, Devyn. Personal interview. 23 Nov. 2020.

Ojeda, Sal. Personal interview. 27 Sept. 2020.

Payne, Piper. Personal interview. 29 May 2020.

Thakkar, Akash. "Successful Freelancing in Game Audio." *YouTube*, uploaded by GDC. 9 Jan. 2019. youtu.be/93ggs7hwJeU.

Vericolli, Catherine. Personal interview. 29 May 2020.

Portions of this chapter are from:

Tucker, April. "Estimates for Post-Production Sound." *April Tucker*, www.apriltucker.com/estimates-for-post-production-sound/.

Tucker, April. "How to Bid on a Project." *SoundGirls*, soundgirls.org/how-to-bid-on-a-project/.

12

Accounting and Financial Planning for Your Business

Where is your money coming from, and where is it going? Whether you are an employee or freelance, tracking this information can help you plan for an emergency or slow months of work. What happens if you have an injury? What happens if your rent goes up or you can't find work? A bad financial situation can stop an audio career before it even gets off the ground.

Accounting is the process of storing, organizing, and analyzing financial transactions. In simpler terms, it is understanding the flow of money in and out of your business. An **accountant** is someone who analyzes and advises regarding financial data.

Freelance professionals typically have ebbs and flows of work (and income) throughout the year, and different areas of the industry can have ebbs and flows at different times. Learning to survive through these times comes down to tracking and understanding your "numbers," or the finances of your business.

Freelancers and business owners need to have systems in place for:

- **Time tracking**: Keeping track of hours worked
- **Invoicing**: Communication with clients about how much to pay you, when it is due, and where to send it
- **Bookkeeping**: Keeping track of when invoices have been paid, expenses (spending), and receipts
- **Accounting**: Analyzing financial information over time, and tax preparation (or preparing for an accountant)

DOI: 10.4324/9781003050346-14

Featured professional

Andrea Espinoza is an audio engineer, tour manager, licensed financial educator, and advisor. Andrea spent eight years in touring audio, working her way up to monitor mixer and tour manager. In 2020, she pivoted into working as a financial advisor.

Time Tracking

Time tracking is keeping track of hours worked (both unpaid and paid). By tracking the time you spend on different tasks, you can have an accurate picture of the total time spent on a gig or type of work. This includes billable time (hours you charge for) and tasks you may not always be paid for (such as travel time to gigs, phone calls, or research or training to prepare for a gig).

Time tracking is used for your own analysis/use and may also be used for invoicing, or billing clients for your time worked. By time tracking, you begin to see trends in your work – how quickly you can complete a task, or what tasks require unpaid hours. If you do not know if it takes an hour or a day to do a task, it can be difficult to set a rate or to know if an offered set rate is fair.

Time tracking can be as simple as a note or a spreadsheet, or there are websites and apps specifically for time tracking (some are free). Invoicing software may include time tracking as a feature, also.

An example (from a real project) of time tracking for music production of a short film:

Task	Date	Time	Hours
Session Prep-Misc.	June 9	1 pm–4 pm	3
Recording	June 10	5:30 am–5 pm	11.5
Editing/Mixing	June 11	5 pm–10 pm	5
Review with Composer	June 12	5 pm–10 pm	5
Output/Delivery	June 13	1 pm–2 pm	1

Invoicing

An **invoice** is a list of services provided (and/or goods sent) with a cost for each. Services can include audio-related tasks (such as engineering or sound editing) and non-audio tasks, such as fees for rental equipment. An invoice only includes billable tasks. For example, you typically would not bill for the time it takes to write an invoice.

An invoice should always include the 5 W's:

Who: Your name (or name of your company), who the payment should be made out to; the company or person to whom the invoice is being sent

What: Tasks (services provided) and the costs

When: The dates of the work performed, the date of the invoice, and the date payment is due

Where: Where your invoice is being sent, and where the payment should be sent to

Why: The name of the project you were working on

Other items that may be included on an invoice:

- Purchase order (PO) number (a unique number assigned by a vendor for your work)
- Taxes owed (varies by location)
- Discounts
- Any preferred credit or title for your work on the project
- Overdue fees (late payment fee)

Mastering engineer Piper Payne suggests including your preferred credit on the invoice, such as, "I really appreciate you crediting me as a mastering engineer on this. Please have the credit read: Mastered by Piper Payne."

Invoicing Systems

Low tech: Word processing software (including Google Docs and Microsoft Word) have free templates available to create an invoice. While these are free options, they also require manually sending files and keeping track of what invoices have been paid or not. When using word processing software for invoices, use a clearly named document title that includes your name and invoice number, and attach it as a pdf. For example: "April Tucker Invoice 109.pdf".

Most efficient for professionals: Invoices can be created and sent out through online services (or apps), and many provide basic features for free. Sites like waveapps.com and invoicely.com are free invoicing systems that create

INVOICE

Invoice No:

Invoice Date:

Due Date:

PO Number:

FROM

Your company name

Address

Phone

Email

TO

Client's company name

Address

Phone

Email

Description	Amount	Quantity	Total
			$0.00
			$0.00
			$0.00
			$0.00
			$0.00
		Subtotal	$0.00
		Discount	$0.00
		Amount Due	**$0.00**

Payment instructions:

Please credit as:

Figure 12.1 Sample invoice.

estimates, invoices, and keep track of when payments are due or received. Some allow payments through the invoicing software (for a fee).

Most in-depth, but paid services: QuickBooks is a popular small business software option that offers time tracking, estimates, invoices, expense tracking, and data accessible by an accountant. While it is the most robust option, there is also a monthly fee.

Pro Tip

[When working for free], send them an invoice discounted to zero. At the very least, when they come back again (to make another song or another EP) after you've wowed them with your free record, now they understand how much it's going to cost this time.

– Catherine Vericolli

When to Invoice

Invoicing, or the process of sending invoices, should be performed regularly. An invoice should be sent within 30 days of the work being performed. Not all clients pay invoices when they are received, so it can take some time to receive payment. Sal Ojeda shares, "The first few gigs that I did it took more than six months for them to pay me, and it was [a major label]."

Tommy Edwards, who worked in music production before pivoting into audio technology, adds:

> It's kind of counterintuitive, but what I've noticed (both as a musician but also as an audio engineer and producer) is when you're working with independent artists, in my experience, I got paid sooner . . . When you're dealing with an established record company or film studio, it can [take longer] to actually get paid . . . You have to plan for it being out. It's not going to be immediate.

Andrea Espinoza suggests putting three dates in your calendar anytime you are hired onto a job: your start date, your end date, and billing date. The latest billing date is two weeks past the end date, but that date represents the last possible date to reasonably send an invoice. "It'll take you maybe three minutes to put into your phone. Then, set reminders and alerts for the last two days," says Andrea.

Regardless of the invoice system used, it is important to schedule time regularly to:

- Create and send invoices
- Verify that the payment received matches the amount on the invoice sent
- Document which invoices have been paid
- Send courtesy notices to clients whose invoice is overdue

Invoicing should be done when you are mentally fresh because mistakes can cost you money. The person processing your invoice may not catch a missing item or that your math is incorrect.

Bookkeeping

A **bookkeeper** is someone who documents financial transactions. Bookkeeping includes keeping track of invoices and payments received, business spending, and organizing receipts. If you want to buy anything to support your business (from audio gear to transportation costs), it depends on **cash flow**, or money coming in and going out of a business. If you do not know how much you have earned (versus spent), or who owes you money (and how much), it becomes impossible to make educated decisions about what you can afford or what gigs to take.

Accurate bookkeeping is also crucial for taxes. What you owe in taxes and what tax breaks you receive depend on knowing how much you spent (and on what). Doing it yourself is not complicated and can help you know what is going on with your business financially throughout the year. Bookkeeping should be checked at least every other week but can be done as often as needed. Physical receipts (or scanned receipts) need to be stored and organized for tax purposes (the time period to keep receipts varies by country).

Tracking invoices and payments is a bookkeeping task and can be tracked manually or through invoicing software (such as waveapp.com or QuickBooks). The idea is to have a place to keep track of whether or not you have sent invoices, when invoices are due, following up if they haven't been paid, and keeping track of when they were paid.

Invoice Tracking Spreadsheet (example)

Status	Date	Number	Customer	Total	Due
Unpaid	2021-9-8	516	Maya Cooper Sound	100	100
Paid	2021-7-7	515	Landon Jackson	250	0
Unpaid	2021-6-5	514	Safara Club	600	600
Paid	2021-4-3	513	Landon Jackson	250	0
Paid	2021-2-2	512	Fizzy Festival	300	0

Another aspect of bookkeeping is having a place where all of your transactions (purchases) are combined and categorized. This can be a document, spreadsheet, or software designed for bookkeeping. QuickBooks is a popular paid option for small business bookkeeping (and accounting) and can download transactions from banks and credit cards for categorizing. Mint.com is free and similarly designed for personal use. All transactions will be categorized – for example, cables, audio equipment, software/plugins, etc.

Business Transactions (example)

Date	Cost	Item	Category
2021-July 28	65	Online course subscription	Education
2021-June 20	100	Plugin upgrade	Audio Software
2021-July 20	5	Leilana's Cafe	Dining Out (networking)
2021-July 14	7	Leilana's Cafe	Dining Out (networking)
2021-July 10	20	Website fee	Website

Andrea Espinoza suggests reviewing and categorizing transactions weekly or within two weeks of doing a gig. By doing it regularly, taxes will go much more smoothly, and you will not be sorting through year-old paperwork and receipts with questions about what you spent money on.

Bookkeeping for Taxes

One good reason to stay on top of bookkeeping is that business expenses can be tax-deductible. This means the taxes you owe could be lower because you bought a guitar, video games, or spent money on audio plugins for your business. While tax law can vary widely between countries, an accountant can advise on expenses that qualify toward taxes and the tax laws for the country where you live. For example, in the USA, nearly any expenses related to operating a business can be itemized (reducing what you owe in taxes). This tax benefit only exists for freelancers, not employees, and the expense has to be justified as supporting your business.

General examples of business expenses include:

- Business meals (meals with clients or colleagues discussing work, attending networking events)
- Concert and movie tickets
- Trade show attendance and related travel
- Research (movie, concert, or theater tickets; music and video game purchases; streaming services)
- Educational materials (books, courses, training programs)
- Computers used for work
- Audio software and hardware
- Home expenses required to run a business (internet, cell phone, furniture for a home studio or work space)
- Transportation to and from a gig (including car expenses)
- Meals on the job
- Insurance

Separating Business and Personal Expenses

For an established small business, business income and business-related spending should ideally be completely separate from personal spending. In reality, for emerging professionals, personal finances and business finances can overlap. The pool of money you use to buy groceries is probably the same pool you are using to buy headphones. However, for bookkeeping purposes, it is ideal to keep business spending separate from personal spending such as using a separate bank account (even a secondary personal account), or a credit card used only for business expenses. That way, when you go through a bank statement (or credit card statement) looking at each item, you can easily tell what transactions (such as meals out) were spending with friends and family and which were with potential clients or colleagues.

When you receive payment for work, it can come in many different ways, including by check, cash, or person-to-person (P2P) services (i.e., Venmo, PayPal, ApplePay). When you use a P2P service for business and personal use, it can be difficult to separate personal money from business income. Is a friend paying you for personal reasons, or are they paying you for work? For bookkeeping, one way to keep these separated is to transfer business income out of a P2P service and into your designated business account.

> **Pro Tip: Patrushkha Mierzwa**
>
> *For those working long days or away from home for periods of time:*
>
> Set up online banking services and use Auto Pay every chance you can. Request e-bills so you won't have to be home to pick up your important mail and because they won't get lost. Use an accounting software program that coordinates with your bank and accountant so you can download transactions rather than having to input them manually. Request digital receipts and subscriptions rather than paperwork that could be lost or damaged. To prevent excessive expenses, barter with roommates for things you'll need to have handled while on a show.
>
> (265)

Accounting

An **accountant** is someone who analyzes and advises regarding financial data. Most accounting needs for an emerging professional will not be complicated, particularly if you do your bookkeeping regularly (and keep it organized). An accountant is not required for doing taxes, but an accountant to a business is like a music mixer to a song – they will know how to tweak and finesse to make the result better for you. An accountant may save you more money than the cost of their service.

A **financial plan** is an overview of your financial situation, goals, and strategies to reach goals (both short-term and long-term). The main reason to have a financial plan is to ensure that you do not find yourself in financial trouble. There are talented emerging professionals who have had to abandon career aspirations or pivot out of the industry because of financial problems such as having no money to pay for rent, transportation, or being overwhelmed by student loan debt.

Goals can ensure you are saving enough money to pay the bills from month to month, to sustain yourself during periods of being out of work, and to survive if you have something unexpected happen. For freelancers, a plan can ensure you can pay your taxes, as well.

Normally, a financial plan is designated for personal finances *or* business finances, but we will look at these combined since business and personal spending tend to overlap for emerging professionals. We are lucky to work in an industry where the things we do for fun (like going to a concert or watching movies) can also be a valid business expense. Financial goals can help you save to buy a piece of gear, attend a conference, pay for education, or move to find more work opportunities. Financial goals can be both personal and business-oriented, like saving to buy a home you plan to use as a workspace/studio.

"Paying yourself first" is a process to save money toward your goals and priorities. Andrea Espinoza explains:

> When you get your first paycheck, your first thought is, "I want to buy this, I want to buy that," right? Because we're taught to pay the bills, [buy] what you need, and then maybe save. People always say, "I don't have any money to save." [Paying yourself first] is shifting the paradigm so that saving is your first thought. I get paid this [amount] but really, I'm only getting paid 75% because 25% automatically gets saved. It's forming habits, and it's rethinking what we've been taught.

The foundation of a financial plan is a **budget**, or a plan on how to spend (or save) your money month to month. A budget is arranged by categories (or "buckets," as Andrea Espinoza refers to them). In personal finance, buckets can be expenses such as rent, utilities (water, electricity, etc.), and groceries. Business-related buckets may be audio equipment, research, business-related utilities (cell phone, internet), and more.

A budget is usually calculated per month even when income changes from month to month. For example, a sound engineer who tours some months and does not work others can still set a monthly budget, which will involve saving money to cover living expenses during lower income months.

Figure 12.2 Buckets are used to separate categories for a budget.

To start, create a list of all the categories you spend money on or would like to save for, for example:

- Taxes – Taxes for self-employment income
- Housing – rent or mortgage, related insurance
- Utilities – electric, phone, internet, television, etc.
- Groceries
- Emergency fund – money set aside for unexpected emergency expenses
- Living fund – money saved for expected periods of unemployment (or low income)
- Transportation – public transportation, bicycle, car, insurance, gas, repairs/upgrades
- Retirement savings and/or life insurance
- Business-related insurance
- Loans/debt – credit card, student loans, etc.
- Medical – health insurance, out-of-pocket medical expenses, dental
- Business-related dining out
- Personal dining out
- Research/education – research attendance at industry events, concerts, training programs and services (virtual or in-person), recurring services or products (streaming subscriptions, video game or music purchases)
- Business expenses – monthly cost for services related to the business, such as website hosting, file transfer services, scheduling or accounting software, accountant
- Audio expenses – necessary expenses such as software licenses, equipment, repairs, upgrades
- Audio wish list – items you would like to purchase for your business (but are not required)
- Miscellaneous business expenses – low-cost items needed to do work, such as Sharpies, gaff tape, office supplies, etc.
- Personal care – haircuts, clothes
- Household – furniture, decorations, appliances

There are many other categories you could create based on your personal priorities and interests, including musical instruments, babysitter/daycare, pets, health and fitness, car payments. Andrea Espinoza also suggests having a "fun" bucket:

> Give yourself rewards for doing the work, for saving the money. Like, I have an "eating out" budget. I give myself a reward, like, "I'm going to go to this restaurant" . . . When you're receiving a paycheck, you're not making

any money, and you're not doing what you want to be doing, it can be discouraging. But that is when experiences and fun come into place. You need to have some sort of balance and some sort of reminders.

Tips for budget categories:

- Have separate categories for personal and business spending. For example, business dining out (related to networking, client meetings, research, etc.) can be a separate category from personal dining out.
- If there are goals you are working toward (buying something that will take time to save for) or categories where you tend to overspend, these can be separated into their own budget items.
- A budget should be as useful to you as possible, so use however it will work best for you.

The goal of a budget is to start building a routine and habits. Andrea Espinoza explains:

> I wouldn't dump all your money into your emergency fund because the whole premise is to pay yourself first and to account for your taxes. So, I would work it out [for] what you can do consistently because consistency builds habits. Dumping your money in on one lump sum is not going to motivate you to do it consistently because now you're in the mindset, "Well, I did one lump sum," and now you're not thinking about it. It needs to be something that is a part of your daily or weekly [routine] (or whatever frequency).

Some buckets will have a goal or target amounts, such as an emergency fund, living fund, or audio wish list. Once a bucket is filled (reaches your goal amount), the money can be allocated to another bucket. For example, once your emergency fund goal is met, your monthly contribution could start going toward your living fund or toward a piece of equipment you would like to buy.

Starting a Budget

A budget can be created with a spreadsheet, using an app, or a website (for example, youneedabudget.com or mint.com).

To create a budget with a spreadsheet, start by compiling recent spending and earnings. The following example will be for three months of spending, but it can be done over a longer period of time.

Income

Add up three prior months of income. Divide this number by three to calculate average monthly income. For example (using arbitrary numbers):

	January	February	March		
Client 1	1000	700	1000		
Client 2	250	250	250		
Client 3	650	0	400		
Total:	1900	950	1650	4500	Three-month total income
				1500	Monthly average

Expenses

1. Download or print three months of all statements – credit cards, bank accounts, P2P (Venmo, PayPal, etc.). Some accounts will have the option to download transactions that can be opened in a spreadsheet.
2. Categorize all transactions using your list of categories (if not done already). If you are already using Mint or QuickBooks and have been categorizing transactions, your categories will come across in the downloaded file.
3. Sum the total of each category and divide by three (the number of months). This gives you an average monthly value for each category.
4. Add each of the category totals together (to see total expenses per month).

Sample Expense Spreadsheet

Category	January	February	March	Three Month Total	Monthly Average
Rent	800	800	800	2400	800
Groceries	150	172	200	522	174
Utilities	200	200	200	600	200
Emergency Fund	50	50	50	150	50
Transportation	85	85	85	255	85
Research	0	100	20	120	40
Dining Out	25	80	30	135	45
				Total Expenses	1394

Spending Analysis

1. **Find problem areas.** Once you have your spending categorized and totaled, some problems are easy to spot. You may be spending more than you are earning, or some categories may stand out as seeming excessively high.
2. **Compensate with lifestyle changes.** If your expenses exceed your income, there are two ways to solve this: increase income or lower spending. Can you cancel any services or recurring expenses? Find a cheaper place to live? Do you need to find a higher-paying job and save money temporarily? If you have debt, can you refinance or lower the interest rate or payment? There can be different strategies to meet the same goal.
3. **Set new budget amounts.** Budgets are not permanent and may not even be long-term. If it is not possible to meet your budget consistently, it can be readjusted.

Sample Budget

Category	Budget
Rent	800
Groceries	174
Utilities	200
Emergency Fund	75
Transportation	85
Research	80
Dining Out	45
Audio wish list	41
Total	1500

Review

In this example (which is simplified from a real-life expense list and budget), this person is earning slightly more than their spending average (1,500 versus 1,394). Because there was money left over, they increased their monthly contribution to an emergency fund and also created a category for "Audio wish list." Over time, the money saved into that bucket will be spent on items from that list.

In some months, spending may go over in a category but as long as a budget is reviewed (and adjusted) from time to time, this should not be an ongoing problem. In this instance, there would not need to be any lifestyle changes to make this budget work. However, if income was lower (say, 1,300 instead of 1,500), this person would need to change their month-to-month spending habits.

Using a Budget to Set Rates: Nathan Lively

Nathan Lively of Sound Design Live used his budget to help set his rates. When Nathan lived in the San Francisco Bay Area, his full budget was US$3,500 a month, or US$875 week, and he wanted to work between two and five days a week.

> My day rate should never be lower than US$175 per day, but I should be shooting for a US$450/day rate so that I can be working two days per week . . . Why would I only want to work two days per week? . . . A lot of times, a day could be anywhere from eight to 14 to 18 hours for some of these things. So, once you've worked two 14-hour days in a row that are non-stop, maybe you don't want to work any more days that week . . . or if I wanted to take some smaller gigs that maybe like a concert venue, I could do it as long as I were still earning at least US$175 per day.

Budget Tips

Andrea Espinoza suggests setting a budget in this order of importance:

1. **Taxes.** One of the most important buckets in your budget (for freelance/self-employed) is taxes. In the UK and the USA, a good rule of thumb is to save 25–30% of your earnings for tax expenses, but in the UK (2020–2021), the tax rate is zero up to a certain amount.
2. **Essential living expenses.** This includes rent, utilities, groceries, etc.
3. **Loan payments**
4. **Building an emergency fund.** An emergency fund is for unplanned events, such as an unexpected car repair or a medical expense. Andrea suggests building an emergency fund with three months of living expenses (and some extra as a cushion).
5. **Living fund.** A living fund is ideal for someone whose income is inconsistent or anticipates periods of time without work. This is different from

an emergency fund because these months of lower income (or no income) are known to happen and are planned for.

6.   **Other priorities**, including retirement, paying off loans early, or buying equipment. These are based on your priorities and goals.

Including budgets for equipment, computers, and entertainment spending can be fun, but priorities change over time. There are buckets that established professionals suggest contributing to that may not be at the forefront for emerging professionals. Insurance (from disability to renters) can help protect your career from a major catastrophe. Retirement funds, investments, and life insurance can help provide options down the road even by contributing a small amount in the early years of your career.

Pro Tip: Michelle Sabolchick Pettinato

Start saving for retirement as early as you possibly can. If you work as a freelance engineer, there are no pension funds, no 401Ks, no retirement savings other than what you save yourself. Open a Roth IRA and contribute the maximum amount every year . . .

The earlier you start saving, the more your money can grow. Touring can be very grueling and hard on the body. Most people don't want to be living this lifestyle in their sixties, but if you don't save for retirement, you may not have a choice. It's hard to think so far into the future when you are in your twenties, and it's also not much fun putting a large amount of money into savings when you'd rather be doing something more exciting with it, but it is something you will never regret. Most people don't realize how important this is until it's too late.

(57)

Scott Adamson suggests looking into long-term disability insurance for freelancers:

It's a little pricey for someone that's kind of starting out, but something worth considering. It's part of the risk of being a freelancer. If you slip and fall and break your hip, nothing is going to cover you. So, it does just go hand in hand with all financial planning and life planning as a freelancer. You have to take all those things into consideration and understand that that's kind of the business you're getting into.

Andrea Espinoza adds about investing:

> I often say after you have your emergency fund, and maybe if you want a living fund (so you don't touch your emergency fund), you should be really investing your money. You should be putting your money in a vehicle that's going to get you a higher return [than basic bank options with low-interest rates]. There are so many other vehicles out there that will give you a better return on your money . . . There are Robo advisor accounts (like Betterment) . . . Your money's already earning more than it would at a bank and in an account that you cannot easily access.

Buying Equipment

One challenge in getting started in the audio industry can be having the equipment and tools necessary to do the work. A budget could include a bucket for audio equipment to set aside money for those purchases.

What if you have some extra money (like a gift, tax refund, or unexpected gig), or you're trying to evaluate where in your priority list to put buying equipment? Andrea Espinoza suggests:

> I often ask: Do I need it versus do I want it? That helps me a lot. Like, I really want headphones at the moment, but I don't need them so I'm not going to buy them. Will they help my career? Maybe, maybe not. But if it's something that is vital and essential to your career, absolutely make a fund for it, make a list and then start to put money away for it.

Sometimes a business purchase is worth **leveraging**, or spending money to increase income. If a piece of equipment will increase your profits, bring in more clients, help you land work opportunities you could not prior, or help you work faster, a purchase may be worth the expense. There are many *bad* reasons to buy gear: to have something new, image, to "keep up" with the latest, or because it's a tax write-off.

For example, if a studio has a stereo setup but a client has a project that needs to mix in 5.1 (with the prospect of more work and paying a decent rate), it may be worth leveraging the cost of the upgrade to land the work (if it does not compromise your ability to pay basic living expenses). If the upgrade is only for a single gig that is low-paying (and no future 5.1 work expected), it may not be worth the expense. While it is desirable to learn a new skill and offer more services, it is not an immediate need.

Shaun Farley explains, "Until you have disposable income, only buy those pieces of software/gear/sound libraries you need for a specific gig. Approach each expense as an investment: How will this help me make money in the next six months?"

Andrea Espinoza adds:

> If you need to go into debt of any kind, always have an exit strategy. How are you going to come out of this? What are you going to do? What's your plan in place? It's fine if you're in this debt for [a set] period of time, you have a plan, and you're already working on it.

Buying Gear: Sal Ojeda

Sal Ojeda had a goal early in his career to be able to work from home with his own clients. Sal started his career working for others (including a full-time job in post-production) and worked toward his goal by saving money for gear and keeping his personal expenses as low as possible.

> Since I already had these steady gigs, I was saving up and buying gear. Early on, I bought a lot of gear used. I bought my speakers new, and eventually, I bought my 5.1 system, [computer], and started buying plugins and a nice interface. I didn't go to Hollywood and rent an expensive apartment. My car was old. I didn't go and buy a new car . . . First, pay the bills and reinvest the rest. The whole idea of living under your means and having that goal of just reinvesting in your business – that's how I've been able to make it work.
>
> My I/O has been with me for almost 10 years. That one paid off many, many times – and that's the brain of your system. You have to do the math and see, okay, how many shows do I have to mix for this to be profitable? If you really know what you're getting, most of it pays for itself after the first or second project that you do. Also, I'm not buying the most expensive plugins.
>
> [My advice is] get a job somewhere, something steady that you can pay the bills while you start getting your clients and your gear.

Tips Regarding USA Income Taxes

Estimated Taxes: If you expect to owe more than US$1,000 of self-employment taxes, you are required to pay **estimated taxes**, or quarterly tax payments. Instead of having Tax Day once a year, you are expected to pay a portion four times (April 15, June 15, September 15, and January 15 of the following year). Self-employment tax includes taxes for Social Security and Medicare and can total over 20% of your total income. If you do not pay estimated taxes, there will be a penalty applied when you file your yearly tax return.

1099 Forms: When you work as a contractor for another US company, they are required to give you a 1099 tax form if you have invoiced for over US$600 of work (in 2022; future years may be different). If the amount is less, they do not need to send you a form. It is possible to receive multiple 1099 forms if you are contracting for multiple companies. You will be asked to fill out a W-9 form (which you can also find online) to gather the information needed for the 1099 form.

If someone you worked for did not provide you with a 1099 for taxes (but should have), you can still declare the income on your taxes without the form. While it may be tempting to leave off income ("under the table") to lower your tax liability, it is in your best interest to be honest and accurate about your earnings. Your tax documents can be used for other purposes in the future (such as a home loan or business loan), and a lender may ask to see prior years of tax returns.

EIN Number: An Employer Identification Number (EIN) is a way of protecting your social security number, especially because you may be giving it out more often as a freelancer than as an employee. An EIN is free, and the process can quickly be done on the IRS's website (irs.gov).

Depreciation: If you purchase any high-value items, keep this information separate for your accountant. Your accountant may want to **depreciate** these items, which means you get a tax benefit for the purchase over several years (versus just the current year). An accountant can calculate to see which route will be more beneficial for your taxes.

References

Adamson, Scott. Personal interview. 14 May 2020.

Edwards, Tommy. Personal interview. 29 Oct. 2021.

Espinoza, Andrea. Personal interview. 28 May 2020.

Farley, Shaun. Email interview. Conducted by April Tucker, 15 July 2020.

Lively, Nathan. "From the Sound Up Lesson 3: Financial and time needs." *Sound Design Live*, 22 March 2017. *SoundCloud*, soundcloud.com/sounddesignlive/sets/from-the-sound-up-how-to.

Mierzwa, Patrushkha. *Behind the Sound Cart: A Veteran's Guide to Sound on the Set*, Ulano Sound Services, Inc., 2021. p. 265.

Ojeda, Sal. Personal interview. 27 Sept. 2020.

Payne, Piper. Personal interview. 29 May 2020.

Sabolchick Pettinato, Michelle. "Michelle Sabolchick Pettinato." *Get On Tour: A Sound Engineer's Guide*, edited by Nathan Lively, self-published, 2018. p. 57.

Vericolli, Catherine. Personal interview. 29 May 2020.

Part 3
Career Paths

13
Live Sound and Theater

Live entertainment has energy and a rush unlike any other discipline. Marc Thériault, who has toured with Celine Dion for over 25 years, says, "If Celine says she's got an issue on stage, I'm jumping on stage. 20,000 people and there's something broken. Can you fix it now? Seconds are like minutes. This is pressure."

Live sound and theater work exists worldwide with a variety of opportunities to learn and practice skills. Live sound is also a gateway for many audio industry professionals to learn basic sound skills. "I got into live sound because I could get stagehand work pretty easily. I got some jobs at different churches, then interning at a studio, and then it went from there," says Kelly Kramarik, who now works in sound for picture.

Alesia Hendley started her career in live sound, which provided a foundation she later used to work in AV and IT (with an audio specialty). Alesia says:

> I did load ins/load outs for years. You have to put in the time to gain that experience. At that point in my career, nobody was going to hire me [without] experience. Ultimately, that load in and load out didn't get to [my dream of] touring with Beyoncé, but that load in and load out got me to an in-house [position].

Despite their similarities, theater and live sound are separate disciplines, require different skills, and have different career path options. Because of this, live sound experience does not necessarily transfer to theater work (especially musical theater). Broadway theatrical mixer Julie M. Sloan explains:

> Having done corporate, live music, and theater sound – for me, the actual execution of mixing theater is the hardest because you're working with the worst of all possible worlds. A lot of times when we get people from other parts of the industry (like rock and roll live touring) and they see what

DOI: 10.4324/9781003050346-16

we're dealing with – like, your entire cast goes offstage and they're changing clothes while they're singing, and they're moving scenery onstage and all that – they're appalled. You don't have those same types of issues when you mix concerts.

(Augustine et al.)

There are touring opportunities in both live sound and theater, but it is not a requirement to work in either field. Both disciplines have some similar entry-level opportunities. **Stagehand** is an entry-level position that provides physical labor for shows (including setup and tear down). Meegan Holmes says even though stagehand work involves lifting, "[It] doesn't mean you have to be big. It just means that you have to know how to use your body [well] . . . and be willing to ask for help."

Live sound engineer Michael Lawrence adds:

It's no longer, "we're going to go hire the biggest, strongest guys we can find." It's now about who has the skill . . . We're at a great time where if you're interested in this stuff and you study and learn these skills, you're going to find a home in this field.

Sound companies are another common source of opportunities. Julie M. Sloan explains:

In nearly every city, there are rental houses that provide gear and personnel to do corporate and club gigs. Larger sound shops will provide support for bigger concerts and events. You can start getting experience working at one of those shops, whether it's part-time, overhire [temporary], or a salaried position . . . It's amazing hands-on experience.

(Augustine et al.)

Nathan Lively (engineer and creator of *Sound Design Live*) suggests the industry is showing a shift toward professionals who have a variety of skills:

In the next few years, the highest demand in touring will be for generalists. Everyone wants to specialize in mixing front of house or sound system optimization, but the highest demand is for people who can mix monitors, tour manage, operate a lift, and drive a truck. It's the same with local gigs. If you can set up a projector and operate a video switcher, you'll get a lot more work than if you can only take audio gigs.

("Get on Tour" 25–26)

Jason Reynolds, a live sound engineer who also does tour production management, recommends:

Carry multiple skills. So, try to learn tour management or production management so you can do more than one thing and not just be a fader pusher. If you have those multiple skills . . . then you're likely to be more marketable.

Live Sound for Entertainment

Live sound skills can be broadly applied more than any other discipline in the audio industry. Live sound work can exist anywhere from concerts to weddings, cruise ships, business meetings, senior centers, or schools. Live sound can be a sustainable career in locations not known for entertainment work.

There are generally two paths into live entertainment as a career. Monitor mixer Karrie Keyes explains:

> One path is for people who want to be exclusively engineers that are mixing the band. They start in clubs or work with bands that are touring in a van. That is more difficult to get to a consistent, professional touring level. The other clear path would be to be working for a sound company or a venue.

For those getting started, it is possible to be pursuing both paths simultaneously. Front of house mixer Scott Adamson explains:

> There are small audio production companies everywhere. Some of them might be a guy with a van and he's got a little PA in the back, and he probably needs someone to help him out. If it's your very first job, maybe that's a good way to get in. Maybe you're doing sound for a band, [meeting] musicians, and those bands are going to be doing other shows. That's how to get into the whole infrastructure.

For those working in clubs and for bands, there are opportunities to learn a variety of skills. One person may handle everything sound-related – from backline tech (setting up instruments and related gear), monitors, front of house, and more. An entry-level opportunity for larger shows and venues is stagehand. While stagehand work is less hands-on with audio, the tradeoff is being able to network with established professionals and seeing how larger scale shows work. Working for a sound company can be a pathway into touring; however, entry-level work may take place at a warehouse or tech shop (not at the show).

Live sound work is predominantly freelance. Scott Adamson explains:

> Even if you do happen to be on payroll working for artists or a production company, the work, in general, does tend to be freelance. That is important for people to understand in live sound. There's a whole host of problems that go along with that, and part of that is financial planning and understanding how to manage that.

Getting Started in Live Sound: Michelle Sabolchick Pettinato

Michelle Sabolchick Pettinato has worked in touring live sound for 30 years, spending on average 250 days a year on the road. Michelle is co-founder of SoundGirls and the creator of Mixing Music Live.

Start by learning as much and getting as much experience wherever you are at the moment.

If you're in school: Does your school have an active AV department or music production program? Does your school have a sound system that it uses for events such as band concerts or theatrical performances? Get involved and learn how to use the equipment.

Already have some experience: Approach local sound companies, venues, or local bands for a job. Explain to them your interest in live sound and ask if they will give you the opportunity to work and learn from them. Be prepared to work as an unpaid intern until you have enough experience. Gaining knowledge and experience – whether it is shop work, loading trucks, or working a show – are all valuable assets. Be available and open to working every opportunity that is offered to you. Proving that you are reliable is as valuable as your mixing abilities.

No experience: Are you a fast learner? Can you learn on the job? Are there any opportunities where you live? . . . If you have no experience, you will need to find an opportunity to learn on the job. Inquire with local sound companies or sound engineers if they will take you under their wing. Often bands cannot afford a sound engineer but desire to work with someone capable, who they can trust, and who knows their material. Approach local bands and see if they are open to this. Expect to start out working for free. Consider the experience you acquire as payment.

Is there a local club or live music venue with an in-house sound system in your area? Explain to them your interest in live sound and ask if they will give you the opportunity to work and learn from their house engineer. Be prepared to work as an unpaid intern until you have enough knowledge.

> If you live in a major [US] city (such as LA, NYC, Atlanta, Chicago, Nashville), you have many options available to you: sound companies, stagehand labor companies, IATSE [union work], equipment rental companies, and rehearsal spaces.
>
> ("How Get")
>
> *From soundgirls.org*

Live Sound Companies

While the live sound industry is known for its mixing jobs, technical jobs exist and are in demand. Chris Leonard explains:

> It's not all about climbing the pyramid to front of house. That's a glorious job, and it's the rock star underwritten position. But, there's nothing wrong with aspiring to be a PA tech. There's nothing wrong with aspiring to be a systems engineer.
>
> (Leonard et al.)

Live sound companies provide crew and equipment for gigs (local and touring) and also have jobs maintaining and preparing equipment for gigs. Scott Adamson explains:

> There are definitely things you can do that are more than just touring as a front of house engineer for a band. A lot of people that I know that are specialists are highly in demand . . . if you want to be a master at setting up a PA system (system engineer is what we call that) – that's a really, really great job . . . wireless audio RF (radio frequency technician) is a great way to get more work.

RF tech Chris Bushong adds:

> We don't need more engineers. We need RF coordinators or some of these niche positions that aren't glamorous and aren't sexy positions but are very important and beneficial to production. It's one of the reasons I got into RF because nobody really wanted to deal with it. I knew I could get more work.

Marc Thériault adds of RF:

> Many shows have a lot of wireless now – microphones, instruments, and monitor mixes. Even if you're the best engineer on the planet, if [signal] doesn't go through or if it's dropping out . . . If you've got bad RF, you've got shit sound. The job is extremely important.

Sound companies may have internships and entry-level positions. However, opportunities may be primarily in-house (shop work), not out on gigs. Michael Lawrence explains:

> I started at the local production company helping out in the shop. Sometimes it's sorting cables or soldering cables. Sometimes it's boring stuff, like, here's 400 XLR cables and we need them to be color-coded with electrical tape by length. That's not exciting. But, being in the shop, around the people doing the work, around the gear, being able to go up to a console and spend some time on the console, being able to spend some time seeing how a power distro goes together, learning how to pack a truck, hearing the language, looking at the advancing documents – that's the best place you can be.

> A lot of companies that I know would be happy to take someone on as an intern or summer help to have an extra set of hands around the shop . . . just go down to your local shop and say, "Hey, I'm really interested in this. Can I help?"

Meegan Holmes is Global Sales Manager for Eighth Day Sound, one of the largest tour sound companies in the United States. Meegan says even though entry-level employees work in their warehouse, the job requires customer service skills:

> We have band engineers that come in and [setup] their equipment . . . It is extremely important how these [people] are treated. They are guests in our warehouse . . . If that engineer is displeased with anything that happens in our warehouse, the engineer can go to the production manager and say, "They totally didn't take care of us. Nobody helped me. I needed assistance. This wasn't working, and they did a terrible job. They were rude." That could lose the account right there.

One advantage of working for a live sound company is the room for growth and training on the job. Chris Bushong explains:

> You get known within a company, and people are like, "I really like that new person. They work really hard." That gets up to department managers who are then saying, "So and so is going on tour. Do you want to do this?" Rather than trying to meet the right people in an ocean, now you're in a smaller pond. You're just trying to get in good with one or two people rather than everybody.

Sound companies come in different sizes, and the work can vary for emerging professionals. Karrie Keyes explains:

> Once you get into [a large sound company], they want you in the shop for one or two years working so that you learn how the systems go together

and you know how to repair everything. It's great, and you need to know that stuff, but once you get out of the shop, you're still going to start as the third or fourth person on a show. You're probably not going to be mixing . . . The small sound companies don't have a crew list of 300–500 people that they can call and they trust to go do a gig. They have maybe 20 people that they can call – if that . . . If you're interning there and they get busy, they're going to start sending you out on gigs. They probably still wouldn't send you out by yourself . . . but it's going to be hands-on, you're going to be doing everything, and you're going to learn.

Working for a large sound company may get you on a large tour faster, but likely in a specialized tech role to start. "You could be on an arena and stadium tour pretty quickly . . . but you're going to be flying PA, setting up gear," says Karrie Keyes.

Chris Bushong says of working as a PA tech without prior experience:

I did a major 40-show arena tour and I had never flown the rig before. [You have to] understand the theory and the principles behind it and reading standards. I know a lot of 22-year-olds who are PA techs on tours.

PA techs work under a system engineer, who is responsible for the PA system (installation, tear down, maintaining, and optimizing). System engineer is not an entry-level position, but some sound companies will pay for training or certification. Chris Bushong suggests one benefit of starting as a PA tech on tour is learning from the system engineer:

If you're smart, then you're engaging with the system engineer and [asking questions]. You're getting off the bus an hour early and walking the room with them as they take those measurements and plug it all into the software and watch what they do. A lot of young [techs] will just sleep to the last minute and get up and [only do their job].

Concert sound mixer Robert Scovill adds:

A great deal of the new generation of truly great systems engineers I have met and worked with are coming out of schools and climbing the ranks of sound company rosters very quickly. And they do so because of the depth of their knowledge and exposure to the latest technologies.

(72)

Theme Parks

Theme parks are another opportunity for jobs and career paths in live sound. Large theme parks may have opportunities for both engineers and assistants (A1s and A2s) and sound designers (who design and oversee system installations). Small theme parks may seek generalists who can do sound and other tasks (like lighting). Kevin McCoy says of working summers at Valleyfair, an amusement park in Minnesota (USA), early in his career:

> For one show, I ran sound and lights. Another gig was outdoors and just sound. That was next to a roller coaster, so every 45 seconds the train would go up the hill and I'd hear "clang, clang, clang." There were musical review shows, so dancing and singing to as many songs as they could cram into a half-hour show. I would do six shows a day. I probably learned more about live sound at that job than I did at many of my other jobs just from having to deal with the challenges. Also, I was given a lot of freedom and responsibility to do things the way I wanted to do things there. It was super fun. I loved it.

Gabrielle Fisher, Audio Designer at Disneyland (USA), says of the variety of shows and events at the park:

> There are a lot of shows that are mainstays, and those are consistent. There are also all these big events that switch [around]. We set one up, we rehearse overnight [during] graveyard shifts, and we do the show in the morning. Then, we tear down and on to the next one. It's a lot of fun, and there's a lot of chance for collaboration.

In a theme park, audio professionals are expected to interact at a high professional level. Gabrielle Fisher explains:

> There's an emphasis on people skills and soft skills . . . You could be interacting with guests anytime, other departments, and others across the park. It's just a general care and way that you should be able to work with others.

House of Worship Live Sound

A house of worship is a space where people come together for religious services and can be a source of work even for those who are not religious. Some houses of worship have budgets for sound equipment, crew, and may have better equipment than local entertainment venues. Engineer Jason Reynolds says of working at churches versus clubs, "Why go fight to work on bad gear just for [low pay]? If you're in this scene, there are tons of churches that would love pro or semi-pro people."

For those who are already involved with a house of worship, volunteering can be a great way to learn and gain skills. Jonathan Hubel, who started in live sound at his church before moving into podcasting and radio, says:

Being actively involved as a volunteer at churches is a great place to learn and gain hands-on experience, often being taught and mentored by professionals and experts in the field. It also greatly depends on the size of the church, as small congregations will be dealing with very limited budgets for equipment, and are almost exclusively volunteer-led. Large churches often have at least part-time paid AV staff (and some full-time), but are often looking for volunteers to help out as well. Even a tiny church that has two microphones and a small PA system is a great place to get your hands on equipment and slowly learn the ropes.

The dynamics of working at a house of worship are different from entertainment venues particularly because workers can be a combination of volunteers and professionals. Engineer Hiro Goto explains:

Usually, it's people attending church who want to help. I'll get a 16-year-old and they really want to volunteer. I'll train them, but the language you use is different. In live sound for church, you have to stay calm, and also you want to be more transparent. They always have last-minute changes. Anything could change, and anything could go wrong. You have to think calmly and then troubleshoot quickly.

Samantha Potter (audio engineer/consultant) adds:

People take church [activities] a lot more personally. It's like an extension of their home. People are forgiving, but if you're a huge jerk, these people (who mostly are volunteers) are not going to want to be around you. That's a big deal.

These opportunities may be better for learning than a long-term career plan. Jonathan Hubel shares:

Churches are absolutely an excellent place to learn! Can you build a sustainable career doing sound in churches? It's possible, but it's going to be tough. I would almost guarantee that you would have to find a very large church in order to ensure full-time employment – and those positions are going to be in high demand. If that's your goal, then start small, volunteering wherever you can, hope to maybe work your way into a part-time position, and eventually *maybe* become a full-time technician.

House of worship work can lead to work outside the church, as well. Jason Reynolds says:

When I was working on a lot of gospel shows, the production company took notice of the work that I was doing and the quality of work. It's like any other gig – you never know who's sitting in the congregation that week. You never know when somebody is going to be sitting in the audience and say, "Hey, you're pretty good at this." For me, that's how it went, and then somebody took notice.

It is possible to pursue opportunities in live sound for entertainment and houses of worship simultaneously. "I don't know if I'd say it's easy," says Jason Reynolds. "I tour professionally but also work in churches, so it's definitely doable. The skills are transferable because at the end of the day, good audio is good audio."

Music Cruises: Andrew King

Andrew King is a broadcast mixer for sports who also does live sound for music cruises.

I do concerts on cruise ships. We call them charter music cruises, so basically, a production company will charter an entire cruise ship and we take over the whole thing. We'll set up maybe five stages on the cruise, and do concerts all week long. We've done as many as 100 concerts in one week on a cruise. Most of them are five- or seven-day cruises, but we might do three of them in a row. They give you a room to stay in and they pay all your travel expenses. We don't get a per diem, but there's free food on the ship. It's pretty good.

If you're a part of a ship's crew (working for a cruise line), you're doing a lot of mundane things, like, we're doing karaoke tonight, we've got dueling pianos tomorrow, there's a presentation in the conference room, that kind of stuff. Whereas I don't have to deal with that. Our entire team is probably 30–35 people – that includes backline people. There are probably 10 or so audio engineers.

Typically, the sound systems on the cruise ships are not that great. We always bring in our own PAs and everything because what the ship has is not sufficient for a concert. A lot of it is similar equipment to what you'd have at a local venue, but there's the complexity of working on a cruise ship. Usually, you'd pull into a loading dock and unload all your gear right behind the stage and you're ready to go. But here, the cases either have to be pushed onto the ship on a ramp (which they almost never allow us to do), or you have big metal cages which get picked up by a forklift and taken to the ship. You unload those, and you've got to push the cases all over the ship. It's amazing how big those things are and how small the elevators are.

If you're working at the stage by the pool, you're going to have a very busy week. Usually, I'm working on the smaller stages, and we might only have shows in the evenings or maybe just a speaking panel during the day.

Live Sound for Jazz: Claudia Engelhart

Claudia Engelhart is a touring sound engineer and tour manager. She has worked as Bill Frisell's front of house mixer since 1989. Claudia has also toured with Wayne Shorter, Herbie Hancock, and David Byrne, among others.

For those interested in pursuing live sound for jazz as a career, there are two typical routes: work in live sound for multiple genres of music (including jazz); Or, work in jazz but also take on roles beyond sound (such as tour managing). Claudia suggests the jazz community is more personable and more accessible than other types of music:

> For the most part, when you're going to a jazz club, you're going to be in an intimate situation. You're going to be much more in touch talking to people or being able to approach a person. Whereas a big rock stadium tour, you're never going to get close to any of the artists, ever . . . A lot of [those] artists don't even really know the people that are putting all their shows together. Their crews are so big and they're just separated. But in jazz, you're one-on-one all the time. You're stuck in a van for three hours all squished together, or you're sitting in a lobby at an airport together . . . It's very personal.

One unique aspect of jazz touring (versus other genres of music) is touring happens year-round and in a wide variety of venues, from small theaters to large music festivals. Assisting jobs do exist, but opportunities are rare. Claudia explains:

> Financially, it's hard now. Traveling, airport, airplanes, hotels, all that stuff, tour buses – it's expensive to travel that way. It's mainly the really well-established jazz artists that can carry any extra people . . . If you're getting started in jazz, you've got to be flexible. You might have to volunteer or you might have to work cheaply and eventually it'll get better as you get more established with that artist.

Finding Work in Live Sound

- **Slow periods in live sound can vary by musical genre and location.** For example, in the United States, New York is slow in January and February, but Florida is slow in August. Los Angeles is busy year-round. Major jazz musicians may tour year around.
- **Connect with people who do the booking.** For example, hotel AV companies have a scheduling manager or labor booker, some venues have a production manager, and tours will have a tour manager or artist management. A production company is a business that plans and oversees an event, show, or tour and could be an independent company or part of a venue.
- **Get a personal recommendation.** When seeking work for a sound company, a referral or personal recommendation can put you ahead of candidates without one. (See Chapter 7 for more.)

Live sound can be more direct than other disciplines asking for work. Scott Adamson explains:

> If someone's on tour, who knows when you will see them again? It could be a year. If you have someone in the room and are [having a] conversation, be nice, but definitely make it known, like, "Hey, if you need somebody, I'm available."

Pro Tip: Nathan Lively

If I were starting over or moving to a new city, I would start by contacting all of the AV companies in town to get some cash flow ASAP and meet new people. I don't want to be desperate for money while pursuing my dream job. Corporate events and hotel AV have high turnover rates, lower competition, and nice people who are willing to help you out. It's not hard to get a job and get trained for free. Other entry-level positions that always have openings are automotive and residential system installation.

At the same time, I would find 10 people in town who are doing exactly the job I want to be doing and get dates with them where I would find out as much as I could about them and their work, ask for advice and contacts, and permission to follow up in the future. I would also hire a business coach and get an accountability partner.

("Get on Tour" 25–26)

For stagehand work, overhire (temporary) work can be a foot in the door. Stagehand work can also help fill in the gaps for established professionals. "If you're a touring engineer, your tour ends, and you have two months off before your next tour starts, picking up stagehand work is a great way to fill in that hole," says Karrie Keyes. In some locations in the United States and Canada, stagehand work may come through a union or stagehand company. Stagehand work can also be found beyond theaters. Andrew King, who works music cruises, shares, "We bring in local crew for the load in and load out, and sometimes they'll stay on the cruise."

Live sound opportunities often come by word of mouth and may not have a formal hiring process. "It's all about contacts. It's very rare that someone gets hired without a direct recommendation," says Michelle Sabolchick Pettinato ("Tips").

A resume is not necessary early on, but can be beneficial to have and keep up to date. Scott Adamson explains:

> Your first job touring, probably first couple of tours – No one's going to care about a resume. No one's going to look. At a certain point, if you're expanding your touring career, you'll start to deal with managers of bands. Sometimes tour managers will ask for a resume, but usually they'll just ask for personal recommendations from people. But, band managers are the people that always want to see the resume. They'll want to see all your experience.

Karrie Keyes suggests seeking runner positions:

> Hit up local promoters. They're not going to hire you to do sound, but they could be hiring you for a runner position. That's a great way to get your foot in the door. You see how the big productions work . . . [Runners] pick up whatever is needed. So, say I've run out of batteries. I go put it on the runner's list . . . There's someone in each location [on tour]. The promoter is hiring them, and they have a rental van.

Corporate live sound work can be found at conferences, business meetings, government or political events, and beyond. Corporate work can be consistent, pay well, and supplement income from live entertainment. Nathan Lively explains:

> Corporate is actually a great place to start if you want to start making some money and are not interested in working for free for a few years until you get your own clients . . . There's no one who really starts out or has [corporate work as a] goal, but there are plenty of people who end up making a career out of that . . . It pays pretty well. There's never a lack of

meetings . . . If you're good at it and you don't mind the type of work that it is, it's a great way to make some money.

("Sound")

Scott Adamson says even though music work is more satisfying and is what drives his career, he also does corporate gigs, including sessions at the General Assembly of the United Nations. Scott explains:

> Corporate work is a great way to pay bills and it's absolutely nothing to sneeze at. Work is work, and putting bread on the table is an absolutely essential thing to do in life. There's nothing wrong with corporate work whatsoever. It actually can be a really, really well-paying job.

There are opportunities beyond traditional paths for those thinking creatively. Nathan Lively suggests:

> I think probably lots of sound engineers (when they're first starting out) are thinking about contacting the big concert venues and contacting artists, but they don't really think about contacting law firms, for example. They could offer law firms their services for live events, or for webcasts, or for PowerPoint presentations and meetings, or whatever it might be.

("Career")

Working at music festivals can be a way to build experience and also network. Claudia Engelhart (front of house mixer for jazz artist Bill Frisell) explains:

> There are hundreds of jobs you could do in a jazz festival, and in the meantime, you're meeting everybody, too . . . Maybe [someone] got sick or something and boom, "Here's my phone number, call me." This is how these opportunities happen all the time. Being in the right place at the right time is true . . .
>
> There's a viola player, Eyvind Kang, who's played with Bill Frisell for years. Bill met him because Eyvind was our driver at a jazz festival. Bill didn't know that he was this monster musician. Eyvind drove us around and suddenly you're talking to one another. You get to know the person a little bit from the 10 hours that you're around them, and boom – he worked with Bill. This is how these things happen.

How Performing Can Help Land Live Sound Gigs: Hiro Goto

Hiro Goto (originally from Tokyo, Japan) is a violinist who performs in a variety of genres from orchestras to rock and country bands in Los Angeles (USA). Hiro also works in live sound for clubs, restaurants, houses of worship, and more.

Playing shows are a good opportunity to meet new people, and being on stage helps get me work. I play in a lot of venues and with different groups. If there's downtime, I always talk to the engineers and start carrying on a conversation. They want to talk about audio because they're passionate. Plus, musicians can be demanding even though the engineer is doing their job right. Sometimes the engineer needs confirmation that they're doing a good job.

You don't see violinists who know audio often in the genres of music I play, so that has helped me. I have a gigantic rack of real gear and an amp. It sounds good and engineers are impressed. They hear it and ask, "How do you get that tone or the effects?" I explain the gear.

It's all about performing on time, especially at hotels and venues. But, some engineers show up at the last minute instead of giving themselves enough time to setup. I'll ask, "Do you need any help setting up?" . . . It looks pretty obvious when they're hustling. When they're packing everything up after the show, I ask, "Do you need help?" . . . They always say, "No one asked me that before." I wouldn't offer if I was just in the audience – only if I'm performing.

One gig, I was playing at a hotel and the engineer was having a hard time with the DIs and was short on inputs. I suggested a way he could do it, and it helped. Later, he asked if I had a business card and said, "I need more engineers like you to help troubleshoot." Since then, he's hired me for many hotel gigs. He has a pool of engineers, and sometimes they bail on a gig, believe it or not. They'll call me last minute. "We need someone to work poolside today," or "Can you engineer tonight?"

Jam Bands: Jack Trifiro

Jack Trifiro is the sound engineer and tour/production manager for Victor Wooten, and also the monitor engineer for Béla Fleck and the Flecktones. Jack can trace his career in touring live sound back to his interest in jam bands, where he started by doing lighting for fun while attending his friends' shows. For those seeking work with jam bands, Jack suggests finding venues where jam bands come through, looking to your local scene for bands you like, getting to know band members, and offering help.

My musical interest goes back to me being a fan. I followed the Grateful Dead around and I used to tape all of their shows. It's funny because companies that I run into nowadays – like a production company in Baltimore is owned by a guy that I knew from the taper section at Grateful Dead shows. You run across the same people everywhere you go.

Being interested in the Grateful Dead got me interested in local jam bands. I met a local jam band at a festival and worked for them. At a festival (working for that jam band), I met other bands, and I met a production company [who hired me] . . . I was setting up PAs as a systems technician doing concerts and festivals. I was being exposed to a bunch of different artists. I used to do Gathering the Vibes [a four-day festival that celebrated the Grateful Dead]. I did that for eight or nine years as the front of house and systems technician. So, that festival alone probably exposed me to 25 acts over three or four days.

Working for the production company, I went to the Discover Jazz Festival in Burlington [Vermont, US], and I started meeting jazz artists. The jam band scene is how I got introduced to Victor . . . Victor and his side projects allowed me to hook up with other artists.

Touring

Michelle Sabolchick Pettinato says of touring:

For me, the best part of touring is that I get to travel the world and experience different cultures, visit exotic places, and meet all kinds of people, all with someone else footing the bill *and* paying me to do what I love! I've been to Red Square in Moscow, Tiananmen Square and the Forbidden City in Beijing, had a private tour of the US Navy's Memphis submarine in Bahrain, had lunch with a Maharaja in India, watched the Super Bowl with the Rolling Stones in Brazil, climbed the Sydney Harbour Bridge in Australia, and had many other wonderful experiences.

I also love the camaraderie. There is a degree of teamwork necessary to make the show happen every night that lends itself to building some really close bonds and relationships with the people you work with. You are, in a sense, a mini-community traveling the world.

(53)

Jack Trifiro says of touring life, "It's a totally different world. Hours, awake time, sleep time – down to what you eat and when you eat. Everything changes. It's completely different from 'normal' life."

Touring can be a short-term plan or a long-term career path (and lifestyle). Scott Adamson shares:

> There's often life after touring. You can think of it as a short-term thing or a phase. Some people are lifers . . . but it's not always the case. Even though there are people that have long successful careers touring, a lot of people do it when they're younger, then they move on to other things when they're older.

Tours can vary based on size, budget, or where you live. Aleš Štefančič explains:

> Coming from a minuscule country of Slovenia, Europe, touring is quite different than in other areas. Since it only takes about four hours of driving to cross the country, we do not usually "go on tour" in terms of packing up our gear, kissing our loved ones goodbye, and hitting the road for months on end. Ninety-nine percent of the time I have the option of driving home and sleeping in my own bed, then heading out for the next gig the following morning . . . So, my touring experience might not apply as much to productions that pack everything in a van/bus/plane and take off as it does to regional touring experiences specific to our small country.
>
> (15)

Building relationships with bands can be a foot in the door to touring. Scott Adamson explains:

> A band is going to bring their friend [on tour]; Or, they're going to bring the person who does the sound at the local bar that they play at (at home), who expressed interest, and can put on a smile and be nice to people.

The entry path to getting on tour may not be through mixing. Artists may need help for anything from driving to tour managing to merchandise. Karrie Keyes suggests backline tech may be an easier path than mixing to get on tour:

> You need someone to tune guitars during the show. A lot of backline techs can do drums, guitars, and keyboards. Usually, when you're starting out, you need to know how to do all that stuff . . . you're going to be setting it all up by yourself.

Major artists and tours generally do not take interns or volunteers because of insurance, liability, and the costs/logistics of travel. Claudia Engelhart explains:

> It's unfortunate, but it's finances in the end. It's how much it costs to fly somebody, hotel rooms, food, every single aspect of touring is an

expense . . . The logistics of it are what inhibits being able to bring more people out . . . unless you're with a small enough group or you're on a van tour or something (and there's extra room in the van), but you still have got to pay for a hotel room for that extra person. There are always expenses that go along.

While touring can be fun and adventurous, workdays can be long. Marc Thériault shares:

If you think you're going on the road because you're going to have fun and party – you might, but the main thing is not that. You still have to work. When you have a day off, you can do whatever you want, but you have to be sharp and you have to be on top of your game. If you go on a tour, you drink too much, and you miss your call, you might [be sent] back home . . . Don't drink with the band. You work a lot. A tour like us – for the PA guy, the show call is at 7 am, then you finish at 2 am. Then it's the next day, the next city.

Australian monitor mixer Becky Pell adds:

Being on the road is not an environment that is conducive to healthy living. We work long hours and the food can be very hit and miss. There's not a lot of downtime, there's not a lot of time to rest, and you may not be getting enough sleep. Let's be honest: depending on the band and the crew that you're with, there are differing levels of partying going on. Some crews can be pretty clean versus others. If you're going to have longevity, it is important to do what you can to take care of yourself. I'm not saying you can't ever go out and party and have fun, but there's a balance to be had.

(Leonard et al.)

Attitude and temperament are important for touring crews. Guitar tech Claire Murphy explains:

You're with 10 people [and sometimes] up to 100 people, and you're in a small space, and you have to get along with people. You can be the geekiest person, know everything about anything but if nobody likes you, you're not going to last long. If you have a bad attitude, you're not going to last long. I've seen it where someone doesn't know that much, but they're keen, they have a good attitude, and they stay on the tour. It doesn't matter that they don't know anything because they're going to learn.

(Keyes et al.)

Karrie Keyes suggests avoiding touring if you do not enjoy traveling or being around people all the time:

If you're in a van, you're going to be around people 24/7 because you're probably sharing a room. So, you've got to be able to deal with people. If you like being by yourself, working solitary, and people bug you, that might not be the best for you.

Scott Adamson suggests trying a tour if you are given the opportunity:

> If you really think that you're going to love it, and you're excited to get away from home for a month, and it sounds fun to you to travel and you like music, you're going to have fun doing it. Everyone else that does that stuff likes the same things. You tend to find similarly-minded people in those environments . . . Committing to a month or six weeks [at first]? That's a long time. Maybe that's not the best first thing to do. But try a weekend or try a week. Try something small, something short, and you'll know pretty quickly if you like it or not.

Tips to Getting on Tour: Claire Murphy

Claire Murphy is a guitar tech who started touring in the UK before moving to the United States.

Save enough money to take 6–12 months off. You do not want to quit at Christmas and have your first two months looking for work [be] the quietest times in the music industry . . . Summer is crazy and if you've put in the groundwork in the months leading up to summer, you should have no problem getting work all summer long.

Touring opportunities can come from offering a little of your time for free – even outside of sound work. Go to a show where you know the support band has a tour coming up and might need extra crew. Go to the merchandise table and meet the band and ask if they need an extra helping hand (for free probably. How could they turn that down?). You will gain so much experience and if they blow up, they'll probably take you with them. You do this successfully a couple of times and you've gained so much experience and made so many contacts. This is the most direct route to getting on tour . . . You don't even need to buy a ticket to the show. Go to the venue around 2 pm on the day of the show and wait for the support band to arrive. Ask if they need a hand loading their gear in! If you get on well with them and you have the means available (train ticket, cheap hotel room) you could ask if you could meet them at the next show and help them again, as you're trying to gain touring experience.

Get on the list for van-hire companies. Sometimes bands will hire vans but none of the band members can drive. None of them are old

> enough to drive, or in the case of a UK tour, they're American and don't feel comfortable driving a manual/stick shift van.
>
> Once I did [my] first tour over the summer in 2010, the band asked me to do their next tour. That's how it works. Generally, once you're in, you're in. Of course, one band alone isn't going to keep you employed for your entire career.
>
> (ch. 1–2)
>
> From "*Girl on the Road: How to Break into Touring from a Female Perspective*"

Tour Managing

It has become increasingly common for touring sound mixers (at all levels) to take on additional job roles like tour managing. Claire Murphy, who worked as a backline tech and tour manager, says of the overlapping roles:

> Doing two jobs at once is exhausting and something will always have to give. But, it is a great way to learn in the short term and will get you hired on more tours, which is the aim at the beginning.
>
> (ch. 1)

Prior experience as a tour manager is not needed to take on the job. Karrie Keyes explains:

> They never say, "Are you also a tour manager?" They always say, "Can you tour manage?" You might have never tour managed before, but they didn't ask you [that]. They asked if you *could* tour manage. So, you say yes, and then you start talking to tour managers or you go and watch the six hours of how to tour manage on the SoundGirls YouTube channel.

Claudia Engelhart, who does tour managing along with live sound, explains her job:

> [I'm] finding lost luggage, making sure everybody has food, dealing with whatever drama happens, dealing with the promoters, making sure the backline is cool, setting up the stage, making sure that dressing rooms are clean and that there's food in back, getting paid at the end of the night . . . I do pretty much everything except play on stage . . . I do all the itineraries. I advance all the show. I spend more time on the phone and online writing emails and typing out itineraries before any tour. The hours that go into organizing a tour, even before I get out there and actually do the fun stuff, are administrative.

The skills needed for tour management are very different from audio engin-eering. Jack Trifiro says of those who would *not* make a good tour manager:

> Someone with a short temper or who panics under pressure. Somebody that makes a lot of excuses . . . It's more of a workload than actual skill. It's being able to handle everything all at once . . . As far as the actual skills to tour manage, I knew all of that already coming into it, as far as spreadsheets and being able to book hotels and flights. You pick up a few tricks here and there as you're going along, but having the basic set of skills already is key.

Jason Reynolds recommends not taking on these additional roles for free:

> Even if you're negotiating more money, it still works out cheaper for the band. It's cheaper than carrying two separate people (it's one less hotel room, one less per diem, one less plane ticket). It is difficult to tour manage and mix the show, but you find a way to make it work.

The Value of Formal Audio Education: Eric Ferguson

Eric Ferguson is an assistant professor in the live sound production program at NESCom at Husson University in Bangor, Maine (USA). The program is currently one of very few bachelor's degrees in the United States focused on live sound. Eric finds the program's graduates are highly employable because of their formal education. Graduates have gone on to work with sound companies, venues, cruise ships, in AV, and broadcast mixing (including sports mixer Andrew King). Eric shares:

> I see my students graduate, go get a job, and within one year work as a PA tech [for a major tour]. In our program, the goal is to help them get a job. I feel really good about putting people in this [degree] track because they can buy houses and get married before they're 30.

The degree program introduces students to a wide variety of job paths in the live sound industry. Eric explains:

> I have a fair number of students that take my system tech class, and they do it because it's one of the badass classes in the program. Will it make them a system tech? No, but it will open their eyes up to career opportunities.

The unique advantages to a live sound degree program (versus one focused on music production with some live sound training) include students getting hands-on with equipment (beyond operating), and relevant, real-world work opportunities as stagehands. Eric explains:

> In the studio, they don't get behind the gear. They don't plug in the console. The tipping, the lifting, the plugging in – the signal flow you learn by actually plugging things in – that never happens in the studio . . .
>
> [After working stagehand gigs], they'll come to me and say, 'I saw this on a show. What do you think of that front of house engineer's approach? Does that make sense? What's the tech behind that?'. . . I will interpret that.

Eric, who has worked in both live sound and music production, suggests live sound today has more stable entry-level opportunities than music production:

> Whether it's a big sound company or the AV tech side, working for a cruise ship or for a venue, you can work for a company that can protect you while you're starting your career out. Mentorship is a key part of it. Somebody can help you along and protect you while you make mistakes. A company can do that for you. The recording industry doesn't have that anymore. The big studios are still there, but they're fighting to stay alive. Most music production is done in a really small, non-corporate/non-business/independent way now.

Theater

While the technology may be the same between theater and live sound, theater is its own art and discipline. Head Audio Heather Augustine says:

> I am a theater person first and a sound person second – where I got into theater just because I love theater itself. I don't think I would enjoy mixing bands or doing corporate as much as I enjoy doing theater just because I love having that story.
>
> (Augustine et al.)

Additionally, musical theater uses an extremely fast-paced style of mixing called line-by-line mixing, which is raising and lowering faders immediately for each character. The level is pre-determined by the sound designer, leaving the A1 with a book of line-by-line instructions on how to mix a show. This process cannot be automated because of performance and timing differences that happen live. Head Audio Kevin McCoy says of mixing, "There are definitely times in the show when I feel like it's me and the actor and the conductor and we're doing a little three-way dance with each other."

There are two general career paths in theater sound: sound design and show operations. A **sound designer** is the lead sound person for the entire sound crew of a show. Sound designers (and their teams – associate and assistant sound designers) oversee everything sound-related for a show, from sound effects, microphones, speaker choices, and the mix/balance of the show itself. Tony Award-winning sound designer Kai Harada says of the path to becoming a designer:

> Historically, some current designers have come up as assistants, some as mixers, some as a combination. Some start downtown, some start regionally, some get their chops on the road, some start off doing studio work, some start and remain in New York.

Sound designers may be staff positions or hired specifically for a show. Working with different sound designers can be a great opportunity for learning and networking. Becca Stoll explains, "If you work in a regional theater like Dallas Theater Center, then New York or reputable Chicago designers are coming in to design those shows, and you are working with them."

Sound operators are the crew responsible for sound needs during a show performance. The **Head Audio** (A1, No. 1) mixes the show live and is one of the few roles who follow a show non-stop during a performance. An **A2** (No. 2) typically assists actors with microphones and adjustments backstage during the show (some shows may have multiple assistants – A3, A4, etc.). A2s may have the opportunity to switch roles with their A1 for some shows. Kevin McCoy explains:

> Adriana Brannon is the A2 with me on this production of *Hamilton*, but we actually cross-train on each other's tracks. Five shows a week I do the front of house mixing track, and the other three shows she does it and I do the backstage track . . .
>
> I like the way that we do it in theater because it forces [the A1 and A2] to have an understanding of the bigger picture, and it makes us both better at our jobs. If I'm backstage and there's a problem with a microphone, I don't need to consult any paperwork to know when that microphone is going to be used because I have the whole show in my head (because I've mixed it). Also vice versa – if I'm front of house and I have a problem with a microphone, the backstage tracks know when they'll be able to get over to deal with it.

Theater is an extremely team-oriented endeavor. Kevin McCoy says of who he reports to on a show:

> I report to the stage manager, to the company manager, and to the head carpenter, because the head carpenter runs the crew. The company manager signs my checks. The stage manager runs everything to do with the show, so they can give me notes about how I'm mixing the show. Then, the general manager is the person ultimately who has hired me and signed my contract, so I definitely report to them. Then, I report to the technical production manager because they're in charge of hiring all technical people. I report into the sound design side, so I report to the production sound person because they're responsible for sound crew, and then I report to the sound supervisor, because they can fly out and [give notes] and listen to me mix the show. Then, I report to the sound designer.
>
> So, I have eight different people that I can directly report to in different ways – all of whom could easily put the wheels in motion to fire me if they needed to. Even though the org chart doesn't reflect any of that, that's the reality of it. I definitely feel I [mostly] report to the sound designer because it's their name on the sound design that I'm reproducing every night. Also, from a practical point of view, they are probably the person who will give me my next job.

Full-time theater jobs exist but may have limited room for growth without moving to another theater or relocating. Becca Stoll said of one regional theater she worked for:

> I had put in four years moving up from being the audio apprentice at the second stage to head audio at the main stage. There were no more promotions available. I did not want to succeed my boss and run the department, and he also was not leaving.

For those pursuing musical theater, some musical knowledge is helpful. Kevin McCoy explains:

> It's definitely beneficial to understand music, and having a general vocabulary about musical instruments is helpful. Also, having a general idea of what it's like to sing on a stage is great, but none of that is required. It's not super common for anyone to use musical language directly with me. It's not required that I could read music, but if someone knows that I can read music, they might change the way they communicate with me.

There are a number of different types of productions and theaters, and opportunities will vary from unpaid or low-paid to full-time positions. Some theaters produce shows (a **producing house**) where they hire staff for all roles. Some theaters host touring shows (**receiving house** or **roadhouse**) and may only hire stagehands. Some theaters are producing and receiving houses. Kevin McCoy says of the small producing houses he worked for early in his career:

> They had maybe a couple of hundred seats, so they weren't getting tours. But, they were producing professional shows with paid cast and paid crew. It wasn't hard to find the jobs, but it was hard to find enough jobs to make a living. You'd find theaters like those in any major city in the US.

Commercial productions are set out to be profitable, whereas **not-for-profit** has a primary goal of cultural/artistic enrichment for a community. For example, in Glasgow, Scotland, local theater is a popular pastime. Educator Clare Hibberd (who has worked in the West End) explains:

> There's a lot of very good theater made in Scotland. It's not big musicals, but there is all sorts of amazing stuff that happens . . . NCS National Theater, Scotland, Scottish ballet . . . is very much leading to the needs of the community. That's got to be more important than making big money out of big shows.

Commercial theater companies are primarily based on Broadway in New York (USA) and the West End in London (UK) and can have a show run in a theater, on tour, or both. Not-for-profit productions can be found worldwide and have varying budgets. **Community theater** can be a collaboration between non-professionals and professional theater artists, or can be all-volunteer. These opportunities can lead to a career in theater. For example, Tony Award-winning sound designer Jessica Paz started in community theater.

Regional theaters are theater companies (in the United States) that may have full-time sound positions (like A1 and A2) but may be unionized. **Summer stock** is seasonal work in the United States (beneficial for building credits, relationships, and experience). **Fringe** theater (or "free theater") can be a part of a festival, experimental in style, and may be low or no pay.

Touring opportunities also exist in theater. Kevin McCoy explains:

> It can definitely provide a decent living for people at the highest level. Entry-level touring work can be not awesome in terms of pay and the hours get harder. Just because you have had one pink contract [union] job doesn't mean that you're in for life. You still have to maintain those contacts.

Finding Work in Theater Touring: Kevin McCoy

Kevin McCoy (Minneapolis, Minnesota, USA) is Head Audio for Hamilton ("And Peggy" Company). He was Head Audio on touring productions of Motown the Musical, Cabaret the Musical, and Once the Musical.

A2s are often selected by the A1 of the tour. The A1 will often present an option for an A2 to the designer to approve. Usually that option carries a lot of weight. So as A2, knowing people who are going on tour as A1 is a way to get those jobs . . . If you network with me, I might be getting calls from other A1 who are working at lower level shows, or more entry-level shows that are interested in having an A2 who doesn't have any touring experience. In that case, I'm going to say, "I'd really love to work with Fred, but Fred has no touring experience. But Fred might be perfect for your job where the pay isn't what someone with touring experience would expect."

I worked as an A2 connected with an A1, and we were a good team. The people who hired the A1 saw that we were a good team, and so they moved us as a team to another show, which was great. I've also worked with A1s that I didn't click with and so I knew that wasn't going to continue after that show. The relationship between the A1 and A2 is really important for the happiness of both [people] . . .

For smaller tours, the more entry-level touring jobs with production companies, sometimes you can just apply directly. For A1, they will hire their people and present them to the designer, like, "Here's your A1." The designer could veto them but often they won't, because the designer might not have a list of people who are willing to work for those lower salaries. For a designer who doesn't have any A2s in mind, they might say, "We have this stack of resumes. Here's a list for you to choose from" . . .

Whether you want to be the A1 or the A2, you really need to eventually connect to the designer, and the designer usually has a say over who gets hired on the audio crew. I think officially that the final say is up to the general manager or maybe the technical production manager, but usually the designer's choice is the one that wins out. So figuring out how to connect to the designer is often key for anyone wanting to get a job on a tour.

How to Network and Find Opportunities in Theater

Relationship building is important for finding work in theater. "You do have to talk to people to figure out if there are any opportunities. You have to be boots on the ground . . . You have to go out and meet people. You have to go see plays," says Avril Martinez.

Theatrical mixer Becca Stoll adds:

> My boss is always going to be a sound designer. So, I'm out here trying to meet sound designers and show them why I, as a mixer, am a good match for them . . . I'm the one representing [their] product every night.

Shadowing a professional can be one way to start building a relationship. Heather Augustine shares:

> I know I am always happy to have people sit with me – Not so much if they just walk up during pre-show like, "can I come sit with you in two minutes?" . . . Even just going up to the board during intermission or after the show and saying "Hey, can I come back tomorrow or another day and sit with you?" that would probably be a really good opportunity to get to see what people are doing . . .
>
> If you're in a city that has a theater that does touring shows, ask if you can shadow, even for a load-in day or something like that. Or, you could essentially be overhire [when extra workers are needed] just to get your feet wet, to see what's happening to get an up-close and personal look. . . Then, if there are any theaters around you, just try to start working with them. Try to start doing shows and building relationships with the community or regional theaters, or anything that's around you, and then go from there.
>
> (Augustine et al.)

Industry conferences are opportunities for networking and job seeking. The United States Institute for Theatre Technology, Inc. (USITT) has a yearly conference that has a job fair for regional theater and summer stock, and some companies interview at the conference. Students can volunteer in exchange for a free pass to the conference. In the UK, the ABTT Theatre Show (Association of British Theatre Technicians) is an opportunity for networking, and the organization runs events throughout the year. "If you can afford to go to conferences, go to conferences. USITT is a very big conference for networking. If you can't go to conferences, don't stress about that," says Kevin McCoy.

Theater job listings can be found publicly. "Job postings seem to appear on social media first and then some will make their way to an official website. Always worth checking both," says Pete Reed.

Becca Stoll says she likes to help those who take initiative to meet her:

> Anytime that somebody finds me, reaches out to me, shadows me – if I have a call that I can put you on . . . [somewhere] I can take a chance on someone, I absolutely will . . . I would love to have you on it because I can get you in a door.

As with any discipline, it is important to be honest about your job experience. Pete Reed shares:

> *Do not lie* on your CV. I cannot tell you the number of times a CV came through and it said they did the No. 2 plot on a show in the West End. The West End is small – very small. Everyone knows each other or has heard of each other. If you have worked as a dep [sub] on a show, do not claim you have mixed it. You will not get a call – unless it is an angry call to say take that off your CV!

Pro Tip: Networking in Theater

If you go to shows, go talk to the people at the soundboard. You'll find maybe one in five of them are going to be a total jerk and shoo you off. Hopefully, four or five of them will be cool and talk to you. I love it when students come up to me and talk to me at the soundboard. Obviously, use your judgment. Don't talk to them when they're busy, and don't be obnoxious. It's okay to go up and just say "Hi, I'm a student. I love this stuff. Can I just talk to you for a minute?" I've given backstage tours before just from someone talking to me. It's not something I can always do, so don't be offended if the person doesn't offer you that.

For introverts who lurk by the soundboard, the sound person might notice it . . . [I can] go up and talk to people easily, and I understand that other people don't have that same socialization. So, if I see someone lurking off in the corner, I'll try and approach them. That's also advice to more experienced members of the community – to look for those people.

Sometimes it's easier for an introvert to talk to someone online than it is in person. I'm very active on Twitter. People can tweet at me or DM me on Twitter (@kmccoy) and I'm always happy to talk to people that way.

– Kevin McCoy

Sending Resumes: Jessica Paz

Jessica Paz is a sound designer for Broadway and was the first woman to win a Tony Award for Sound Design (for Hadestown).

> The culture around hiring designers is changing. Basically people told me, "Don't send your resumes to places because they only hire people they know, or people from word of mouth," etc. I think that that's changing now because there is this seismic shift in the industry to create more diverse opportunities and it's about equity, etc. So, I think that a lot of theaters are, in fact, looking for new people. Going back to the same people over and over again is sort of a big no-no now.
>
> In this day and age, sending your resume is a good thing to do. Having a website, put your stuff up on a website, put your resume on a website . . . I don't believe you have to be a union member . . . email directors that you want to work with, whose work you like, and see if they would have a cup of coffee with you. I get people emailing, students emailing me all the time . . . and I make time for them. So, I'm certain that there are directors and producers (and all in all) who will respond. If they don't have time right now, we'll put you on their list to get to when they do have time.

From SoundGirls: Ask the Experts

Unions

The advantage of working as a union member is the employment protection for workers.

A **union contract** is an agreement between union members and the facilities or productions they work for. A union contract can be extremely detailed, down to who can touch mic stands or even use a ladder, and includes minimum pay rates and rules about overtime. In the United States and Canada, sound technicians and stagehands fall under the International Alliance of Theatrical Stage Employees labor union (IATSE).

Whether a production, venue, or a tour is union can vary. Kai Harada says of different arrangements:

> Not all venues are unionized, but all Broadway shows are unionized . . . Regional theatres may be under union jurisdiction, or they may not be.

Designers working in smaller markets may not need to be union members, but designers working on Broadway must be union members.

Even those with professional experience may have to start in an entry-level position (such as stagehand). One way to get started is **overhire work** (temporary help when needed). Broadway A1 Julie M. Sloan explains:

> Whatever city you're in, find out about getting on the IATSE overhire list, because most cities don't have a large enough membership to staff all of the calls and they do use overhire. You can still work overhire without a union card. You're not obviously at the top of anybody's list, but you're going to start to see what shows look like, how shows get put in . . . the calls for a touring show tend to be very large the first couple of days, which is when they really, really need extra hands.
>
> (Augustine et al.)

Kai Harada adds:

> If you are in an area with a strong IATSE presence, and a roadhouse that hosts theater as well as live music, this is a great way to learn, by being part of the local IA crew, seeing how a particular show is structured versus another, and meeting the road crew (some of whom can be nice). It may not be the most intellectual of work, but a lot can be learned, and in many cities, it can be lucrative . . . Having an IATSE card can also be useful making your way into New York and onto working on a show.

The requirements to join a union vary by location and can be found by contacting the local (the branch in that region). Kevin McCoy suggests, "Not having union membership should never be a reason for you not to apply for a job. So, apply for the jobs whether or not you have a union membership card or not."

The union for creative industries in the UK (including the West End) is BECTU (Broadcasting, Entertainment, Communications and Theatre Union). BECTU offers support to members (such as financial and legal services), and in some cases can negotiate with employers.

References

Adamson, Scott. Personal interview. 14 May 2020.

Augustine, Heather, et al. "Ask the Experts – Mixing for Broadway and Theatre." *YouTube*, uploaded by SoundGirls. 24 Feb. 2021. youtu.be/ifBC_ErEDCs.

Bushong, Chris. Personal interview. 27 Sept. 2020.

Engelhart, Claudia. Personal interview. 8 June 2020.

Ferguson, Eric. Personal interview. 2 Dec. 2020.

Fisher, Gabrielle. Personal interview. 18 Aug. 2020.

Goto, Hiro. Personal interview. 13 Feb. 2021.

Harada, Kai. "Kai's Sound Handbook: Advice to Bright-Eyed, Bushy-Tailed Youth." *Harada Sound*. www.harada-sound.com/sound/handbook/advice.html. Accessed 28 April 2020.

Hendley, Alesia. Personal interview. 17 Dec. 2020.

Hibberd, Clare. Personal interview. 7 May 2020 and 15 May 2021.

Holmes, Meegan. Personal interview. *What Makes You Stand Out*, SoundGirls, 20 May 2020, youtu.be/wYamiOK8Y6A. Accessed 21 May 2020.

Hubel, Jonathan. Email interview. Conducted by April Tucker, 22 Oct. 2021.

Keyes, Karrie. Personal interview. 19 June 2020.

Keyes, Karrie, et al. "Career Paths in Live Sound." *YouTube*, uploaded by SoundGirls, 1 Oct. 2019, youtu.be/dbBxX9PddhI.

King, Andrew. Personal interview. 15 Dec. 2020.

Kramarik, Kelly. Personal interview. 18 June 2020.

Lawrence, Michael. "Getting an Audio Education, Landing the First Gig, and More." *Signal to Noise Podcast*, ep. 12, ProSoundWeb, 6 Sept. 2019, www.prosoundweb. com/podcasts/signal-to-noise/.

Leonard, Chris, et al. "Becky Pell, Monitor Engineer & Yoga Therapist." *Signal to Noise Podcast*, ep. 41, ProSoundWeb, 15 April 2020, www.prosoundweb.com/podcasts/ signal-to-noise/.

Lively, Nathan. "Nathan Lively." *Get On Tour: A Sound Engineer's Guide*, edited by Nathan Lively, self-published, 2018. pp. 25–26.

Lively, Nathan. "Sound Engineer's Path Webinar." *Sound Design Live*, 16 Feb. 2015. www.sounddesignlive.com/sound-engineers-path-webinar/.

Lively, Nathan. "Career Advice For Freelance Designers." *Sound Design Live*, 16 July 2013. *SoundCloud*, soundcloud.com/sounddesignlive/career-advice-for-freelance.

Martinez, Avril. Personal interview. 28 Aug. 2020.

McCoy, Kevin. Personal interview. 22 Dec. 2020.

Murphy, Claire. *Girl On The Road*. Kindle ed., Self-published, 2020.

Paz, Jessica. "Ask the Experts – Sound Design for Theatre." *YouTube*, uploaded by SoundGirls. 5 June 2021. youtu.be/rFMLCU1g19k.

Potter, Samantha. "Veteran Audio Professional (And STN Co-Host) Chris Leonard." *Church Sound Podcast*, ep. 11, ProSoundWeb, 16 July 2020, www.prosoundweb.com/podcasts/church-sound-podcast/.

Reed, Pete. Email interview. Conducted by April Tucker, 24 Nov. 2020.

Reynolds, Jason. Personal interview. 27 July 2020.

Sabolchick Pettinato, Michelle. "Michelle Sabolchick Pettinato." *Get On Tour: A Sound Engineer's Guide,* edited by Nathan Lively, self-published, 2018. p. 53.

Sabolchick Pettinato, Michelle. "How to Get Started in Audio." *SoundGirls*, soundgirls.org/how-to-get-started-in-audio/.

Sabolchick Pettinato, Michelle. "Tips On Writing Your Resume." *SoundGirls*, soundgirls.org/tips-on-writing-your-resume/.

Scovill, Robert. "Robert Scovill." *Get On Tour: A Sound Engineer's Guide,* edited by Nathan Lively, self-published, 2018. p. 72.

Štefančič, Aleš. "Aleš Štefančič." *Get On Tour: A Sound Engineer's Guide,* edited by Nathan Lively, self-published, 2018. p. 17.

Stoll, Becca. Personal interview. 27 April 2020.

Thériault, Marc. Personal interview. 4 Dec. 2020.

Trifiro, Jack. Personal interview. 4 June 2020.

14
Music Production

In the 20th century, most music production work was reserved for those who came up the ranks of a recording studio or record label. Today, the industry has shifted toward DIY, and even a home recording can have millions of streams or win a Grammy. "The gatekeepers are gone, and that brings everything great and that brings everything terrible. It brings its own set of challenges," says music producer John McLucas ("Interview").

One of those challenges is building a financially stable career. "Until the late 1990s, records still cost US$200,000 or $300,000 or $400,000. Now, you're lucky if you get $2,000 or $3,000 or $4,000," says mastering engineer Piper Payne. Susan Rogers (who engineered and produced major artists in the 1980s and 1990s) says of the industry shift:

> There's definitely money in the music business. Always has been, always will, and there always will be. But, the pyramid keeps changing its shape. There have always been those few at the top who made the most money. I know from making records in that middle tier of the pyramid (signed artists, people with record deals) that people weren't getting rich, but they were living their lives. They were making their mortgage payments. They were okay. But now the pyramid is more and more narrow, it seems like, and there's just that top percentage who are really getting wealthy. Then, everybody else is down in that lower layer where they kind of need to have a day job.

Jeanne Montalvo Lucar is a Grammy-nominated engineer who works in music production and spoken word audio in New York City. Jeanne says if she were to only pursue Latin music (her special interest):

> I wouldn't make any money. I wouldn't have any money. I would be on the street. I have literally just started finally getting to do more Latin music in the last two or three years [after 14 years in the audio industry] . . . my main

DOI: 10.4324/9781003050346-17

[music] client is not sustainable . . . I always joke that podcasting pays for my music habit because that's what it is.

Producer/engineer Nick Tipp, who works in live recording and studio recording across many genres of music, adds:

> If you work really hard and dedicate yourself to something and just keep listening and take every audio job, you'll find somewhere you can make money. It may not be what you thought, but you're still building the skills for your music production work.

Historically, there has been a divide between musical artists and those serving artists "behind the glass" in the studio. Music production careers used to be centered around using audio skills in service of other people's music. These career opportunities still exist, particularly in music that involves large ensembles (such as orchestras), acoustic ensembles (bluegrass/country, jazz), and live event recording. Other roles, even those who do their work alone, can require a high level of client services and communication skills. Marc Thériault, Chief Mastering Engineer at Le Lab Mastering in Montreal (Canada), explains:

> With mastering, you need to know how to talk to artists and deal with their insecurity. Mastering is the last step before it goes out on the market, so they're all nervous. It's like having a baby and it's the last day [before you leave the hospital]. You need to show them that you're very confident in what you do. "I've given a baby to you, and I know it's going to be fine."
>
> It has to be technical first. You need to fix issues . . . There's a technical side you need to know to clean it up. Then you can be artistic . . . The way you're being creative is how you communicate with the artists and make sure you have the right vision of what they want to hear. It's not what you think is good. It's what makes them have an emotion. Before I touch a button, I want to make sure that I understand what the project needs because you end up doing too much stuff, and it's not the right thing.

Mariana Hutten (Toronto, Canada) says of recording versus mixing and mastering:

> A recording engineer has to be 100% attentive to somebody else's mental state and requests for eight hours straight – "I need water. I need this. Can I get a bit more of that? Can I get a bit more of this?" . . .
>
> I [like] working with people, but there's something that I love about showing up to the mastering studio and it's just me and the gear. It's just being able to do things on my own terms and on my time. Like, being able to listen to the music, make a choice, take a break, and then go back. It's a complete connection between me and the music.

There's something about mastering that absolutely does not give any stress in my life. A lot of the time when I'm recording or doing more production work . . . the interpersonal relationship aspect takes a stronger stance. With mixing and mastering, I'm thinking about purely the music.

Susan Rogers adds:

I'm the kind of person who's going to do her best work when you put me alone in a room and shut the door, which is why I was a good mixer . . . when I'm alone, creativity flows. For other people, that's a choking, stultifying environment. They need to be with other people in order for their ideas to flow.

Music production can be emotionally charged at times and can require navigating different personalities and difficult situations. Chris Bushong, who worked in music production before pivoting into live sound, shares:

One thing I found working in the studio with bands – you're almost more of a psychologist than you are an engineer (more so with rock bands than session musicians). Like, the singer and the guitar player get into a fight . . . The producers are off dealing with the grand vision and trying to mediate a problem over here but then you see something going off the rails on the other side . . . Now you have to mediate that so you can get the singer in the vocal booth to do what you need to be done.

The modern audio industry is also full of emerging professionals learning audio skills for their own music. Susan Rogers refers to them as "sound sculptors" who take the traditional skills of music production and combine them with a modern element of music composition. Educator Eric Ferguson calls this a "non-traditional musician." The term "music producer" has expanded to include artists creating (or contributing to) their own works.

For non-traditional musicians, following the "traditional" path of working for studios or being in-service of others may not align with their interests and goals, which may fall closer to that of a composer or songwriter than a recording or mixing engineer. "If you're going to be a non-traditional musician, you really should be learning music business, licensing, and royalties," says Eric.

The Risk of Producing Your Own Music: Susan Rogers

Susan Rogers is a professor in the music production and engineering department at the Berklee College of Music (USA). Susan is known for her engineering and producing work in the 1980s and 1990s, including engineering for Prince from 1983 to 1987.

In my day, all the analog engineers were working toward high fidelity, and we were working toward capturing realism because it was so damn hard to do (especially with analog being so nonlinear). But in today's world, that little silicone box – it's been an entire game-changer. It's made it so easy. Push of a button – you've got a perfect kick drum sound. Today, the goal is not to express reality the way that it was in my era. The goal for the engineers now is to express ideas. We're finally entering into music what they entered into in the visual arts [world] 150 years ago – which is now we can express abstraction [in music] . . .

I see students at the end of the pipeline (I usually teach the 400-level classes when they're about to graduate). By the end, up to about 40% of them will want to be on the artist's track. They think, "Now that I know about engineering and production, I want to do my own art-istry." They are not the strongest of the artists. What I'm thinking is they were never strongly pulled into engineering and production or mixing in the first place but they thought it might give them some job security. They have enough semesters and recognize there's no job security, so might as well go back and do the thing that they came to this college for, which is to pursue their artistry.

I strongly encourage them to pursue [their artistry] first. They can add the more technical things later should they choose, but it's really, really hard – much harder – to go from being an engineer to being an artist than it is to go the other way around.

For the students today who are really fast with Pro Tools, who are great with editing and sound replacement and things like that, they'll get work. If they do their jobs well enough, they have a chance of moving up the ladder and getting to do more and more artistic and creative work. But for those who don't want to do the technical work and want to enter as artists (either composing artists, producing artists, whatever), they are going to have to be self-taught. They're going to have to get out there and compete in the marketplace of ideas. It's brutal.

The Two Paths of Music Production

In music production (whether being in service of others or as a non-traditional musician), it is possible to build a career by two paths: working for yourself (finding your own clients), or by working for others. Each path can require different targets for networking and finding clients (see Chapters 7 and 8 for more). There are challenges and advantages to each path:

Working for others	Working for yourself
Someone else is responsible for your schedule and workload, but you may not have much choice of your work or schedule	You are responsible for finding all your own clients but can approach whoever you would like. You are managing your own schedule
Opportunities to observe professionals and their workflows (possibly getting paid to learn)	You are responsible for teaching yourself on your own time (unpaid)
Access to better equipment than you may own (or be able to afford) and technical support	You have to purchase, learn to use, troubleshoot, and maintain all your own equipment
Entry level positions may be doing menial tasks outside the studio	The same menial tasks are necessary, but it is related to your own business (such as keeping your studio clean)
Opportunities to network and build relationships on the job/at work	You have to find your own watering holes and participate. Networking and relationship-building happens on your own time
It may be a slower path (at first) to the roles you want but will likely have training and support	You can work whatever roles you choose, but some opportunities may be low/no pay, and there's a risk of being in over your head
Working with higher caliber artists than you may reach on your own	You have to seek out artists willing to work with you and build a relationship with them on your own

Working for a Studio

Educator and recording engineer Rob Jaczko explains the value of working for a studio:

> There is something to investing your time as you're paying your dues. You're building a resume and you're building relationships . . . Quite frankly, the independent landscape is dicey. It's very difficult unless you're super successful in your lane, you have a track record, and you have a lot of deep relationships to keep filling your plate [all year] and pay the bills.

Even those that are Platinum-award winning and Grammy Award-winning engineers and producers that we know – it's a daily struggle to make sure the next gig is coming in. If you are lucky to be part of a studio system, you're probably [reluctant] to want to leave that. On the flip side, of course, that makes it more difficult for the next generation coming up, because there are fewer opportunities to advance. The folks that are ahead of you are sort of impacting your progress. They're staying put, and they're not [moving on] as they once did. So, it's complicated.

Even those with a bachelor's or master's degree in music production may be expected to start at the bottom. This is partly due to the onboarding required to learn professional studio etiquette and workflow. Chris Bushong says of country music recording in Nashville (USA):

Studios in Nashville [are] very punctual just because of the union. If you're doing three sessions, then it's got to start at 9 am for that first session, because they have to have their hour break before the 1 pm session starts .. . On the country side, it's more of a business, and people realize that. It's all about get the job done, get out, and move on to the next thing. It's really organized and very regimented, in some ways. Very scheduled as opposed to, "Hey, we're going to lock ourselves in the studio for six months and see what happens."

Entry-level studio jobs (such as intern, runner, or assistant) can be highly competitive and may not be publicly advertised. Tina Morris, studio manager of the historic Village Studios (Los Angeles), receives 5–10 inquiries a *day* from job seekers. A personal recommendation can be the difference between going into a resume pile and getting a call back.

Studios can be more selective about who they hire because there are so many job seekers. A studio may seek out people who best fit into the work culture (both the studio's culture and their clients) versus the most qualified person. Entry-level work is often related to client services or supporting clients in having a pleasurable experience. Tina Morris explains, "We're here to serve the clients, and they're our guests. It's like a hotel. We're here to provide a service, and the service is not providing friends to hang out with."

Composer/producer Giosuè Greco worked as a runner at two major Los Angeles recording studios during his emerging years. He says of studio work:

I noticed that the people who were more inclined to become engineers (from runner to engineer) were people that had great interpersonal skills more than technical skills. In retrospect, I'm realizing that having technical skills in such a very well-equipped studio – it's not super important. If something breaks down, you always have support. There's such a support system of people that come and help you out. But dealing with the client in the most accommodating and professional way – that is what's super important.

Producer/engineer Lenise Bent explains the importance of entry-level positions during recording sessions:

> Every one of those positions in the studio is absolutely valuable . . . If it's 10 o'clock at night and I know I'm going to four in the morning, I can't leave the console and I need that cup of coffee, all I have to do is look at that intern, and point at the cup. [They were] the most important person in the room at that time because I couldn't stop everything and say "Hey, you know, I'm going to go get some coffee. You want some coffee?" That's not my job. I can't do that because I have to make sure that client is happy and the session is going on.

Ghazi Hourani, who regularly worked 12–18 hour days at the Record Plant in Los Angeles (working his way up from runner to engineer) adds of the importance of runners, "When you're in a room all day, sometimes all you have to look forward to is that meal." Ghazi says of his runner experience, "The only time you get to touch gear is when you're cleaning it. I didn't get to touch anything for six months." However, by sticking through it (and getting promoted to assistant), his credit list filled with major artists of the time, including Lady Gaga, Shakira, Snoop Dog, T-Pain, and many more. "When you have those moments that are awesome, it's awesome," says Ghazi.

Not all recording studios will have these limitations, and a small studio may offer hands-on opportunities much earlier than working for major studios. Educator Eric Ferguson suggests when taking a studio job, "Don't ever take a job where you're not allowed to touch the gear. It's a deal-breaker."

Working on Your Own

For some, working on your own may be a better fit than working for a studio. Producer John McLucas made this decision early in his career, in part after cleaning an outdoor refrigerator as part of an internship. John explains:

> I thought I could use my time better and not wait, not do the "hope and pray" method, which is kind of what I feel like in a studio environment. You work hard, then one day, "Oh, Bobby head engineer has the flu. Guess what, buddy? It's your time. Step up to the plate. Show us." Then you have this epic heroic moment. There are plenty of people who've had that, and that's great. I've just seen a better path for me and have never looked back. I've been in studio environments, but it's only when I want to. I'll run Pro

Tools for engineers in their seventies who don't like running Pro Tools, and it's great because they're legends. Even in 2014, I thought, "I can work with artists from a Facebook group." I would just rather do that.

("Interview")

Working on your own can provide opportunities outside traditional paths (such as assistant or engineer). Giosuè Greco says of his recording studio experiences, "There is a lot of time invested in that studio, but there is not a lot of headroom. You can't really go above being an engineer in a lot of those facilities."

Music mixer/producer Jeff Gross, who has had a home studio over 20 years, says, "I always hated the hierarchy of the studio. It just never resonated with me, and that's kind of why I just never went to get a job at a music studio." However, being a business owner adds its own set of challenges. "Your job is not engineer only. Your job is babysitter, your job is runner, your job is therapist, and your job is Dad," says Jeff.

One route to finding clients is offering education. John McLucas explains:

> When I started producing, a lot of people on my calls were like, "I want to start learning how to do this too, but I can't, so I'm hiring a producer." I was like, why don't I just take three hours and I'll show you some cool shit? I show them what they want to know. They get to learn a ton, but I know that showing them something for two hours doesn't put me out of a job. I feel like it's helped with my really good retention. It actually helps them communicate, it helps them make better demos, or they actually come up with a sick idea all on their own. Like, they'll chop up some crazy thing after being inspired by what we learned together. So, I never look at it as a bad thing. I just answered the call that the market tells us . . . But if some-body comes to you and says, "I need help, I'm in pain, please help me," and you can help them in some way, just help them.
>
> ("Interview")

In addition to one-on-one consulting, John McLucas creates educational content. John's clients have found him through his podcast, content on YouTube, TikTok, and a Facebook group he created. "Once they're in the group, I'm able to really talk with them a lot more. I try to keep it as [artist driven] as possible and not go too gear/techie," says John ("Interview").

Mixing/mastering engineer Mariana Hutten says of education:

> One technique that I do is sharing information and teaching these young producers to do stuff. I actually do get a lot of work [from it]. It feels like you'll lose work if you're giving out information but actually, for me, it's been the opposite. I give out information in workshops, and then it sells

my expertise instead of losing work. There are people who are actually trying to become better in engineering, but still would send me their stuff for me to do. There are a few detailed secrets that I have that I don't give up. I think every engineer does that. You don't give away the secret sauce.

Mariana adds that some clients are prepared to hire even if they initially are only seeking education:

I take advantage of the fact that people in the hip hop scene like the glamorized aspect of hiring an engineer. It's weird, because there's a lot of DIY. But there's also a bit of "I paid somebody to do this."

Relationship building (and maintaining) is crucial for finding opportunities in music production. "With this service, this specific industry, people don't need [your services] all the time. So, you can be nurturing but they just aren't ready to record," says John McLucas ("Interview"). Musicians can be a major source of networking. Piper Payne explains, "Musicians are also tied to so many other musicians. They're literally your calling cards for you." Catherine Vericolli adds, "Every person in the band has another f—ing band."

Connecting with Artists: Langston Masingale

Langston Masingale is a studio owner, music producer, engineer, and product designer in Syracuse, New York (USA).

If you're freelancing, or you have a facility and you really want to get more bands in, then you've got to go out and chase the work down. The phone's not going to ring, nobody's going to come to you (not in the beginning). You've got to establish yourself . . . Go talk to people, introduce yourself, and pass your name around. Relationship building is a process . . .

If you're dealing with different types of artists, you also have to understand their cultural situations. You can't just open a studio and expect gospel musicians to just pile in there. They're from faith-based communities, and you probably will have to connect with them somehow through those pathways. That's why a lot of times, you'll meet [religious] musicians that work with Christian engineers who happened to go to a church that's in their network. Some of that is part of the way this works . . . Same thing with rappers. Like, what type of rappers are you trying to get? . . . Experimental stuff, adult contemporary, and so on? Hip hop is just as diverse as any other type of music . . . If you're not looking at it from that angle (of cultural competency and culturally relevant practices) then you're going to hit a brick wall. You have to make yourself available to people the best way that you can . . .

I've done a lot of rock [producing]. If you don't go see the band live, how do you develop a knack for what you think they might sound like when you try to do a cool iteration of them in the studio? That's why you get a lot of rock records that sound very under-produced because that person who's engineering their record is so enamored with the live sound that they're bringing from the tracking room floor they forget to actually produce the record . . . When you go there live and get the energy of the audience, and you get to understand how they move and flow through songs – they might be doing certain things live that you might want to remind them about when they come into the studio . . .

Come [to a show], come twice, bring a friend . . . That's part of showing people that you actually care about what happens to them. Even if you don't land them, other people are watching you. Just like [social media] where people don't "like" your post, people saw it. There are other people watching you get down, and that's how you force the universe to create something out of nothing.

Offering your services to other professionals can lead to opportunities. Music producer Nick Tipp prefers to hire others so he can focus more on producing:

The kind of thing that requires the heavy lifting, like cleaning up tracks, [Izotope] RX and editing and all that stuff, well, that's not something I'm interested in doing. It's something that my assistants can do fine. It's not that hard to teach these things.

Niching, or having a specialty (especially a style of music), can help tap into markets of potential clients. Eric Ferguson explains:

There are so many things in audio and you can find those spaces. I worked in smooth jazz for years. I [mix] Thai records, and I've got this little scene over there. The clients I work with are in alignment with me in age, and they work in really good studios with really good session players. It's good music, like full orchestras and top rate, and I get to mix great stuff.

When working on your own, having a demo/portfolio of your work is crucial. Catherine Vericolli says about building a portfolio when you do not have one:

You don't have to go out and find a band. It's so overwhelming to think, "I have to convince this band or artist to make a record with me." Don't do that. Find somebody who already knows you, who knows your struggle, who is on your team, and who's willing to put some time in so that you can get something going . . . if you can't afford [a studio], then use what you have at home, use your own musicianship, use your friends' musicianship, use your family's musicianship.

Jeff Gross suggests finding solo artists versus bands to simplify the process, and to team up with professionals when a project is beyond your ability:

You might not have the knowledge, and this person has a studio. Bring them artists that you like and say, "What if we did this together?" Then you can learn some stuff. This other person has the gear and the knowledge, you'll get a lot of information and experience out of it, and you'll get a demo out of it. I have people approach me with that kind of thing. "I've got this project. You want to do it?" They don't have the equipment, skills, or the education to operate it.

Networking at the URM Summit: John McLucas

Music producer and content creator John McLucas shares how to network and find opportunities at the Unstoppable Recording Machine Summit, a yearly conference for rock and metal producers.

URM is a hard rock and metal-focused audio education platform. Audio engineers [are] mostly in those two genres and the vast majority are guitarists as well. Most of us start as musicians then we find this love for audio and will transition over, so that's why there's a lot of core strength in riff-writing and song structure. So, going to an audio convention [like the URM Summit] and saying "I'll help you write riffs" isn't super helpful. Anything that most of the people are already good at is not going to be a thing that they're going to want to work with you on. Think of something complimentary, or something that they're not as specialized in, something you have a talent in or interest in that you could develop [to] become the definitive person for that one thing.

You don't have to come in with a complementary skill for you to thrive in that environment, but it's really smart to take a moment and think about what can you do that these people are going to need and can be valuable to their clients and the final product?

If you're bringing a new skill to the table or working to build that reputation, forget the idea of getting money. It's much more powerful to do the work for the social brand power or building yourself up as an authority than for [little money] . . .

As you're having conversations, anytime somebody mentions it or it comes up naturally, you talk about what they are struggling with and then you say what you do. Write their name down. You [can say], "I'm not looking for money right now. I'm just looking to build a portfolio and build a reputation." It's so powerful because you're going to be doing free work for all of the right people instead of just cold DMing, cold emailing, cold-calling random bands or producers and not having a great relationship. You're going to have plenty of in-person, really intimate, strong foundational relationships and then you can build on that. If you do it for 30–40 people, you will be an absolute authority in what you're doing.

From *The Modern Music Creator Podcast*

Working With Low Budgets

Music production as a career can be highly rewarding creatively but challenging financially. "The most talented musicians that I've ever known have no money when they get started. So, they're not going to be able to pay you probably what you really should be getting paid," says Catherine Vericolli.

It can take creative solutions to help artists do projects within budgets they can afford. Mixing/mastering engineer Mariana Hutten shares:

I know that my market is people with [a certain] kind of budget and place in their career. So, I charge an amount that's fairly cheap, but then I automate a lot of my processes. That's sort of the drawback – a lot of the industry had to lower its costs because we're not making that much money, but then we're doing all these things that are slightly automated and less special, in a way. I have all sorts of presets for different genres, and I have channel presets. Of course, I tweak it. I'm at the point that your average hip hop track, I can mix it in like two hours.

Music producer and engineer Nick Tipp says of helping his clients with lower-budget projects:

You record it all in one day. You have the artist do all the editing. You have them do all the vocal tuning. You teach them how to edit. They're paying you to teach them and they're paying you to record them. I provided them with the iLoks. Their budget for post suddenly [drops].

Jeff Gross says of taking **points** (a royalty percentage), "I'll take them as an engineer if they don't really have the money. People assume if you're working for free for them that there isn't any value in that." When asked how often those arrangements pay off, Jeff said, "How many people are successful in the music industry?"

Working for Free to Build a Portfolio: John McLucas

John McLucas built his business in music production through content creation and utilizing social media to start building relationships.

Have genuine connections. If you genuinely, deeply care, then offer yourself for free with no strings attached. You don't try to convert it into anything. You just say, "I love what you do. I love your band. You should come into the studio." People think [things have] changed a lot, but the principles of content marketing – doing free work when you don't have paid work – it hasn't changed. It used to be done at concerts. Now it's in the DM or on the phone.

I'll send a little video clip message like, "Thanks for reaching out. Let me check out your stuff. If I feel like it might be a good fit, let's totally jump on a call," and send it to them. He's not terrible or discussing horrific topics, and I like his voice. So then, "I can shoot you a link. Go ahead and book a time. Looking forward to chatting." Then we'll actually just sit down and have a conversation. It's no different than when you bring [someone] into the studio to sit down and just kick it and listen to some tracks you've worked on. The only difference is that it's on a different platform now.

Start with the people you know, offer yourself for free, and be invested in them. Start with the people you already know who are likely to let you just work with them to build yourself up and build that portfolio. The old guard still talks about "you've got to have a good portfolio." Yeah, you've still got to have a good portfolio. I don't get hired if they don't like my work.

("Interview")

Buying Gear

Many professionals interviewed strongly suggested not buying recording equipment when getting started in music production, and instead investing in quality equipment over time. Catherine Vericolli explains:

> You already are going to be inexperienced – do not be broke and inexperienced. If you're broke and inexperienced, you don't have the money to get experience, which means you have to get another job. Save up some money and get something good. You know what's even better than that? Go to a studio where they have that stuff. Find a client that you can work with who's going to pay for you to be in there to use it. That's how the industry works. Building a recording studio right now and spending a lot of money on gear is not a good business plan.

Chris Bushong adds, "It's a flaw I see with a lot of people starting out. What about when it breaks? I'll let somebody else buy the gear and I'll rent it. When it breaks, I don't have to deal with it."

Piper Payne suggests saving for when it is a necessity to have equipment:

> At some point, you will have to invest in the equipment and be self-sufficient in the event that somebody dies or you get fired or you broke something and you're no longer allowed at the studio. Like, shit happens. So, you have to be squirreling away a little bit of money, if you can.

Importance of Credits

Jett Galindo, mastering engineer at The Bakery (Los Angeles), says, "Credits are a currency in this business." However, when you work on a project, it is not guaranteed you will receive public credit for your work. To help receive credit, Piper Payne suggests:

Send out a credit request early. If you see a record you are working on is close to finishing, reach out. "When we think they might be working on graphics, we send out the request for 'we really appreciate any credit [for working on the project]. If you are going to credit us, here's how we need to look,'" says Piper.

Include your preferred credit on your invoice.

Request physical copies of albums (in any format). This is necessary so you can submit to allmusic.com, an online catalog of music albums that includes credits from the liner notes. Piper suggests asking for additional copies for personal archive and to give to clients.

Jett Galindo adds:

> For independent releases (those that aren't tied to record labels), you can easily reach out to the artist/client to ensure that you are properly credited on platforms where their music is distributed. There are several websites out there that allow indie artists to easily list down song credits (e.g., Bandcamp, Discogs, Genius, Jaxsta, Muso.AI, etc.). Even the act of crediting you on social media platforms adds value to your work. Your credits on indie releases are just as important, so don't take those for granted!
>
> There may come a time when projects (usually high-profile label releases) don't list down all the credits behind a production. Knowing your substantial role in the project (e.g., assistant engineer, production assistant, etc.), don't hesitate to mention that credit on your personal portfolio. It's a credit you've rightfully earned for yourself. Just be aware that credits are meant to highlight your involvement, so avoid claiming credit on projects you didn't contribute to.

Film Score Composer/Electronic Musician: Giosu è Greco

Giosuè Greco was raised in Italy, where he attended the Conservatory of Music in Vibo Valentia studying saxophone performance. He received a scholarship to attend the Berklee College of Music (USA), and decided to attend even though his English skills were poor at the time. After graduating from Berklee in 2013 (with a degree in music production and engineering), Giosuè moved across the US to Los Angeles with a goal of writing and creating music. "When I moved to LA, I was a little bit more inclined to find a job rather than developing connections because I needed a job to pay rent," says Giosuè.

Giosuè landed a runner job at The Village, one of LA's historic and well-known recording studios. Giosuè says working at The Village provided some opportunities to get feedback on his own music:

> There are a couple of producers who are really dear to me, and I showed them my music. One told me that it was good. I didn't take that as a thing, like, now I'm going to start working with this giant producer. I knew that was never the context in which he was telling me that. But when someone like that tells you that you don't suck, it's a thing. Then you kind of build your strength from other people's feedback.

When I walked into it, the idea was that I was doing that part-time. The problem with doing one of those jobs part-time is that you end up working 16 hours and you have no energy to write ... I didn't consider that aspect. I was really starstruck by the opportunity. I'm glad that I did it. I knew that I didn't like it, but at least I tried it. I didn't want to have any regrets. I didn't want to have any missed opportunities. If you don't try you don't know what it's going to be.

Giosuè later took a runner position at another major studio (Larabee Studios) but found the studio path was not a good fit for his personal priorities, particularly since he entered the industry at an older age than his peers. Giosuè says of the studio jobs:

Starting from the bottom, I accepted them with a lot of humbleness, but then I just kind of realized that I could do something else. I could actually spend my time doing something else that might be a little bit more meaningful for me and possibly for artists. So, that's why I quit [that path].

As a freelancer, Giosuè took a variety of audio gigs to support himself financially. Giosuè tested audio software (an opportunity he found through Berklee's job board. He also recorded ADR for a small post-production studio. Giosuè says:

Those gigs paid relatively well for the number of hours that I put in. [Music] writing doesn't pay quite as much as you would think – meaning, you put so much time in writing a score, even a score for a documentary, and then when you actually calculate the amount of money per hour that you get, it's like, US$5. It's a very long run game.

Within five years of moving to Los Angeles, Giosuè had built a sustainable career as a film score composer and electronic musician. During that time, he worked on multiple television series and composed for a documentary short film, *Period End of Sentence*, which later won a prestigious Academy Award. Giosuè does not actively promote his music online and instead finds work through his professional relationships:

I spend zero time on social, and it shows because I don't have a following. But at the same time, I'm a music professional. A lot of these artists that put all their energy (financial and time) into becoming artists might not be able to live off music the way I do it ... I don't do any performance. People come to me, "Can you produce this thing?" This album, this soundtrack, and I do it.

Looking back, Giosuè would have approached his start in Los Angeles differently:

> I moved to LA without much money saved up. That wasn't the smartest idea, but at the same time, those were the circumstances. What I would do now, if I had an amount saved (like US$16,000) and I moved to LA, I would probably try and write more music with people. So, developing my personal connections to writers, whether it's in the pop realm or in the film scoring realm. Then, I would probably start and push the boundaries in the sense of trying to get some of my music placed – developing relationships with A&R, trying to get my music placed on lists, talking to different publishers, trying to get a publishing deal a little earlier.

> I would probably write more music. But at the same time, you need to have some finances set aside to do that. Because if you decide to write, you might not be making money for months and months and months. Some people work for years and years, write songs for years and years, and then they finally get a huge placement – that's when they're able to jumpstart their career. Or, that might never happen.

Giosuè relies on his formal music and engineering background to be versatile in a competitive space:

> There's a lot of noise, especially right now with social media. Even a 16-year-old from my hometown [in Italy] can get 16 million followers in 24 hours. At the same time, I do think that the best music always sticks out.

> There is always going to be demand for content creation based on whatever trend is going on. The sample packs that you buy are trend-based because music is a fashion. In three or four years, it's going to sound [dated], and we're going to need different samples.

> Making electronic music is very easy, as long as you're following a trend. Shifting amongst trends becomes really difficult. When a client comes around and asks can you produce a song [outside your niche], you need to be able to change the style. You need to be able to come up with something new, and that's what they're paying for. So, the dubstep producers that have 100,000 followers on Instagram . . . it's very unlikely that that person is going to make a long-lasting career. If you can jump into different waves, jumping in and out of different trends, that's your skill. That's what people are paying for.

Giosuè says his work is not for those seeking to be in the public eye:

> If you want to do scoring and writing and be an engineer and do a Kontact library, if you want to do all these things and on top of that have fame/internet fame, that's not the job for you. This is one of the compromises . . . You can potentially lose a very big stream of revenue coming from one place because you're busy building all these other things. People are not going to recognize you on the street and be like, "You're my favorite." No one's going to do that with me. I don't care, but some people do care.

Building a career requires a balance between pursuing your interests and recognizing the limitations of what you are trying to pursue. Giosuè explains:

> If you need to make money for whatever reason – because you need to pay rent immediately, or you're moving to a new city – you have to get whatever comes your way. But at the same time, you've got to keep in mind that if what you want to actually do is be a successful songwriter or artist, there are some hard deadlines in your life. For example, I don't want to be ageist, but if you want to be a pop star, there are some target ages that are perfect. If you miss that window, then there would be less of a chance for you to be a successful pop artist.

> So, maybe you don't want to take that engineering gig that takes you 90 hours a week and you don't have time to develop your own art or to do promotions or play gigs. In the end, do you want to be a starving artist for X years, or do you want to start making money right away but you might be missing a lot of opportunities? There is no right answer.

> You have to find what you're good at, and that has to be certified to some degree by people around you. You cannot think that you're good at guitars if nobody tells you you're good at guitar. Then just build on that [reality].

> It's easy to say (now that I'm doing it) that I knew all along I would end up being a composer. The reality is, I didn't know that I was going to end up becoming a composer. I knew that I wanted to write music and possibly not die of hunger. I think everyone knows that to a degree.

Film Scoring (Recording and Mixing for Sound for Picture)

For those with a classical music background or experience who are seeking other opportunities, film scoring (music for picture, including television, web, and movies) may be an area of interest. Skills like reading scores, knowing how to record and mix acoustic instruments, and the general aesthetic of classical music are all assets when recording, editing, and mixing music for picture. Score mixer Agnes "Aji" Manalo (Manila, Philippines) explains:

> To be able to work well as a scoring recording engineer/mixer, a strong background in the classical music genre, in classical music instrument training at a highly advanced level, and performance experience are needed to fully understand the film score composer's intentions and music directions.

Work in music for picture often comes through relationships with film composers and directors. Film scoring work may not be exclusively acoustic, and low-budget projects may require mixing a combination of acoustic recording and synths. It is not necessary to own software synths, but it is beneficial to be familiar with composing software and terminology (for the sake of communicating with composers). Other important skills are knowing how to use click tracks, tempo maps, and understanding timecode.

Classical Music Production

Classical music production is a unique niche where formal education and professional experience are respected and valued, and work is generally paid. There are professional training opportunities and professionals are generally welcoming of those wanting to learn the craft. At the same time, classical music has some barriers to entry, including some background in classical music (as a listener and/or performer) and the ability to read music and musical scores. Jeff Dudzick, who worked for the Boston Symphony Orchestra (at Tanglewood) and restoration of the Met Opera archive in his emerging years, says of the industry:

> It's a small group of people. You can't hide. It would be like me going into pop. I feel like it would be very obvious that I don't belong there and I don't know what I'm talking about (either with talent or technical

people). It just wouldn't work. If you go into classical, you've got to have some decent familiarity with classical music . . . You can't hack your way through. You've got to know some stuff.

Classical music can be a source of sustainable work in locations not known for classical music. Students of all ages perform in concerts and recitals (and seek out recordings) and need quality demo recordings (for seasonal camps, scholarships, and school admissions). As an emerging professional, music producer Nick Tipp utilized his classical music (and acoustic recording) skills to record local school ensembles:

> I would interact with all the high school jazz band directors around southern California. I would offer, "We could come down and we could record each band's performance at your jazz competition. We have [recordings] ready to go when the parents are walking out the door." I was also giving a kickback to the school band program and advertising that.

Classical music recording includes capturing live performances (concert recording, archival recording) or session recording (in a concert hall or recording studio). A venue can hire engineers and assistants directly, and gigs may involve recording, live sound, or both (with the venue's equipment). An ensemble or artist could hire an engineer directly (usually freelance). Location recording requires a complete recording system to be brought into a space by the engineer. Mastering engineer Piper Payne, who learned classical music production through two audio degrees, utilized these skills as a freelance side hustle early in her career. Piper explains:

> When I first moved to California, I didn't have any income. I knew literally the one person who brought me there. What I did have was a couple of decent mics, a good little recording system, and a laptop, and I had my little portfolio of location and archival recordings to fall back on. So, I went out and I found some smaller ensembles that had a small budget that needed archival recording done – not necessarily full-blown sessions, but they needed to document their performance. I would say, "For $1,000, I'll record the entire fall season (seven or eight performances), and I'll give you a [recording] at the end of every one." I had to hire an assistant once in a while, or I had to pass off a gig to a colleague if I had competing schedule needs, but that really sustained me and it paid my bills for many years.

Foreign language skills are not a requirement to work in classical music production. James Clemens-Seely explains:

> I've worked on classical productions where I don't speak the language [of the music], and that hasn't been a problem. They need someone else to help with pronunciation stuff or whatever. I do speak classical music, so no one in the room instinctively mistrusts me.

Coming into Classical from the Outside: Jeanne Montalvo Lucar

Jeanne Montalvo Lucar is a Grammy-nominated audio engineer and radio producer. Jeanne learned classical music recording skills through an internship at the Tanglewood Music Festival (USA) and as an audio assistant at the Banff Centre for Arts and Creativity (Canada). In New York City, she was an engineer at the Manhattan School of Music and worked on the Metropolitan Opera restoration project with Long Tail Audio LLC.

I landed in classical music totally randomly. I didn't go to the classical music hub schools that feed into [professional training] programs . . . I didn't know that classical music was such a niche. I landed in it and I enjoyed it. I grew up on classical music and I'm a classically-trained pianist. I had the skills to be able to do it, but it was definitely something that I had to learn because I didn't go to school for it . . .

My advice is to learn to edit . . . One of the things that surprised me when I started classical is the amount of editing. I thought that classical music was the untouched realm where people still played . . . I was so taken aback by that and I didn't expect it . . . The systems that they use in classical (like Pyramix) are user-friendly.

Reading scores is a big one – not just reading a score of one instrument but reading an actual orchestral score. When we would QC music and it was 20th century or Prokofiev or Stravinsky, I'd get lost in the score because that stuff is all over the map. You have to make sure that there's no music missing.

[Working at a music school] is good experience . . . Orchestra recordings, Afro-Cuban jazz recordings – those were huge gigs, tons and tons of mics, tons of networking/routing. I learned visually how to make something look nice on the stage because they were performances, too. That went into my other gigs. If there was ever a camera, I knew that the stuff needed to look nice . . . Those jobs are live so you have to be on your toes, too. There's a ton to learn from those kinds of schools.

Classical music really helped me in my future endeavors [especially] the other kinds of music that I wanted to record and do. Learning how to mic an acoustic instrument beautifully and capture what you're hearing in the room, too, has been really helpful for the kind of music that I've been recording now. All of that stuff can adapt.

Specialization in classical music occurs over time. For example, opera work goes hand in hand with classical music recording, and any experience in classical music can help build experience toward working in opera. Joe Hannigan, the owner of Weston Sound & Video (USA), says of pursuing opera:

> If you're not living in a large metro area already, you should consider moving to the areas with universities and schools that specialize in opera coaching and performing. In other words, go where the operas are. You may find slim pickings at first, but if you're able to stick with it (while doing other related work along the way, like recitals, chamber music, oratorios, and choral performances) you may eventually be able to specialize. It's definitely a niche market, but it's there if you know where to find it. Building relationships is key for this, just like the rest of the classical world.

Where to Find Classical Music Audio Opportunities

- Network with local musicians and professional engineers working in classical music. Musicians may be able to give an introduction or referral to engineers they know.
- Intern with (or assist) local recording engineers who specialize in classical music. They may need help with remote recordings, recording sessions, or editing.
- Work at seasonal music festivals. Classical music festivals may have seasonal work for engineers, assistants, and interns. Pay may be low/minimal but housing may be included. Positions may be listed online (or found through networking/relationships).
- Reach out to local concert halls and venues (including college/university) that hold classical music concerts and ask if they need recording or live sound engineers.
- Be open-minded to opportunities. A children's piano recital or beginner's orchestra concert may not be a high priority for a busy professional recording engineer, but they are likely getting calls for those gigs. With a good relationship, they may be able to recommend you and provide equipment.
- Formal audio education can be beneficial. Studying at a music school with a strong classical music program will provide many opportunities for recordings and sessions, meeting musicians, and forming relationships with people who may be able to help throughout your career.
- Video skills are becoming increasingly valuable to classical music engineers (who may be asked to film a concert or audition tape).

Technology and Classical Music: James Clemens-Seely

James Clemens-Seely is Senior Recording Engineer at the Banff Centre for Arts and Creativity (Banff, Alberta, Canada). James has worked with garage bands, electronic artists, award-winning artists and producers, and nearly every professional orchestra in Canada.

One of the things about classical music that I find most different from other genres is the duration of historical practice that hasn't included technology. If I'm making a rock record or an electronic dance music record, those genres evolved with technology, and are mediated by technology . . . Whereas in classical music, there are centuries of people making music in a room for people in the room. When I show up with a bunch of microphones and cables and computers, even if the person making that classical music grew up in an age of cables, computers, and microphones (and may use a cell phone to record themselves), it's still a fundamentally [unnatural] thing to have the technology there. That requires a little extra sensitivity.

Because of that, one of the biggest prerequisites for finding success in classical music production is being able to speak that classical music language at a pretty high level. If you can't speak the language, you're going to be struggling all the time to help musicians feel heard. It should be possible to make clear to a classical musician that you understand them . . .

Most of the time it involves making eye contact. Like, when you're putting a mic right over their music stand and they look at it with alarm, you say to them, "Don't worry, 95% of the sound [will be] those mics way out there further than the conductor. This one is just to add a little presence in detail." . . . Or you say, "Wow, that's an interesting-looking flute because it's made of glass." Just showing a little bit about that mode of music-making puts them at ease.

By definition, if you show up with microphones and cables, you're an alien . . . It's very quick for them to think that you are an enemy and to start pointing the mics at the ceiling or actively start sabotaging things, in some cases . . . Anything about recordings seems like an invasion because it's so not integral to classical music . . . You just show them, "No, I came from where you came from, and this is my way of being musical with you."

It's still the rare classical musician who makes a living off of their recorded catalog. Classical musicians are still like, "I play 40 shows a year and make one record a year." The shows are obviously the part of their life that matters. They're used to no technology or minimal technology. It takes a lot of sensitivity to set them at ease within that context.

References

Bent, Lenise. "SoundGirls Career Paths in Recording Arts Part 3." *YouTube*, uploaded by SoundGirls. 30 May 2019. youtu.be/d6BJNkSM4O0.

Bushong, Chris. Personal interview. 27 Sept. 2020.

Clemens-Seely, James. Personal interview. 20 Dec. 2020.

Dudzick, Jeff. Personal interview. 1 Oct. 2021.

Ferguson, Eric. Personal interview. 2 Dec. 2020.

Galindo, Jett. "Using Quote in Textbook." Received by the author. 10 Aug. 2021.

Greco, Giosuè. Interview with April Tucker and Ryan Tucker. 28 Sept. 2020.

Gross, Jeff. Interview with April Tucker and Ryan Tucker. 14 Jan. 2021.

Hannigan, Joe. "If a student was interested in learning opera." *Facebook*, 20 May 2020. www.facebook.com/groups/2165880873644305/.

Hourani, Ghazi. Personal interview. 12 Nov. 2020.

Hutten, Mariana. Personal interview. 19 Nov. 2020.

Jaczko, Rob. Personal interview. 15 Oct. 2020.

Lucar, Jeanne Montalvo. Personal interview. 12 Oct. 2021.

Manalo, Agnes "Aji". Email interview. Conducted by April Tucker, 3 Oct. 2021.

Masingale, Langston. Personal interview. 14 Sept. 2021.

McLucas, John. Personal interview. 25 Jan. and 8 Feb. 2021.

McLucas, John. "How I Rebranded My Friends Audio Business By Accident – Value-First Thinking." *The Modern Music Creator Podcast*, ep. 53, 15 July 2019, pod.co/the-modern-music-creator.

Morris, Tina. Personal interview. *What Makes You Stand Out*, SoundGirls, 20 May 2020, youtu.be/wYamiOK8Y6A. Accessed 21 May 2020.

Payne, Piper. Personal interview. 29 May 2020.

Rogers, Susan. Personal interview. 12 Oct. 2020.

Thériault, Marc. Personal interview. 4 Dec. 2020.

Tipp, Nick. Personal interview. 22 Sept. 2020.

Vericolli, Catherine. Personal interview. 29 May 2020.

15
Interactive Media (Game Audio)

The game industry has grown from a few companies in the 1970s to nearly three billion gamers in 2021, and a market of over US$200 billion is predicted in 2023 (Wijman). At first glance, the game industry appears massive compared to the music or film industry. According to Comscore, the film industry made US$42.5 billion worldwide in 2019, and IFPI shows the total revenue of the music industry was US$21.6 billion in 2020. However, game industry figures include gaming software *and* hardware, plus a broader variety of applications including games for mobile devices, arcade games, and slot machines.

Video games are considered interactive media, whereas television, film, and animation are strictly linear media (where audio content is identical at every playback). While the sound requirements may seem similar, they are different industries with different workflows, norms, skills, and relationships required. Sound designer/composer Emily Meo explains:

> It is possible to think about it as the audio industry and then there is a gaming part, but that's never really how I thought about it. It's the gaming industry, and then there is an audio part within the gaming industry.

Audio Lead Jack Menhorn adds of trying to work in both areas, "Film/television and games are completely different universes of people, so keeping up those contacts requires double duty." For those specializing in Foley or ADR/voiceover recording, there may be opportunities to work in both disciplines.

The game industry varies from designing games over the course of a weekend (called a "hackathon" or "game jam") to AAA (pronounced "Triple-A") games created with the industry's highest budgets and production value. In AAA, teams are large, and projects can take multiple years from start to release. Jobs are highly competitive, sometimes having hundreds of applicants for an audio position. The game industry also has professional opportunities working

DOI: 10.4324/9781003050346-18

on smaller budget games (called "mid-core"), and self-published and/or self-financed games (referred to as "casual-core" or "indie").

Having a passion for video games is crucial for career success. Lead Audio Designer Will Morton explains:

> I have been in interviews where applicants have turned up and demonstrated they know absolutely *nothing* about games. It's not only that they don't know about the games the company they are interviewing with creates (bad enough) – but they don't know anything about games at all. Amazingly, some people I have seen in interviews said things like "Nah, I don't really play video games, that's more my nephew's thing." Game developers are usually gamers, or at the very least they are ex-gamers, so you really want to be excited about the same things that the people interviewing you are.

Mariana Hutten, who worked in video games before pivoting to music production, adds, "When you work for one of those companies, that's their life. They're the same as music people – they spend the day working on the game, then they go home and they play more games."

Working in interactive media requires learning language/terminology specific to game developers (also called "game devs"), and also learning the process of making a game. "You work with engineers, animators, level designers, game designers, concept artists, and VFX [visual effects] artists. I had to learn what everyone else did – about what all the roles were, and how games work as a whole," says Audio Director Kristen Quinn.

Composer/sound designer Harry Mack adds:

> I've played games all my life and I've had music as part of my life. But, [it's different] to put it all together, see how it works in a project, learning the jargon, and knowing who to talk to. The artists, the animator, the producer, the engineer – who's going to answer my question? Who should I not be bugging? How long does it take?

Implementation is a skill that exists in interactive media but not linear media. **Implementation** is a technical process of placing sounds into a game, such as choosing what sounds will trigger in an environment, or manipulating how it will sound. Commonly used implementation software can be downloaded for free, but it takes experience to learn how to do implementation well. VR sound designer Megan Frazier explains:

> It isn't going to be a five-hour blitz in an afternoon. It's something that you have to take on over the course of a couple of years. There is a need to understand direct implementation right now because you [may] be the

only audio person and the only consultant for the entire pipeline. You're not going to have anyone else on your team.

Sound designer Javier Zúmer adds:

> The biggest studios usually have an audio programmer – a person that only programs for audio. In the real world, smaller studios don't have that. So, usually the sound designer knows some coding and programming, and you always get the help from the programming team. For me, for example, I didn't know much programming, or Unity (which is the integration [software]). I knew a little because of my freelancing, but nothing compared to what I know now after eight hours every day trying to make it work. So, to me right now, programming is pretty important to do my job, and to make things sound cool and interesting.

Organization and documentation are crucial in interactive media, and often it falls on the audio team to develop and manage. "It depends on the company, but sometimes we [audio] are kind of our own producers. We're organizing our own asset lists, we're going to all the meetings, we're tracking down all the information," says Kristin Quinn.

Tiziana Mazzucco says of video game localization (dubbing games into different languages), "It can be complex dealing with thousands of files recorded by 50 or 60 different voices. In this case, precision and organization are a must."

The game audio industry used to be primarily employee-driven but has shifted to contract workers. Today, jobs can be in-house (working on-site for audio companies or game developers) or remote (working off-site). In-house work could be employee or contract. According to a 2019 survey of game audio professionals, about half of the respondents were employees and half were freelancers. 18% were employees who also worked freelance (Schmidt).

Jack Menhorn says of contract positions:

> Many companies will hire entry-level jobs on contract, and if the employee does well in the position, transition to full-time. This is a sort of "probation" where the company tests out the employee and see if they flourish or not. That way, if it isn't working out the company has an easier time letting the employee go . . . I did not get a full-time job in game audio until I was 30 years old. I know people who got a job right out of college.

It is also possible for tasks to be outsourced, or hiring others to do the work. Sound designer and vocal artist Bonnie Bogovich explains:

> I subcontract out to other colleagues who have the strengths in the areas that [I'm not as] quick . . . I'm not so good at this, but I know three people

who are really good and can rock it in half the time. One thing you learn when you're on a team is production doesn't see people as much as it sees dollar signs. No matter what happens, there's a budget for a game that has to be shipped. It breaks down to hours, how expensive those hours are, and how quickly can things get done. If it takes you 20 hours to do one task but if you contracted someone else with more skill it would take two, show them that information and those charts. It really helps a lot when pitching the idea of hiring on help . . . As a team player, you want to help get the game done on time.

Paths in Interactive Media

Sound Designer (audio designer) – creates and manipulates sounds for interactive media. Sound designers can be entry-level, mid-level, or senior positions. Sound designers may also be expected to compose music, implement sounds into the game, and handle other areas of sound (possibly recording original sounds or editing).

Composer – writes music; may also be a sound designer.

Technical Sound Designer (audio implementer, audio programmer) – makes audio assets (such as sound effects, dialog, voiceover, or music) run in-game through implementation.

Audio Lead (sound lead, audio director, lead sound designer, audio manager) – oversees sound needs and the sound team for a project. This is a senior-level position.

Other related paths include dialog recording, dialog director, Foley mixer, and Foley artist. These tasks are done in-house by some companies or outsourced.

The responsibilities of a game audio position will vary from company to company. The term "sound designer" is used in many disciplines from theater, sound for picture, spoken word, and more. In the field of interactive media, "sound designer" can be a blanket title to cover everything required for audio in a game. Technical sound designer Avril Martinez explains:

> The bigger the company, the more the lines are drawn between jobs. The smaller the company, the less the lines are drawn. Sometimes sound designers will also be implementers. Sometimes they'll even be composers and make their own music. There are people out there who do everything all at once: Sound designer, implementer, and composer. Those are usually

small games. As soon as you get to bigger places like Microsoft, there is space for more structure in the hierarchy.

Bonnie Bogovich says of taking on many roles:

> When I was at Schell Games, I was the only person in the audio department for four years of my five and a half years on staff . . . It wasn't uncommon for me to be assigned to eight projects at once, though you can only focus on so much at one time. Eventually, I did get an intern who we (thankfully) eventually hired on. But we were rarely put on the same projects. That was a smaller-mid-sized studio situation, but in larger studios tasks and staff are more spread out. One [person's] job might just be dialog cleanup, i.e., removing all the spit out of [one voice actor].

While there are key cities known for video game work (such as London, San Francisco, Montreal, Seattle, and Tokyo), it is not necessary to relocate to have a successful game audio career. However, some relocation may be required as opportunities arise. Jack Menhorn says of his career in the United States:

> I started out in North Carolina doing entirely remote game audio work for years, and was patient/persistent until a local job opportunity (two hours away) popped up. It was my goal to move to one of these major cities, and fortunately, I had a developer move me to Seattle. When I was in Raleigh [North Carolina] I was doing just fine, although I was lucky there as it does have more major developers than many other cities. It wouldn't have been required to [relocate to] continue my career.

Interviewees shared the view that diversity has not been a major barrier to entering the field or building a career in interactive media. Kristen Quinn explains:

> Games are a really eclectic group of people. Because we are interdisciplinary by nature means it's really different types of people. It's also people whose communication styles might be extremely different, and the way they process information might be extremely different. Do I think there are problems in the industry? Yes. There are pockets. There are some companies that are maybe problematic, or maybe there are companies that aren't even problematic (that are really great), but have had a couple toxic people who've come in and caused problems. I've had an amazing experience in the games industry, and the games industry has treated me very well – outside of the [long] hours, sometimes.

VR and AR

VR (virtual reality) and AR (augmented reality) are emerging technologies within the interactive media industry. The skillsets required overlap with video game sound. An emerging professional could start building their career directly targeting VR or AR work (and relationships) or start in video games and pivot. Kristen Quinn moved into VR after spending years in video games, and shares:

> It's really not that different in my world. If you've worked on games, you've worked in game engines. You've already gotten used to building sounds based on game objects that are positioned in 5.1 [surround sound], 7.1, or whatever your endpoint is.

Megan Frazier, who worked as a recruiter before pivoting into VR, came into the industry targeting VR work. She says of networking:

> I knew that VR was going to be this little, tiny cluster of people . . . I had done it before as a recruiter where there are these really insular little communities where everybody knows everybody, and everyone's in everybody's business. [It's about] injecting oneself and being visible in those communities . . . You should be networking with everybody in the VR community at large – the artists, the devs, the producers, everybody who's involved . . . You can leverage personal connections a lot easier in VR.

Generally, the principles of building a career in game audio apply to VR and AR. VR and AR communities can be found online and through in-person networking events. There are also hackathons, which are a way to build experience and a demo piece.

Avril Martinez has found the VR community to be very welcoming, especially when it comes to diversity. They explain:

> In VR, there's this really strong community that just wants to support each other and wants the technology to grow and become a good, useful technology. So that's why we're being as friendly as we can. We want this to succeed, and we want good people to be in it. We want to support good people, essentially . . . Working in VR as an audio person there are not only games, but there's also the medical field, engineering, architecture – there are tons of branches that you can jump into. They all need some sort of sound.

Pro Tip

Some of the biggest mistakes I've made are not quite knowing when to keep my mouth shut (in regards to being so excited to be a part of a project), and just going out and saying, "I'm working on this project." Something as simple as confirming [a project] is the big no-no. You can't release any information at all about the game or that you're working on it – anything – until you get a clear from the people you're working for. Even just the project title . . . Don't you want people to know the name of your game? No, not until the PR people are ready to release it.

– Harry Mack

The Hiring Process

Interactive media can have a formal hiring process, and it is possible to land an interview by applying to an advertised job listing (a unique aspect of interactive media vs. other disciplines of the audio industry). However, positions can be very competitive, and small mistakes can remove you from consideration. Mark Kilborn, Audio Designer at Certain Affinity (Austin, Texas, USA), explains:

> Don't send me letters or CVs with typos. Sound implementation is a very detail-oriented job. A typo or a missed semicolon in our game can break the build and hinder the progress of 300 other developers. Correct spelling, punctuation – this stuff matters. If you can't pay attention to it when selling yourself, then how can I trust you not to break the game?
>
> (Hughes)

To apply, a job will generally require a cover letter, application, demo reel/portfolio, and resume or CV. A **demo reel** (or **showreel**) is a collection of short works to show potential employers your skills. A demo reel can demonstrate a specific skill (such as Foley, sound design, or implementation), a role (such as music or sound design), or style (shooter game, vehicles, trivia game, etc.). A **portfolio** is a collection of short works to show a broad and more diverse range of skills. Some professionals use the terms "demo reel" and "portfolio" interchangeably.

Lead Audio Designer Matthew Florianz recommends:

> When a job is posted, your website, CV, and showreel should already be up to date. Some job openings close within a week or even days, when enough

candidates apply . . . Do links work on different browsers? (Try accessing your site in incognito/private mode.) Does it work on mobile?

Ariel Gross, founder of Team Audio and the Audio Mentoring Project, suggests doing research when applying for a job to help your presentation stand out:

> If you haven't played any of a company's games, ideally, you'd make some time to play them . . . If you can't play the game, go consume lots of gameplay videos on YouTube, read articles and reviews about the games, and look for interviews with the people that worked on it.

Ariel also suggests trying to find out who will be reviewing your application:

> If you're applying at an established AAA game developer, it's more likely that first contact with your application will be made by a recruiter or someone from HR (i.e., not the hiring manager). They may filter out candidates that are missing non-negotiable requirements for the position. In these cases, it can be very important to align your resume with the job listing to make sure you end up in the right pile.
>
> If you're applying to an independent company, it can be a little less clear. Sometimes there's a recruiter, other times your materials go straight to the hiring manager(s), other times it may go to a council of stakeholders. In the case of indies, sometimes the posting will say what happens with your stuff, and other times you can figure it out by poking around on LinkedIn or social media.

Javier Zúmer says of cold-contacting, "If I really, really like their product, I will actually do some design for them, like for a video of their game or something. That's a pretty cool trick if you really want to get someone's attention."

Some game devs (including AAA) may use a third party to help find or screen job candidates, and each company may utilize these recruiters differently. Kristen Quinn says of recruiters she works with in the United States:

> Some recruiters might take the first call, and some recruiters might just [find candidates] and have the [person] hiring reach out for the first time. I've worked with recruiters that I really trusted and that I really valued their ear . . . It really just depends on how involved a recruiter wants to be, and do they care about the audio hire experience. It may get past that first recruiter's eyes if I get a recommendation.
>
> If I'm at a bigger company where I get so many applicants, there's a first screening that's not even done by me . . . I've had recruiters do the first phone screen and [look at] how they communicate. It's a casual conversation, like,

what games do you like? Based on communication, or the types of things they said in that call, they might potentially not get past the recruiter call.

The Hiring Process: Matthew Florianz

Matthew Florianz is Lead/Principal Audio Designer at Frontier Developments (Cambridge, UK).

The Process

The application process generally consists of rounds. Getting through the first round is a priority so don't make any obvious mistakes. Examples of mistakes are: applying for the wrong job, not acknowledging the job description, missing crucial information, or defective and unclear links.

A correct application is usually passed on to the audio team, who will decide on a first selection. The first impression (for me) is your showreel, so make it land. Cover letter, CV, and portfolio will be checked in detail after this (with the showreel fresh in mind). From the first selection, a smaller group will be invited for a practical test.

The Test

Tests usually consist of a video requiring new sound and perhaps some technical questions. Read the instructions carefully and make sure that whatever is specifically asked for is as good as can be achieved in a set time limit . . . Some companies have tests that are four minutes [of content] or longer – I don't think it is reasonable to ask for much more than a minute per week (that's already a lot) especially considering people have jobs and families. If the work exceeds time availability, I would opt to concentrate on a smaller part and do this really well than delivering a mediocre whole. Keep untouched parts silent and explain why in a letter.

Ask for clarification when instructions are unclear.

Loudness, tone, specifications are specifically stated if important. If not, listen to games made by the studio, use those as a guide.

Write about your creative process. It allows more insight, but don't write a novel.

First Interview

Interviews sometimes happen before an art test and sometimes after. Expect a Skype/Zoom etc. call for a first interview and pick a time and place that is comfortable . . . Approach this as a casual talk that will almost certainly have a follow-up, usually in the form of a test.

Final Interview

- At this point, you are qualified for the job.

- Final interviews are often on-site, though jobs might offer work from home and interviews as such.

- For those not selected for this final round – audio people will generally be happy to make time and feedback on your test.

- Not hired at this stage? Good tests are remembered. Some people hired are from leaving a good impression on a previous test . . .

- Don't over-dress . . . it could feel uncomfortable when everyone is casually dressed. Just dress and behave like your normal self.

- At the end of an interview – ask if there are any pressing reasons [you may not] be considered for the role. It is the best time to address any concerns.

Applying for Jobs: Javier Zúmer

Javier Zúmer is a sound designer at Pixel Toys (Leamington Spa, UK), where he has also been involved in the hiring process. Javier has worked in the audio industry in Spain, Ireland, and England.

You get loads, hundreds of applicants [for a job posting], but 80% don't have the experience or knowledge. Maybe they did some audio course or tutorial or something, but they don't really have anything that you can touch, that is tangible. It's mostly about the reel – the one-minute video of your best stuff. Most people don't have that, and that's the key. That's how you get the door open pretty much in video games . . . Either they don't submit a reel, or they do and it is very poor. There is a lot of, "Maybe I'll get lucky." People don't realize that looking for a job is a job . . .

Apply to a lot of jobs in video games and try to have the most attractive portfolio and reel possible . . . I didn't apply until I thought I was ready to apply, and I think I lost the opportunity of learning how do you get your CV ready, how do you send out the stuff, how to write a cover letter, how to prepare for interviews . . . Think about job interviews as training. Even if you don't want to work for a studio, even if you want to be a freelancer, it could be nice training to do a technical test or an artistic test for a game since it's very, very good practice. It's a bit bad for recruiters because they get people that can't really do the job, but for you, it's very good training.

Demo Reel/Showreel

When professional credits are lacking, it is common practice to do a **linear redesign** (or sound replacement), replacing the audio from a commercial project with your own. Since this is done without the permission of the game's license holder, it's important to label it clearly as a redesign so it is not confused as claiming credit on a released product. Javier Zúmer suggests:

The reel is the thing that gets you through the first phase – the thing that opens the first door in an application. Do anything you can to have a great reel. There are a lot of people that go to a video game cinematic and they redesigned the sound effects. That's cool. It's a nice thing to do as practice. But, the problem is that everybody's doing that. So ideally, get into a real project where you can really learn.

Matthew Florianz advises for showreels:

- Don't leave any doubt about the consistency of your work – present your best work only.

- Check your video after uploading (I've reviewed mono showreels).

- Clearly explain the work done – visual annotation is recommended.

- When redesigning an existing trailer or gameplay: Choose something exciting!

- Be original and creative. For all job levels, skills are expected to go beyond track-laying or finding sounds in a database.

- Don't save the best for last.

- When linking to your work on a website, point toward the specific and relevant examples.

Matthew Florianz adds, "I once got hired because the studio liked that I had written [in a cover letter] about how I designed and approached audio in my showreel. It gave them a feel for how I might work."

One resource is the web series *Reel Talk* (found on twitch.tv) by Power Up Audio (a game audio team in Vancouver, British Columbia, Canada). *Reel Talk* has been critiquing showreels and websites (submitted for review) since 2015. Javier Zúmer had his reel reviewed and says of the experience:

> It helped me a lot to see: Don't waste anybody's time. Go to the good stuff, be bold, and do something interesting. When you're looking through applicants, you need to surprise. You need to "wow" people. People forget about that. When they send an application, they forget that whoever is looking through those is looking at 100. It's pretty easy to be boring.

Finding Opportunities

Some opportunities will be low pay or unpaid. Jack Menhorn says of working for free, "It's definitely valuable to build up credits and experience on smaller projects. Many of these may be low-paying indie/mobile projects. I caution about getting taken advantage of and working for free." As a rule, Jack only works for free when everyone on the project is working for free.

Ashton Morris adds:

> Many people would say, "But I am looking for any project to get involved in. How can I be picky at all in the first place?" I get it – It's hard to find work in the first place. But don't sell yourself short. If you see a game and you don't understand why anyone would be interested in playing it, then maybe your time would be better spent elsewhere. You will be doing them a favor because they will eventually find someone passionate about their game, and you will produce better work with a game you really believe in.

Kristen Quinn suggests larger companies are more likely to have internship opportunities versus small and indie companies:

> Smaller companies really care, and they want to have internships, but the level of capacity is a lot tighter. Maybe their capacity for being able to mentor and spend more time with an individual isn't as great – because of the amount of responsibility and the number of hats they have to wear at any given point in time. With that said, I know a number of small companies that really value that mentorship, and value wanting to figure out ways to include that.

Javier Zúmer adds about seeking out work and internships in Europe:

> I didn't see many internships – Mostly entry-level or junior sound designer [jobs]. Actually, I had the problem of thinking, "Am I a junior or not?" I had the [audio] experience of someone who is not a junior, but I didn't have any actual studio development experience.

Jack Menhorn says of job seeking:

> There is definitely an ebb and flow of hiring and layoffs in the game industry, but there are always exceptions to that, so keep up the search . . . In the absence of work, making sound replacements and reaching out to small student and indie projects online are a good way to start building yourself up with experience and technique. Work on sound all the time because note: if you succeed in getting a job, you will be working with sound every day, so you might as well get used to it.

A job listing may have a variety of prerequisites, but some professionals recommend applying even if you may be underqualified, but explaining why you might be a good fit. At the same time, it is not a good use of time or energy to apply to every position available. Jack Menhorn suggests:

> If you're pretty new to game audio, it would be wise to limit what you apply for (to a degree). If you only have a few projects under your belt, applying for a senior sound designer role might not be the best use of your or the company's hiring manager's time, for example.

The Catch-22 of AAA Titles: Mark Kilborn

Mark Kilborn is Senior Lead Audio Designer for Certain Affinity (Austin, Texas, USA). He spent 10 years at Raven Software as Audio Director on multiple Call of Duty titles.

> Once you've got experience on a AAA title, a lot more doors open up. But getting that first one is challenging.
>
> Get to know as many people as you can in the field. Don't just get to know them to get work – get to know them to get to know them. Make friends. Be genuine. Most of the people making game audio are amazing people who are well worth knowing, so just enjoy knowing them.
>
> Do whatever smaller work you can – smartphone games, indie games, whatever. You'll get more experience while you continue pounding on the AAA door and, frankly, you might find you prefer working on

the indie stuff. I know plenty of people who are very happy in that world, and some amazing sounding games are made in that sector . . .

So-called "luck" is really just the collision of preparation and opportunity. Develop your skills. Build a portfolio of good work. That's preparation.

Network. Surround yourself with people who are doing what you want to do. That's exposing yourself to opportunity.

At some point, opportunity will appear, and you'll have put yourself in the right place with the right skills to take advantage of it.

(Hughes)

From The Sound Architect

Kristen Quinn suggests being open-minded to opportunities that allow you to meet and work with sound designers:

> We had positions that grew into sound design positions but maybe didn't necessarily start as a sound designer. We hired someone who managed our sound labs, so they managed all the studios and did the troubleshooting . . . When I worked at Riot Games, we had an audio coordinator. They would be responsible for helping with all our purchasing and things like maintaining the studios and making sure the machines were updated. They would also get some design tasks, and eventually, every one of those project coordinators got into a sound design position.
>
> It could be someone who gets brought on to help run and manage the studio, or they manage our sound effects library. Microsoft hired someone that would come in and take all of the recordings we had done on every project, edit them down, and built a really clean sound effects library out of that. I find those jobs to be really good "get your foot in the door" opportunities, because if you can edit and you're good, you can learn tools and methods to become efficient at doing it and speeding up the process. That allows opportunities then for people to see you [and think], this person's adding a lot of value, so I'm going to give them project opportunities.

Indie work can be a way to build up experience and create portfolio items. Avril Martinez shares:

> If you have a good portfolio, website, resume, and cover letter, the sky's the limit. But it is sometimes easier to just go for the small indie jobs because those are the people who need the help. You'll learn through them, too. There's a little more flexibility for learning. Small indie games are probably the easiest and fastest way [to get started].

Javier Zúmer found working for studios was better for learning than freelance opportunities:

> At first, I had a system to contact the studios to get freelance work. But then I decided to try to apply to studios to get that experience of working at a studio with the actual developers. I learned that doing sound design for people as a freelancer made it quite hard to feel connected and to know what they're actually doing. For most people, audio is an afterthought. It was hard to do cool things when I couldn't really talk to the programmers and the designers. So, after doing a lot of networking, I just applied to midsize indie studios.

Pete Reed was working on a West End theater show when he decided to pivot into game audio, building a portfolio working on small indie projects. Pete says of finding opportunities:

> Game devs post [job] positions all over the place. If you like the look of a game in development, you can always ask if they have anyone doing the audio. No harm in asking. At this point, it's about getting experience working in the industry, and you might need to start at the bottom of the ladder. Don't expect to get a AAA company knocking on your door straight away.

Quality assurance testers (QA or QV/quality verification) are entry-level positions playing video games as a job (looking for errors or bugs). QA requires troubleshooting and the ability to communicate issues clearly in writing. QA positions can be a good introduction to game development, but also requires some effort to pivot into audio.

Mixing/mastering engineer Mariana Hutten worked as a QA tester in Montreal and Toronto (Canada) early in her audio career. Mariana says of QA:

> It's certainly possible to end up in sound design [from QA]. The way to do it is as a day job – you do the audio testing (or game testing in general). Then outside of that, you work on indie games doing sound design to build your portfolio. You're not going to go from testing (which is a different skill) to straight up designing. You'll need to have some experience. In that case, when you apply for that position, because you have experience in testing, that's going to be a plus. You're going to have a competitive advantage over the person who came out of sound design school and never worked on a video game, and never actually worked in a team making a game.

Jack Menhorn started his career in QA and found it was a good source of stability:

> Working QA for me was more of a side job for money and general dev experience rather than a way for me to "break in" to game audio. I was

already freelance at the time, so the QA was more about a reliable source of income. That is not to say I did not take the QA job seriously; I was as professional about that role as any other . . . I actually strongly recommend not getting into a field of game dev that isn't your focus just to "get your foot in the door." It is generally frowned upon as people want you to be focused and passionate for the role in which you were hired, rather than using it as a stepping stone.

QA may not be the right fit for everyone. Emily Meo shares:

QA is low pay, it's grueling work, and it's very thankless. It's really rough. It's a shame, because QA is so important to the work that we do . . . If you get in there, and it's just terrible, and if you decide that the cost is too great, then find something else. There's always another way.

Paths to Finding Work: Ashton Morris

Ashton Morris is a composer and sound designer based in Colorado (USA). Ashton shares different ways he has found paid work in game audio.

From best to worst, and in terms of frequency/likelihood and enjoyability, in my experience:

1. Repeat clients (developers you have already worked for and have an existing relationship with)

2. In-person friendships or networking (making friends at a meetup or game jam and eventually working with them)

3. [Cold-Contacting] (personally reaching out to a developer that you really appreciate and asking if they need audio help)

4. Clients that found you through your website or past work ("I really liked the music for this game, are you available?")

5. Referrals (someone you know or have worked with refers you to a developer)

6. Other audio people needing help or having too much work ("Hey, I have a friend working on a cool game, but I don't have time," etc.)

7. Random tweets or develog musings ("This game is coming together! I should really start thinking about audio soon.")

8. Want ads (developers posting ads on forums)

9. Audio-for-hire posts (you post your information on various sites and forums looking for work)

10. Stock music or SFX (sites where you write general music or sound effects to be licensed)

11. Freelance sites (sites where you place a bid for posted work)

Personally, I think numbers 8, 9, 10, and 11 suck! But realistically, what I did (and what a lot of people do) was start from the bottom few and work my way up [the list].

From ashtonmorris.com

Hackathon/Game Jams

One way to quickly gain experience is by participating in **hackathons** or **game jams**, which are gatherings where a game is conceptualized and created in a matter of days. The name "jam" comes from music, where a group of musicians get together to play or perform with no preparation. Game jams and hackathons can be found around the world (with some virtual) and are open to experienced professionals and those new to game development. Emily Meo explains:

> Game jams are an excellent way to build up your portfolio or even just to test out new ideas. A game jam could be in person, or it could be online. It is a set amount of time in which a bunch of devs have to make a game (usually with a theme). A really big one is Global Game Jam. It happens once a year, every year. It's like 72 hours . . . you get a theme at the beginning of it, and everybody divides up into teams. Or if you want to solo dev it, you totally can. There aren't really rules – it's kind of up to you. Hopefully, by the end of it, you have a working prototype game. If you don't have one of those, then at least maybe you made some friends and had a good time.

Megan Frazier started pursuing a career in VR by doing hackathons in Seattle (USA) on weekends while holding a full-time sales job. At the time, she had no industry connections and her audio skills were completely self-taught. Megan shares:

> I faked it until I made it. The first time I ever did it, we were hawking sounds over and it sounded awful. To a certain extent, that's what a hackathon project is . . . I didn't know anything about it, and it was okay. Hackathons

always need sound people, but you should only pick one team because then you're actually invested in it, and they're invested in you . . . and you need to get a portfolio piece out of it. That's a big deal for someone starting out.

Hackathons: Avril Martinez

Avril Martinez is a sound designer and implementer for video games and AR/VR. They started in theatrical sound design and live sound before pivoting into video games. Avril gained skills and experience by participating in hackathons.

Hackathons are how I met my community. That's how I met other people who were interested in the things that I like doing. To participate, you pay the fee and show up with your computer. You show up with your skills, knowledge, and hopefully a lot of sleep (because you're going to need it).

I loved the rush of building something so fast over a weekend with a bunch of people who also didn't know what they were doing . . . That's how I started really getting the skills and working towards making those connections. These are the people that I literally slept on a cold concrete floor next to, on our sixth cup of coffee of the night, at three in the morning, trying to figure out why an arm isn't working . . . It becomes a built-in network of people who know you and your work . . . and we know how we work when we're at our worst.

Just start jumping in. Hackathons are a safe environment to mess up, learn, and meet people, too. These people care about the same thing that you care about. They want to create something amazing. If you feel like you can supplement that, then do it. That's how you get stuff for your portfolio, too.

Show up with Audacity and a pair of headphones and a recorder and go for it. Unity is free – you can download it anywhere . . . One barrier (that was my barrier for a really long time) was not having a VR headset. That was a big blocker for me because you want to test in your environment, but I used to just ask my friends to borrow a headset.

Networking

The interactive media community is full of information online and many places to network. Online "watering holes" include social media (Facebook groups,

LinkedIn, Twitter), Slack (such as AudioVR Slack), Discord, and Reddit. Meetups can be found for different niches, such as independent developers, VR/AR, or game engines.

On Twitter, the hashtags #GameAudio and #GameDev are commonly used. Responding to posts and interacting on Twitter can be a way to start meeting people and building relationships.

Pete Reed found a unique hashtag to connect with his local game audio community:

> One of the biggest things I did was follow #gameaudiodrinks. Twice a year, sound designers would meet in a pub in London to network. There are other events like this throughout the UK (the one in London was the closest to me). It was open to all . . . I just went to meet people.

Javier Zúmer adds:

> Go out and network, and also do whatever you can so you're not just one more person out there. Have a blog, do libraries. Have a great reel or port-folio – things that truly make you shine among everybody else.

For those working remotely, Jack Menhorn says of networking:

> A strong online presence is pretty key for people to know who you are if you are 100% remote. Reaching out to potential clients, to let them know you exist as well as building a rapport with the online game audio and game dev communities, is critical to help people remember you when they have opportunities. There is certainly the ability for living almost anywhere that has a reliable internet connection.

Conferences are also a source for networking and job seeking. Popular in-person conferences are GDC (Game Developers Conference) and GameSoundCon in the United States, and the Develop Conference in the UK. Some in-person conferences have recruiter booths, where you can inquire about positions. GameSoundCon and GDC also hold a "Demo Derby," where participants bring examples of their work and receive feedback from professionals in front of an audience. Will Morton says of game conferences:

> There are plenty of events to try out, and you don't have to limit yourself to the big ones. Get on Twitter and follow game development people and teams as you will often see tweets from developers saying they are going to such-and-such event – conferences, conventions, meet-ups. It's easy to find events, and easy to get a feel for what the events are like (YouTube often has videos from previous events, so you can see what you are letting your-self in for). If you are new to game development conferences, make sure

you do this research early – you will often need to schedule a few days to do a conference, including travel if you don't live near main cities, and there are often a lot of extra details to be sorted out when going to a conference.

Rates and Contracts

Brian Schmidt's GameSoundCon (www.gamesoundcon.com) performs an annual survey of professionals around the world. The survey collects information on compensation, work arrangements (freelance or employee), contract terms, and more. Not surprisingly, rates can vary widely based on the size of the company and location. Salaries in the United States can be significantly higher than in other places in the world, for example.

Akash Thakkar says it is crucial to ask for details about a game before providing a rate:

> You need to know about the team size, the deadlines, what game engines are used, how much work they expect from you, how fast they want it to be delivered, what sorts of contract that you're going to work with, what tools do you need, if you need to buy any – all that sort of stuff before you even so much as think of a [compensation] number for this game.

Rates for game work can be hourly, a flat rate, or charged by the asset. Professionals have their own preferences and reasons for choosing each. For example, Akash Thakkar suggests using a flat rate, when possible:

> This puts the client at ease and makes things so much easier for them. In fact, I've gotten gigs because I use flat fees. I've had developers visibly relaxed when I give them a proposal and they say, "Oh, thank goodness, this is so helpful. No one else was using a flat fee. Let's get to work." So, it is an extremely good way to work.

Jack Menhorn says of rates:

> Billing can be whatever works for you and your client; per asset, per minute, total for the project, per hour, and so on are all valid. I myself prefer hourly as it is easiest for me to track and protects myself from endless revisions and feature/asset creep . . . meaning additional features, feature requirements, and unscoped work added to the project as it goes on. An example being, "Driving around in a car is fun, but how about we also have the car fly and shoot missiles?" Adding in the flying and missiles is unscoped and unagreed upon work . . . If you charge hourly and a client wants revisions, it's as simple as "cool, that's another X hours to do the next revision."

If you charge per asset and a client wants multiple revisions, then you have to make sure you have a contract clause for the number of revisions [included] before charging for a new asset. Then, you have to track [assets].

It all comes down to what you want to make per hour. If you charge per hour, then that's easy. If you charge per project or per asset, then you need to do the math of how many hours it will take you to do the work, how much you want to make per hour, then multiply those two numbers together.

Contracts are common in interactive media work, and it is important to define who owns the sounds you have created. If a project is canceled or work goes unused, who owns it, and what are you allowed to do with it? In an **exclusive** contract, the client owns the work, but you may still be able to use it for demo purposes (depending on the contract). In a **non-exclusive** contract, you own the rights and can re-use the work in other projects. Jack Menhorn says of contracts:

Always, always, always get a contract. Even if it is a very small project that seems a bit informal, a contract is in place to protect both parties involved. I would be hesitant to work with anyone who would refuse a contract. If someone is unprofessional enough to not want to have a contract, then who knows what other ways they will be unprofessional? You can make a case that if the project is just a free project made on the side by friends, then one isn't needed. This gets fuzzy because what happens when that friend decides to make it a "real" project and sell it for money? Do you get a cut? If so, is it a fair slice?

With large companies or recruiters, a contract may be set with little room for negotiation. Avril Martinez explains, "When you're looking at contract jobs, a lot of times it's with a recruiter, so the recruiter actually has the contract already done for you."

Megan Frazier adds of indie clients, "Most indies, unless they are just really, really savvy, aren't going to know [about contracts]. They're going to want you to drive."

Foley for Interactive Media: Jeff Gross

Jeff Gross is a Foley mixer for film, television, and video games in Los Angeles (USA). Jeff has mixed for major studios, including Sony Pictures, Paramount Studios, and Formosa Group.

Foley for video games is very territorial . . . Companies want sounds created specifically for their game, and they pride themselves on this.

You're creating a world, so why would you want to use somebody else's sounds for that? Gaming is very specific, and the sound is so important to them.

There are cinematics, which are the short videos that propel the story forward. The action is predetermined. There's a story, a plot point, and there's dialog. So, it's literally [linear media] – the same as an animated project or cartoon. How the [interactive] game is done is you basically loop the actions. There are usually actions like run, walk, jump, land, stop – there are between 10 and 20 actions needed per character. We'll get a short video of each action, and we loop it in Pro Tools . . . It's repetitive stuff, where you do the action 20–30 times, and then they can have [options]. They'll do that for every character doing every action [on different surfaces], like dirt, water, grass, puddles, mud, snow. It's very repetitive, but I think it's actually fun.

I don't think you can have a career just doing video game Foley. You have to just do Foley. You're not going to get enough work to niche in one thing unless you become amazing at it (and work is not ever guaranteed). Then there's [scheduling] – video games can run the span of a couple of years, and the work comes in bursts. What are you going to do in between those bursts? You can do another video game, but it can bottleneck . . .

It's also beneficial to do both because it gives you more of a feel of what kinds of projects you can tackle. Video games are a whole different way of recording [than film or television] . . . Games are creating a world as you would with a cartoon or an animated project (although some [clients for] animated projects might request more real-world acoustics for the Foley – setting the sounds back into the room and not recording them dry). That's where communication comes into play with the sound supervisor, so you don't hand something over that's not what they want.

The budgets are shrinking in film and television but video games aren't. Clients usually leave you alone for TV shows and movies for the most part. Every once in a while, you get a supervisor who will come down and want to hear a playback of how things are sounding, and maybe they'll give some notes. The game that we're on right now is a high dollar one so we've had somebody on the stage with us every day [supervising] . . . These games have a lot more money for Foley than TV and film. More money becomes more time which translates to more creativity and elaborate sounds.

References

Bogovich, Bonnie. "Ask the Experts – Game Audio." *YouTube*, uploaded by SoundGirls, 14 Aug. 2021, youtu.be/_L_mRsLJtCU.

"Comscore Reports Highest Ever Worldwide Box Office." Comscore, 10 Jan. 2020. Press release.

Florianz, Matthew. "How to Stand Out When Applying for a Job in Game Audio." *LinkedIn*, 25 March, 2018. www.linkedin.com/pulse/applying-job-game-audio-matthew-florianz/. Accessed 14 Aug. 2021.

Frazier, Megan. Personal interview. 11 and 21 Aug. 2020.

Gross, Ariel. "A Big Jumbled Blog About Joining Team Audio." *Ariel Gross*, 26 June 2012, arielgross.com/2012/06/26/a-big-jumbled-blog-about-joining-team-audio/. Accessed 8 Aug. 2021.

Gross, Jeff. Personal interview. 30 Dec. 2021.

Hughes, Sam. "Interview with Mark Kilborn" The Sound Architect. www.thesoundarchitect.co.uk/interviews/interview-with-mark-kilborn/ 29 July 2013. accessed 8 Sept. 21.

Hutten, Mariana. Personal interview. 19 Nov. 2020.

"IFPI issues Global Music Report 2021." IFPI, 23 March 2021. Press release.

Mack, Harry, guest. "Career Advice For Freelance Designers." *Sound Design Live*, 16 July 2013. *SoundCloud*, soundcloud.com/sounddesignlive/career-advice-for-freelance.

Martinez, Avril. Personal interview. 28 Aug. 2020.

Mazzucco, Tiziana. Email interview. Conducted by April Tucker, 30 Sept. 2021.

Meo, Emily. "Ask the Experts – Game Audio." *YouTube*, uploaded by SoundGirls, 14 Aug. 2021, youtu.be/_L_mRsLJtCU.

Menhorn, Jack. Email interview. 30 June and 26 July 2020.

Morris, Ashton. "Freelance Game Audio: Getting Started and Finding Work." *Ashton Morris*, www.ashtonmorris.com/freelance-game-audio-finding-work/. Accessed 1 Aug. 2021.

Morton, Will. "Make Some Noise Getting a Job Creating Sound and Music for Videogames." *Gamasutra*, 8 April 2015. Accessed 8 Aug. 2021.

Quinn, Kristen. Personal interview. 30 Sept. 2020.

Reed, Pete. Email interview. Conducted by April Tucker, 24 Nov. 2020.

Schmidt, Brian. "Game Audio Industry Survey 2019." *Game Sound Con*, 11 March 2020. www.gamesoundcon.com/post/2019/09/10/game-audio-industry-survey-2019. Accessed 8 Sept. 2021.

Thakkar, Akash. "Successful Freelancing in Game Audio." *YouTube*, uploaded by GDC. 9 Jan. 2019. youtu.be/93ggs7hwJeU.

Wijman, Tom. "The Games Market's Bright Future: Player Numbers Will Soar Past 3 Billion Towards 2024 as Yearly Revenues Exceed $200 Billion." *NewZoo*, 1 July 2021. Accessed 8 Sept. 2021.

Zúmer, Javier. Personal interview. 24 April 2021.

16

Sound for Picture (Broadcast, Production, and Post-Production)

Sound for picture is an established industry that is continuing to grow worldwide. According to the Motion Picture Association, the global theatrical, home/mobile entertainment, and paid television markets combined were US$328.2 billion in 2021. Additionally, there were over 1.3 billion subscriptions to online video services worldwide in 2021.

Careers in sound for picture are generally divided into three areas:

Broadcast – capturing sound for live transmission. Work may take place from a set or on location (OB, or outside broadcast). Broadcast job titles include EIC (Engineer in Charge), A1 (sound operator, audio broadcast mixer, broadcast audio engineer), A2, comms (comms tech), RF tech.

Production – capturing sound as filming occurs (not live). Work may take place on set or on location. Job titles include production sound mixer, boom operator (second assistant), sound utility (utility sound technician, UST, sound assistant), playback engineer.

Post-production – enhancing/manipulating production sound, creating, or adding sounds to edited picture. Work generally takes place at a post-production sound studio, post-production facility, or a home studio. Job titles include assistant (mix tech), sound editor, sound designer (sound FX editor), dialog editor, Foley or ADR mixer, Foley artist, sound supervisor, and re-recording mixer.

In all three areas, work may be booked project to project. Stable full-time positions do exist, but are not the norm. It is also common for professionals to work in multiple roles within an area. For example, a broadcast A1 might also work as an A2, or a sound designer could also be a re-recording mixer. Professionals tend to specialize in one of the three areas, but in the modern audio industry, some opportunities may require knowledge in multiple areas.

DOI: 10.4324/9781003050346-19

Doing Production and Post for Online Media: Mabel Leong

Mabel Leong works in production and post-production sound at Night Owl Cinematics (Singapore).

> Night Owl Cinematics is a local food, lifestyle, and comedy production company that originated from a YouTube channel. It has been running for 10 years now and has content on YouTube, Instagram, and TikTok. In short, it's a really busy place! It would make sense for the company to have dedicated staff on-hand who can handle all the audio duties required for the entire production cycle, including post-production and equipment upkeep.

> I had to pick up production sound full-time in this job as it's a one [person] band for every shoot. I wear all the audio hats. I am working in tandem with another sound engineer on the staff . . . We work individually, either alternating post-production or on-set duties weekly or both being on separate sets on certain days.

> Having both of us on hand keeps the machine (that is Night Owl Cinematics) going apace with consistent shoots and uploads, so being active in both production and post is a huge asset. There is the added benefit of being on the set and noting the audio issues that will have to be treated later in post, or being able to advise the crew that we have to redo a scene.

Production and broadcast work can require a significantly different lifestyle than post-production. Some production and broadcast jobs require physical labor, whereas post-production work is generally at a computer (with the exception of a Foley artist). Production and broadcast require closely working with a team in real-time. Post-production has opportunities to work with others (recording ADR, Foley, mixing with clients) or to work alone (some mixing, sound editing). Mehrnaz Mohabati says of the nuances of post-production sound, "We are doing art. I don't think about myself as an engineer more than being an artist. Even when I do dialog editing, I look at it as the music."

In production sound, assistants may be **wiring**, or placing microphones, cables, and wireless packs on talent. Sound assistant Eliza Zolnai says of wiring:

> It's something you have to handle or at least enjoy – just not to be shy. We did a movie with young ballet teenagers (pretty much in ballet clothes) . . . You have a very intimate relationship with the actors by the end because you just get really close to all sorts of body parts.

Post-production hours tend to be daytime or evening, and may have some flexibility in schedule. Patrushkha Mierzwa says of production sound schedules:

> During production, your work can start at any hour of the day or night and continue for as long as the company says; rarely do two consecutive days have the same call and wrap times. Along with widely shifting call times will be the location of your work; some days you may be at the beach at 5 am, other times you may be at a power plant on the other side of town at 9 pm. A job may last one day, one week, or six or more months in multiple countries.
>
> (263)

Camille Kennedy refers to production sound as a "traveling circus," and shares of the work:

> Every day is a little different. You have no idea what you're going to be doing. I might be putting a wire on a goat. Like, I've wired up animals before . . . If you like something different every day, not knowing what's going to happen, or knowing when you're going to go home, this job is for you.

Post-production work is indoors (often with no windows), whereas production and broadcast may require work outdoors (including extreme weather conditions or environments). "I know from experience (especially from work on the film *Shy People*) that I can get motion sickness on boats or aerial craft, so I ask about these elements and am honest about my ability for the work," says Patrushkha Mierzwa (39).

Broadcast sound requires quick learning on the job under tight deadlines. Broadcast A1 Andrew King says of learning new consoles and equipment on the job:

> There's nowhere around here to just say, "Hey, can I go play around on your console for a couple hours?" [The answer would be], "It's on a TV truck that we're taking to Cincinnati," or "There's no time for that." When I step on that truck, I need to know how to use that stuff because I don't have time to spend an hour trying to figure out how to route the auxes properly on a console.

While sound for picture work can be creative, the discipline is generally more business-oriented and has more structured schedules than music production. Mehrnaz Mohabati has worked as a music engineer in Tehran, Iran, and in post-production sound in Los Angeles (USA). Mehrnaz says of the differences:

> In post, you can't be loose. In music, you can because music needs that flow. Back home [in Iran], my husband and I had two studios. At lunchtime, we

asked our clients if they wanted to eat together. We arranged a setup where we could all sit together, talk together, get to know each other more, and have fun. Sometimes we [would play] music together for an hour before we would go back to recording or whatever (it could create collaboration between artists). But in post, that doesn't happen. We work.

Sound designer Aline Bruijns adds:

Most of the time [the direction is], we need to get it done. We don't care how you do it. This is the amount of time that you have. Good luck. It needs to be good, it needs to be tidy, and it needs to be done in time.

Sound for picture requires niche skills that are different from other disciplines. Patrushkha Mierzwa explains the specialized knowledge required to work in production sound:

Typically, the costume people don't know the names of lamps, and the grips don't know about the script supervisor's book. But the sound department must know about lighting to be able to discuss ways in which light can be attenuated to remove microphone shadows or otherwise modified. We need to know how the script supervisor assigns numbers to scenes as much as how to communicate with the costume people over issues that involve the placement of radio microphones on the actors . . . We need to know camera lenses and be able to imagine accurately what the operator is framing so we can adjust our movements . . . We need to tune into the director and into the specifics of the actors' performances, so we know the blocking and when the timing or the dialog is changing. We need to work with the props and art department people to solve noisemaking problems, or create opportunities for planting microphones in the set. Even the locations department will hear from us while we work out issues involving unwanted sounds from a building or location near the shoot. I believe that when we do our job well, the sound department becomes far more involved with and knows far more about filmmaking than most of the departments know about sound.

(3)

The opportunities in production and post-production work are wide – from short films, feature films, television series, commercials, marketing/promotions, web content, and more. Additionally, there is scripted content (also called narrative) or unscripted, such as documentary or reality television. Sound utility/boom operator Camille Kennedy shares:

There are always so many little productions that are always looking for people. You can make a good living doing that, and then maybe move on to commercials or something. There's a whole circuit of mixers that only do commercials . . . With commercials, you'll get more [scheduling] notice. You might get a whole month ahead before you actually have to work.

Pro Tip: Camille Kennedy

*Production sound utility/boom op Camille Kennedy shares who would **not** be a good fit to work in production sound.*

- Anyone that can't stop staring. Some people just stare.
- People that would make inappropriate comments. Not mean, but they think it's a joke.
- People that don't have good hygiene. I always tell my assistants cut your nails. Keep your hands nice because if you're wiring talent . . . I've had actors say somebody's odor was too much, and even costumes said they leave an odor. They're just naturally that way.
- People that can't pay attention or self-motivate . . . There is so much downtime on sets. It can be crazy, crazy, and then a 45-minute lighting setup. Some people think, "nothing's happening" and go away because their attention is gone. We still have a lot to do. You have to really self-motivate. You can't just expect everything to be told to you.

Broadcast Sound

Broadcast sound includes live transmission of events including sports, concerts, news, and more. Sound crews vary in size. Some broadcast crews (including small news crews) may not have a dedicated sound person, instead relying on a cameraperson to cover the role. A large crew could have specialized roles like comms or RF coordinator. Troubleshooting skills, understanding signal flow, and fast learning are crucial skills in broadcast sound. Andrew King says you may be expected to work with unfamiliar setups with no time for training. "There's a bit of a learning curve every time you step on a TV truck because every truck is set up a little bit different," says Andrew.

In some ways, broadcast sound is a crossover between production sound, post-production sound, and live concert sound. Eric Ferguson (who has worked in live sound and as a broadcast mixer) says the characteristics that make someone a good broadcast mixer are the same as a good monitor mixer:

> Really thick skin, really fast response time. Not taking it personally. [You do] all the broadcast mixing and the tech in the truck, so you need to be pretty technically savvy . . . There's an intercom yelling at you all the time. It's really hard. It's definitely the hardest form of mixing I've ever done.

The two main job roles in broadcast sound are A1 and A2. Jeri Palumbo explains the A1 position:

> The A1 is in charge of a broadcast show's final mix-to-air and is usually located in the production truck (on remotes) or an isolated audio booth (in a studio setting). The A1 is in charge of the entire audio crew; creates and advances the paperwork (mult sheets); creates signal flow to all cameras, producers, and production; patches the truck (or studio); patches/dials in communications (comms); and, of course, mixes the show live.

The A1 may be responsible for comms (the intercom system that allows communication with crew around the venue), or comms can be a separate person/role, depending on the gig. Mixing for broadcast can be complicated because of the multitasking required. Andrew King says of broadcast mixing for sports:

> Where the real challenge comes in is you're listening to the show that you're mixing; then, you're also listening to the director and the producer. Then, you might have the instant replay [person] calling you on the intercom, and then you've got an A2 calling you on the radio. It's sad to say, but the sound of the show is almost the last priority for the A1. We want it to be our number one priority, but if the producer can't talk to the play-by-play [person], it doesn't matter. That's why we care about intercoms.

Jeri Palumbo explains the role of an A2:

> The audio A2 has completely different skill sets from an A1 . . . one may think the A2 to A1 progression is a seamless one, [but] the positions couldn't be more different in functionality. If the A1 wires everything inside the truck or control room and mixes the show, the A2 wires everything *outside* the truck, such as the broadcast booth, the field, the basketball court, etc. The A2 does not mix or have a console (unless they are in a sub-mix position) . . . Even though there may be a bit more "freedom" during a show once the audio is set (the A2 being able to move around the venue, versus the A1 being inside a production truck), the A2 position is more labor-intensive. The A2 must have extensive knowledge of booth and field gear.

Andrew King says of helpful qualities in an A2:

> A good A2 is somebody that understands signal flow (how the signal from this microphone gets to the TV truck), and is able to do it in a way that works. Troubleshooting skills are number one because if we're on the air and something's not working, you'd better be able to figure it out pretty

fast . . . As an A1, I don't have time really to help A2s do stuff. If they don't know how to do it then I have to go do it, and it really puts me under some serious pressure. For a Rapids game (our local MLS [soccer] team), just walking down to the field and back to the truck might take 15 minutes of my day. That's my lunchtime, basically, taken away from me.

Finding broadcast work is a matter of networking and recommendations. Andrew King explains:

If you try to look up the top A1s in the industry, all you'll find is maybe some interviews with them. They don't have websites or anything like that . . . I'll get random calls sometimes like, "Hey, this other person gave me your name. Are you available for this show?" That's really how this part of the industry works – word of mouth. Reputation is everything. Your name starts floating around the market. If your reputation is bad, you don't work. That's it.

Location Sound Mixing for News: Pete Bailey

Pete Bailey (Manchester, UK) has been a location sound mixer on Good Morning Britain, *a morning news program for ITV, since 2012. He discusses a typical day working in news and his outside work in production sound.*

Typically, it's a 2 am to 3 am start. We are on location by 5 am, ready for the program to go live at 6 am. News is an ever-changing production. I always have to be prepared for whatever the story is, never knowing if we are going to be outdoors or indoors. It could be floods, extreme weather, sports, politics, court cases . . . or it could be something nicer, like the world's biggest mince pie . . .

I work in a small crew consisting of just me, Geoff the cameraman, and Katy, our correspondent. We used to have our own satellite engineer with a sat truck, but now we have our very own that both me and Geoff operate and set-up each morning. We all meet on location with the satellite truck at 5 am and discuss how we're going to shoot that morning. During set-up, I hand out radio comms to the crew so we can hear the directors/producers in London, and we also do a vision and sound test with the technical director there.

Proudest Accomplishments

My proudest accomplishment has to be working on *Captain America* as a sound assistant. Working on a Marvel film was the peak for me:

I even have my wage slip framed. It's amazing to see what goes into something like that, but boy, were they long days!

My favorite shoot has to be a film I did in Positano, on the Amalfi coast in Italy . . . I was given an entire day to go and wander around the town and the beaches to record wild tracks of the waves crashing up the cliffs and onto the sand, and the bustling markets between the labyrinth of defending steps and alleyways. During the scenes, we shot completely noiseless audio and we even had unscheduled fireworks on our final scene of the shoot down on the breezy beach, where I luckily had my stereo mic already recording. It made me fall in love with sound all over again.

Advice to Those Starting Out

You must be passionate about the job and take pride in your work. As long as you have that interest and attraction to your role, skills can be learned and developed along the way.

I would advise people to network like crazy. Build relationships and friendships with people who are at the same point in their career as you, and those already established in the industry. It's the number-one key to success, as all jobs are word of mouth in this industry. I host a podcast called *Film Industry Pro* where I interview people in all areas of the industry . . . I get all sorts of crazy stories and incredible tips and advice.

Besides that, if you're pursuing a technical role, then buy your own professional kit ASAP. Never buy cheap gear; people won't take you seriously. I started with a really cheap mixer which I used on my first professional production and completely messed it up because my kit wasn't capable of handling the job. I literally threw it out of the window, marched to the bank, got myself a loan, and bought some serious kit, which immediately attracted attention from other professionals when they saw my gear list.

Interview from rycote.com; Pete's website: petebaileysound.com

Production Sound Jobs: Patrushkha Mierzwa

Patrushkha Mierzwa has worked in production sound for over 80 movies and is the author of "Behind the Sound Cart: A Veteran's Guide to Sound on the Set."

> The *production sound mixer* (PSM) is the contractual head of the department . . . and is responsible for recording the dialog and sound elements present during the shooting phase of the project.
>
> The *boom operator* is the on-set sound representative and is responsible for the selection, placement, and operation of microphones (in collaboration with the mixer) and the capturing of the dialog and sound elements present during the shooting phase of the project.
>
> The *utility sound technician* (UST) is the support person for the sound department – able to step into any of the aforementioned positions when needed, as well as setting up, maintaining, and even repairing the equipment, placing wireless microphones on the actors . . . On many film and television projects, the UST is the ambassador of the sound department to the other crafts, the director, the producers, and the company. The UST is often the main connection to the actors . . . [The UST is] taking care of the mixer and boom op, food/beverage runs (it's never beneath me to take care of my department members who cannot leave the set), make phone calls, deep cleaning the gear or truck (the yucky jobs), soldering, organizing, inventorying . . .
>
> There may also be an intern, a Y-16A trainee, a playback operator, and/or video assist operators and video playback operators. During the COVID-19 period, it's common to have an additional person for sanitizing and managing sound equipment used by other departments.
>
> (Mierzwa, 2-4, 288)

Production Sound

Production sound work takes place on location (traveling day-to-day) or on a set; it can be working on a team or alone (a "one-person band"). A team is typically made up of a production sound mixer, boom operator, and a utility sound technician (UST)/sound assistant. Eliza Zolnai says of the terminology used in some countries (including the UK, Hungary, and New Zealand):

We don't really use the word "utility." We separate it as "first assistant sound" and "second assistant sound." First assistant sound mostly means the boom op or the two boom ops. Second assistant sound means probably the person who's either a second boom and wiring; or if it's four, then it's a third boom and wiring. We try to completely separate the wiring person and make it the job for that one person, because it's just a lot.

Both working with a team or alone can help gain experience, but a one-person band would be expected to have (or rent) their own equipment. Patrushkha Mierzwa explains:

In this scenario, a mixer will have a recorder, possibly a mixer, a boom microphone and pole, a couple of wireless microphones, and possibly an IFB system so others can hear. Many smaller productions – including interviews, documentaries, industrials, and some commercials – budget for only [a one-person band].

(84)

Eliza Zolnai says of working on teams and as a one-person band:

The only places where I would be alone are documentaries. That's the only place where I actually [mix]. Most of the time I'm the assistant of other mixers . . . A bit of both is the best [when learning] – watching very experienced people work and at the same time, try to do little jobs on school movies for free, or on short movies, or no-budget movies.

Camille Kennedy adds:

I did the one-person band, booming and mixing by myself . . . I didn't buy or invest in gear, but I did work for a sound company that gave me gear. They made half of whatever money [I made on] the gigs.

One barrier to working on a professional set is a lack of prior experience on a professional set. While film schools and audio programs can offer some experience, the size and scope may not match what happens at a professional level. Shadowing or observing may not be possible due to privacy or safety concerns.

Camille Kennedy learned set etiquette and safety by working in construction, props, and as a grip. This also helped her build relationships with sound departments that later hired her (see story in Chapter 3). "If you don't know set, or the way to work on a set, you're not going to get hired . . . How can you get there? I pushed a broom around a studio and now I'm booming," says Camille. She says of the challenges:

How do you know how to wire a car if you've never been taught how to wire a car properly? . . . You haven't learned the proper mic placements or the proper tools to use for the situation. I'm sure you could go out and figure it out, but the end product is going to suffer because you're using that as your test and your trial. The standards are so much higher with the professional sets. You know everything is going to be close to perfect all the time because the standard is so high.

Mabel Leong worked in post-production before taking a job at Night Owl Cinematics (Singapore) that required her to do production and post. Mabel says of working on set:

There are some extra things you've got to consider compared to a recording studio or post-production-only role. As someone who transitioned from post-production only to a dual production [and post] environment, I had difficulty fitting in initially because I lacked production skills and the "set sensibility" (i.e., not standing by a corner looking like a fool throughout the shoot). There is a lot to be mindful about . . .

Unless the crew you work with is sensitive to who handles equipment, endeavor to learn how to operate or use the non-camera equipment on set to help set up or tear down if you're free. This was a big sticking point for me, because not doing so and standing by the side (despite being available to assist) can make you look "lazy" or "unhelpful" to the rest on set. Some things are not too hard to learn with repetition.

Sound utility (or sound assistant) is typically the first role for an emerging professional working with a sound team, and can turn into a career on its own. Jenny Elsinger (Wilmington, North Carolina, USA) explains:

I really like the freedom I have by staying as a utility. My work is very varied. I do a lot of second boom. I get called to boom or mix as a day player with another mixer's gear. So, I kind of get the best of everything . . . I really enjoy mixing, but you can't do it halfway. It's a huge financial commitment . . .

I never sit down during the day. There's always something to do and I like to stay busy. There's all that fun and problem-solving, fixing stuff, and coming up with creative solutions. I love the freedom, too. The sound mixer is very often trapped with the cart, and the boom operator has to stay on set, and I like to run around kind of behind the scenes. I also like the variety of the other departments that I work with because, as the utility, I'm really the go-between with props, costumes, grips, and electrics, too . . . I'm truly interested in everyone else's job. I'm one of those people who loves to pick your brain and see how you do things. The entire process of filmmaking fascinates me from the ground up.

(Garrett)

Eliza Zolnai adds of being a sound assistant:

> Although I have the qualifications to work as a mixer . . . I just like to be
> on set, and I like to be around the actors, solve problems, talk to everyone,
> and be around everyone. I just enjoy that a little bit more than just being
> behind the mixer, sitting in the corner of the studio all day.

Sound utility work requires physical labor. Patrushkha Mierzwa explains:

> Most people assume they aren't strong enough to hold a boom pole, so they
> immediately write off this job . . . Yes, there will be minutes when the boom
> is a strain, and you'll have to muscle through . . . [A utility sound techni-
> cian] has carts to push up hills, onto ramps, through doorways, and man-
> euver into elevators. There's a good amount of lifting, twisting, pulling,
> and squatting, so having a general fitness plan is a must . . .

> We are one of the few positions that must have all their body parts working
> to be able to do our job, whereas a director of photography can have broken
> arms and broken legs . . . But as USTs, we can't. We can't even have a
> head cold (we'll have trouble hearing). From head to toe, we need to be in
> perfect condition, and that's why it's so very important to have a physical
> training program that you follow. What we do is so physically unnatural
> that we must train consistently so we aren't getting hurt . . . The job of a
> UST requires nearly constant movement; it's common to walk 14,000 or
> more steps in a workday.

> (Mierzwa 21, 144, 240)

A sound assistant may be dedicated to wiring. Camille Kennedy explains:

> [Wiring the cast] is a full-time job especially since things are wireless. When
> you get into period pieces with costumes or space costumes, it's always hard
> to wire . . . You're wiring cars ahead of time. You're planting mics on set.

Eliza Zolnai adds of wiring:

> The bigger stuff you do, the more attention and time the actors
> need . . . after a certain level, if the actors are watching you crawling on the
> ground, putting carpets down and getting completely dirty, then running
> up to them and getting under their dress . . . that can be a little tricky . . .
> Even before COVID I always had hand sanitizer on me and baby wipes. I
> always wipe my hand right in front of the actor – even if I wash my hands a
> second before – just so they feel that I care about it . . . I think that's super,
> super important.

Sound assistants may need to boom op at times and will receive guidance and training. Boom operating does not require having your arms in the air all day. Camille Kennedy says:

> You have to be taught how to use your own muscles and your own skeletal system for your benefit. I work off ladders all the time because I'm short. You can sit holding a pole all day . . . As long as you're getting good sound, who cares how you're doing it?

Boom operating can require some performance to navigate around cameras or to compliment an actor's performance. "Boom operating is really about a duet with the actor, more than focusing on the dialog and movement and beyond anticipating seamlessly whatever is coming, almost being an extension of the actor," says Patrushkha Mierzwa (153).

Production sound mixing on a team is not an entry-level job, and taking on other roles first can be beneficial. Camille Kennedy explains:

> A lot of mixers were boom ops first. They're the ones that really know what's going on because they've swung boom, they've wired people, they know exactly what they're looking for. When you work with mixers that never strung a boom, they can't tell you what to do better, or they don't know. They've never actually pointed a microphone with a boom. So, when you actually have a mixer that really has owned their skill as a boom op, if you're in a really tough situation, you [as a boom op] can call for backup and ask, "if you were doing this, what would you do?" They'll be able to grab the boom and work with you. Those are generally the better mixers.

In production sound, bookings could be for a day, weeks, or months. A sound crew could go from project to project together or separately. Eliza Zolnai (who has worked in production sound in Hungary and New Zealand) explains:

> In Hungary, they would be calling the same [team] every time. My experience with New Zealand is not that long, but there's a bit more wiggling around in teams. It really depends on where the job is – if it's in Wellington or if it's in Auckland . . . In Hungary, it's not production who asks to work with you. It's the mixer who says, "I'm bringing in this person and you should negotiate about prices with that person. That's who I want on my team."

Camille Kennedy says of the challenge of working with different crews as a sound assistant:

> Every mixer has their own gear and their own system . . . So, when you get on set with a new mixer, you've got to learn their gear, the quirks about their gear, where you store it in their carts, all that kind of stuff. When [I

was] jumping around from different mixers, by the end of the month, I had to learn like seven different whole rigs of gear.

Being a Boom Op: Camille Kennedy

[As a boom op], you really read the energy of the room. When we're doing a deep scene, like a deep topic, the crew is silent. No laughing, no joking around because that actor has to be in that headspace of something. How can they get into that state when you have some crew next to you joking around?

[During comedy scenes], I've totally burst out laughing while booming, like I couldn't help it . . . If it happens, it's okay as long as you're not doing it all the time. I've boomed scenes and I've ended up just bawling, crying because it's a scene that was like that. When you're doing the audio for it, it's right in your head. Being right there and feeling that energy with that performance and stuff, you're just super moved.

Then you get into sensitive, intimate scenes. You have to be so respectful during those times. No talking, no goofing around. When I did *Below Her Mouth*, most of the time the actors were all naked all the time. So, you have to be really professional about that.

There are times when there's a really interesting and intricate scene. There's the Steadicam operator or the dolly cam making lots of camera moves, and everyone's performing. The cameras are doing their performance. I'm booming, like dancing with a camera operator doing my performance. The actors are doing their performance. We all know where we're going to be during this specific time so no one crashes into each other. The rehearsal between sound and camera is almost more important than the actual actors themselves. It's a lot of work and a lot of memory work, and you only have like three minutes to figure it out. If I don't, I get run over by the Steadicam operator. They're not going to stop.

Post-Production Sound

Post-production sound (or "post") has a variety of jobs. ADR and VO recording involves working with actors re-recording lines, adding new lines, or recording into another language (called **dubbing** or language localization). Foley involves original recording (typically in a Foley stage by a Foley artist) to enhance details or add coverage for missing sounds (like footsteps or props).

Sound editors clean up sounds provided by a picture editor or add sounds for enhancement. Some sound editors edit all elements and others specialize in one role such as dialog editing, music editing, or sound effects editing/sound design. A **sound designer** focuses on adding sounds (by creating original sounds or using a sound effects library). Sound editors usually work more independently than other sound roles in post.

Mixing is performed by a re-recording mixer (or dubbing mixer), and high-end projects can have two or three mixers. The studio where mixing work is performed is called a dub stage or mix stage. The **sound supervisor** is the lead person of the sound team who oversees the process. Not all projects will have a sound supervisor, and it is considered a senior-level position.

A post-production sound professional may specialize in a single role or have expertise in a variety of roles. In Los Angeles, Oscar nominee Ai-Ling Lee specializes in sound effects (as a sound designer and re-recording mixer) for films. At the start of her career in Singapore (where she grew up), Ai-Ling was expected to cover many roles:

> I got in as an intern in Singapore at an audio post house for radio, TV commercials, music recording, and mixing. They hired me and I worked there for a little over a year. It was a great introduction for me because, in commercial work, it's just you alone with a bunch of backfield clients. We have to liaise [with them], record the voiceover, edit all the music, edit all the sounds, and mix. It's a good opportunity to be thrown into all kinds of areas of sound, to know how things work, and to work with clients. But I knew that I really wanted to work in film, and at the time, there wasn't much film work in Singapore.

Sound designer Aline Bruijns (Netherlands), says of having more specialized roles:

> People hire you for a certain type of sound design. It's not just like you throw in some sounds and it's done. You always have some sort of thing that you like. I get asked a lot for ambient sound design and ambience editing for film. I truly love it because I play around with birds, I play around with something happening in the back, or just to make it really spatial.

Professionals may also work in multiple roles. Karol Urban, whose primary interest is mixing, finds opportunities to learn from other mixers when she works as a dialog editor. "If [other mixers] explain to me how they want to work, I see that as being the gold from this job. I'm learning their way of thinking and working," says Karol ("Interview").

Foley mixer Jeff Gross adds about being versatile in post:

> Some people fall short because they focus too much on one aspect. I focused on both the editing side and the recording side, and it's what I think made my career. A lot of people who are editors don't have to stick a mic in front of something and record it, and a lot of people who record Foley don't edit. We're at that threshold of generations where now people are starting to do more of both . . . If you have a bag of tricks in your back pocket, these experiences and skills, you can move around and become more employable. If you're an editor, learn how to record. If you're a mixer, learn how to edit.

Post-production sound has a variety of work environments that can appeal to different personalities. Some like the rush of tight deadlines that come with working on a television series. Some films have an extended timeframe, which allows a higher level of detail. A professional in post could work on commercials, marketing and promotions (called "promo"), corporate videos, political ads, online content, and more. Mabel Leong, who has worked in commercials, advertising, and online media in Singapore, says of the differences:

> In my working experience, advertising and corporate work demand a more polished veneer to your mix . . . In comparison, being in a production house allowed me to get hands-on with a variety of emotive works and subjects . . . I preferred the internal reviews and directions from my producer colleagues compared to managing clients in one-on-one corporate freelance work, but it also means I am at the mercy of the production house's schedule and internal deadlines . . . If you are confident in handling clients and their expectations on a one-on-one basis, along with the quick turnaround time and the responsibility to work well alone, freelancing with advertising gigs can be a great fit for you.

One common challenge of post is budgets. Foley mixer Jeff Gross explains:

> Sound is an afterthought. It's like, we've got to have the [best] camera, the best lighting, all of this. But the end of the line is sound, and that's what gets skimped on the most. You can watch something and if the picture sucks, but the sound is good, it's forgivable. But if you're watching something where you're struggling to understand what's going on because of poor sound, even if the picture is good, you lose interest.

Dub Mixing: Tiziana Mazzucco

Tiziana Mazzucco is a freelance dubbing engineer in Milan, Italy. Her career started in Rome, working for a studio where she learned post-production sound and dubbing.

Over the years, I have worked on countless audiovisual products, such as TV series, films, documentaries, cartoons, and reality shows. My job is to record the actors and then synchronize the Italian voices on the mouths of the real actors on screen. I have also done a lot of mixes of reality shows and cartoons . . . Italians dub everything, so [source] languages can be from anywhere, especially in the last few years. In fact, there has been an exponential increase in foreign content from all over the world.

I think that a dubbing sound engineer does a purely technical job. There is no space for creativity if the aim is to reproduce someone else's work. Above all . . . A good sound mixer must know how to use and how to choose the right tools to make a mix sound like the original one, but without having any indication from the person who created it. You shouldn't hear differences if you switch from one language to another. There are also many technical specifications to be respected – and a sound mixer must be able to achieve them.

What I like most about this job is its dynamism. It is never repetitive because you deal with different people on different projects every day.

Learning Foley in the Modern Audio Industry: Jeff Gross

Jeff Gross is a Foley mixer for film, television, and video games in Los Angeles (USA). Jeff has mixed for major studios, including Sony Pictures, Paramount Studios, and Formosa Group.

Learning Foley has got to be a hybrid approach [learning alone and from professionals] . . . You have to figure out a way (if you're just starting) to get to somebody who's done it and get some feedback on what you have recorded. They might be able to give you pointers or tell you what's wrong with your recording. When I started doing video games, I needed some guidance on it. I was working with professional Foley artists who have been doing it for 30 years. [Learning from others] is so important because then you have somewhat of a baseline. You have to somehow pull yourself up to the minimum of quality of what professional mix stages expect.

Bad Foley is Foley that sounds thin, or it just doesn't mix in right with production. Foley needs to fit in [naturally]. The problem that nobody tells you about is when you put Foley into the mix and pull the levels down, it can't sound ticky and thin. The idea is to capture things (generally speaking) realistic and weighty. Being aware of your acoustics is important. Take for instance capturing water. You can't just use a bathtub because it's reflective. There's a water pit on a traditional Foley stage. You can make it sound like a bathtub, or you can make it sound like the ocean. How it's constructed can affect your recording greatly . . . If you have no baseline to jump off of, you're shooting in the dark and don't know how it's actually translating on a mix stage. You don't know what you don't know.

If you can, sit on a stage with somebody – with the Foley artist, paying attention to the prop they're using to create a sound; or with a mixer, checking out the tools they're using, watching how judiciously their choices are so that the sound that's being recorded doesn't sound too thin, too sharp, too harsh, too distorted, not distorted enough, or not big enough. You have to learn the rules in order to break the rules . . .

If I was just starting out and I wanted to just jump in (without exterior experience), I would just start with where you live and a Pro Tools rig. A house is going to have natural ambiences, and you're going to have to learn how to move mics to get the right sounds. The problem with a house is bleed from outside noises, but jumping into building a stage is expensive, and there are no guarantees. You can cut hand pats and certain things from libraries, but there's always going to be feet, cloth, or props (like paper movement) that are very specific to the action on the screen, and are not cuttable.

The theory of "if you build it, they will come" doesn't exist anymore . . . I wouldn't open up a Foley studio until you've had several years under your belt of experience doing Foley, getting feedback (however you can get it), and networking with sound supervisors, re-recording mixers, and sound design folks. I would always be asking the dub stage, "How are things sounding? How are things translating?" If the feedback is good, like, "we're using it, it's working, it's playing well," and you've networked enough to know that you can get shows, then I would pull the trigger and build a studio.

Finding Opportunities in Post

Studios (or post facilities) were the primary route to build a post-production sound career in the 20th century. Similar to music studios, it required starting at the bottom in a position such as an intern, runner, or production assistant (PA). This could lead to opportunities such as assistant, mix tech, or learning other sound roles. Kate Finan, supervising sound editor and co-owner of Boom Box Post, shares her experience starting from the bottom:

> I was a secretary at my first job, and I got promoted within one month because I used to come in early and I would stay late every day. I would ask every single day, "Who can I observe today? While I'm eating my lunch, can I sit in on a mix?" When I got told [at the end of a shift], "Go home," I would say, "No, I think I'm going to stay and I'm going to check all of your cables for you."

This pathway to a career still exists and can be an opportunity to learn industry norms and professional etiquette. Mehrnaz Mohabati says of interns:

> An intern is learning how to be organized, punctual, and how to have good manners with clients, which is really important. I had an intern that I had to ask them to leave my room [during a session] because they laughed inappropriately. Or, I had an intern that I asked to print [copies] . . . I'm so glad I checked it. I saw they missed four pages . . . If you can't [make copies correctly for ADR], you can't do this job.

This pathway (learning from others in-person) is still important for passing along certain crafts that cannot be gained through self-teaching. This especially applies to learning skills like Foley and ADR. Foley mixer Jeff Gross shares:

> You can only go so far learning and recording Foley by yourself. You're in a bubble, and that bubble is so small compared to what the bigger world actually is. There is a baseline of understanding what is acceptable and what is not. In a perfect world, you try to at least hear and see what people who have been doing it [a long time] are doing because you're not necessarily going to know what sounds right. The worst thing is if you do a show (and send it to the mix stage), and then the mixers say, "What's [wrong with] this? This doesn't sound [right]."

There can be a ceiling to opportunities at a studio, and may require going freelance for career growth. Sal Ojeda explains:

> I didn't leave my day job (which was at a post house) until I knew that I had enough clients to keep me busy. Even when I did go, I went part-time

[with the post house]. At the beginning, they were my main client. I wasn't staff anymore, and I had to give up all the benefits, but I knew that was the move that I had to make.

In the early 21st century, the post industry had explosive growth of smaller project studios and home studios. This shift partly came about when expensive videotape machines were no longer required to receive materials or deliver mixes. Aline Bruijns says of the modern industry in the Netherlands:

> A lot of freelancers can work from home. They don't need all the rooms anymore – as long as there's a place where they can have viewings with a director which is better located, or has a nicer feel or look to it, than someone's home studio. It's easier for people to work from home, and it's more accepted as well . . . The bigger post-production facilities in the Netherlands get the larger budget productions . . . There are three or four larger production studios. But at the same time, when I started out there were more, and now it's scaling down.

This trend can be seen worldwide, which has made it difficult for emerging professionals to find in-person mentoring, entry-level jobs (including internships), and to learn niche roles like ADR or Foley. Learning and networking opportunities have shifted online. Aline Bruijns explains:

> As much as I would love to help people, I cannot take people into my house. SoundGirls, AES, MPSE – all of those kinds of organizations open up the world a little bit more because we're all situated in our own little cave . . . You cannot walk into someone's office or someone's studio and say, "Hey, what are you doing?"

At the same time, this industry shift has removed the barrier of having to work up the chain at a studio. An emerging professional can take the initiative to learn skills (through courses or online resources) and find projects to practice on their own. Karol Urban advises:

> I find the best way to get to the position of being a mixer is to simply be a mixer. Go find a student film to work on. You don't need someone to give you permission to be a mixer. Be a mixer. [The] people that you're working with will organically become better at what they do, and they will bring you along because you helped formulate what they create. Your filmmakers will take you with them. Your picture editors will continue to come back to you. As that happens, you will be more experienced, and you'll be more desirable to different companies. That's how I found myself in the position of being a mixer.
>
> (qtd. in Gaston-Bird)

The Value of Networking in Post: Shaun Farley

In 2011, Shaun Farley was a sound designer, editor, and re-recording mixer in Washington, DC (USA) while also running the popular industry website *Designing Sound* (with Jack Menhorn and Varun Nair). Shaun attributes the website as to how he met professionals who worked for Skywalker Sound, a worldwide-known sound facility in the San Francisco Bay Area (USA). Shaun says:

> When I started talking with all of these people, I never actually thought I had any chance of working there. So, my conversations weren't colored by a desire to work there. I was just happy to be talking craft with them, and to find I approached things in similar ways.

A couple of years later, Shaun moved to the Bay Area. To prepare, Shaun had been networking and developing relationships there for years, including with his connections at Skywalker Sound. Shaun explains:

> On my own, I spent a lot of my first six months [in the Bay Area] going to every networking event I could find. I'd run into a lot of other professionals repeatedly, and would start introducing people to others I had met . . . People at the networking events were visibly confused by how I knew so many people. That made an impression, as did helping them find the kind of connections they needed. The ability to help people (even when they didn't need my specific skills) made an impression, and I did end up getting some work out of it. Overall though, it was just a matter of surviving until some of my contacts at Skywalker Sound were ready to pull me onto a project. I still had to prove myself when I went in there, but I think it's safe to say I've done so at this point.
>
> Network your butt off, because getting work is still very much a question of *who* you know. When I say network, please understand that I'm saying you should be building relationships. Collecting someone's business card and not speaking with them for two years is not effective networking. Yes, the skills are important (editing, mixing, storytelling). You have to have those. The biggest obstacle you're going to run up against, though, is everyone else who also wants the few jobs that are available.

Tips for Finding Work in Post-Production

- Even a handful of credits (including student work or short films) will make an applicant stand out against those who are curious/interested in post but have no credits.
- Seek out shorter content versus longer to start. For example, short films are a great way to build up credits and experience without the commitment and workload of a feature film.
- Keep your credits up to date on iMDB.com (do not expect others to add them). IMDb.com is a searchable database of entertainment programs, cast, and crew and can act as a public credit list of your work.
- Be open-minded to content. Any post-production sound experience can lead to other opportunities or can turn into a career path on its own (such as dubbing, promo or trailers, reality television, etc.).

Sound designer Shaun Farley suggests:

> Do whatever you can to find projects to work on with people: 48-hour film festival, student projects (especially if you've made connections with film students who are still in school), and look for local filmmaker meetups. If you haven't already, get yourself a DAW and a recorder (whatever you can afford). Pro Tools is still the industry standard for film, but that's not the case for games. You can use Reaper for free on non-commercial projects, and it's cheap for an actual license. You can definitely start with that for film and video if that's all you can afford.

> Handheld recorders are cheap now, too. Get something that has an XLR input so your next step can be to buy a decent mic. Microphones don't lose value or become obsolete in the way recorders do. So, make the leap to a decent mic before you splurge for a top-of-the-line recorder. This will help you build your personal sound library. It's more economical to start building your own than it is to buy a big "general" library that you'll get sick of. If you find you need a sound you can't record or create, then there are boutique sound libraries that can fill those needs without costing you an arm and a leg.

Partnering with other professionals can help land opportunities. Luisa Pinzon, who was the sound supervisor for *Shark Tank Mexico*, says:

> I've seen that, at least here [in Colombia], they prefer to hire a team when the team is already formed rather than hire three people differently that will need to team up and work together. Maybe there are some people that are better than them, but the team already knows each other, and they

know how to work together (workflow and everything). It saves a lot of time for the company. With *Shark Tank Mexico*, they were just like, "This is what needs to get done. You guys figure it out." It was great, and they were really happy.

In post, demos are typically not necessary and rarely asked for. A "good" mix is not necessarily demonstrating clean and pristine audio (because of the lack of control over deadlines or production audio quality). One exception to this is sound design, where an original sound design demo could help show an emerging professional's ability level and creativity.

Post is a relationship-driven industry, so networking and building relationships are crucial for finding opportunities. Karol Urban explains why this is the case in Los Angeles:

> Because there are so many qualified people, I really do think a feeling of comfortability and artistic connection is of paramount importance. Do they feel like they can chat with you? Can they say, "I hate it," or "I love it," or "Do you get me?" I feel like that's extremely important – possibly more important than your referrals and experience.
>
> In LA, I think people are supportive during job changes – not just because they personally wish you well, but also partially because of the gig economy. Many of us in LA move around all the time. When a friend works with a new company, they may have the opportunity to offer you a new door or window to a new opportunity.
>
> ("Interview")

For example, Kelly Kramarik met Matt Waters (re-recording mixer, *Game of Thrones*) as a student and stayed in touch with him. Through a connection of Matt's, Kelly received a recommendation for an audio editor position at Starz, a television network based where she lived (in Denver, Colorado, USA). Kelly landed the job, saying she heard from the manager who interviewed her, "You're young, but you seem like you're really qualified, you can learn quickly, and you also are the only person that had a recommendation from the *Game of Thrones* mixers."

Music Editing and Score Mixing for Sound for Picture

Music editing and score mixing are career paths that overlap between music production and post-production sound. "Music Editor" (as a job title in sound for picture) is a high-end position, often only found on major productions and projects. A music editor is the liaison for music between the composer and the mix stage, and also between the filmmaker/director and the composer. Music editors can be tasked with testing different music tracks (called "temping") or editing a music track to a scene. The job involves interacting with a wide variety of people and departments (composers, directors, producers, picture editors, music publishers, sound supervisors, re-recording mixers, and more). Music editing is a role where it can be highly beneficial to learn the craft under an established professional.

Score mixing is recording and mixing composed music (a score) for sound for picture. Score mixers usually do not interact with the post-production team and instead are part of the music department (composer and/or music editor).

Andrew King, A1 Mixer for Sports Broadcasting

Andrew King was drawn to sound as a career after watching the opening ceremony of the 2008 Beijing Olympics. "I was just blown away by it. I was like, 'That's amazing. I have to do that.' That's what set me on the professional path," he says. Andrew learned live sound while in high school in Colorado (USA) and worked in live sound briefly after graduating. He looked for a degree or vocational program in audio (other than music production) but could only find two. He attended one – the New England School of Communications (NESCom) at Husson University in Bangor, Maine (USA). Andrew says of his time at NESCom:

> It was really a live sound program, but they had a TV truck. I spent a lot of time in that truck. My junior year . . . they basically handed the keys of the TV truck to the students and told us to go do some shows. They'd never done that before, so it was a little scary for them because there were no teachers on-site at all. They would drop off the TV truck and leave.
>
> It allowed for a tremendous amount of experimentation and trying to figure [things] out. How do we do this? What kind of mics do we use, and where do we put them? . . . We had a little bit of guidance and whatever you can find on the internet, but "how to mic a basketball game" is not really a popular topic on the internet . . . We had a place where we could go check out equipment, so I'd go a couple of weeks in advance and reserve all ten shotgun mics they had for the game. I think we produced 15 games that semester – basketball and soccer, mostly. We did a volleyball game, softball, and some football games. I just got hooked on it. I knew it is what I wanted to do.

After graduation, Andrew moved back to Colorado and worked for an IT company that was flexible with him taking time off to also do audio work. Andrew explains:

> That summer (2013), a regional sports TV network here contacted me to see if I wanted to start mixing some shows. I said, "I don't think I'm ready for that. I think I'd better A2 for a while and figure out what's going on." . . . When I left [school], there were a lot of things I didn't know about the broadcast side, so it was quite a learning curve. The first time I stepped on a TV truck in real life, it was like, whoa, there is a lot more going on here than I knew. I wanted to shadow because I wanted to understand it. I'm the type of person that I need to understand how things work. I'm not satisfied with just plugging it in.

Andrew was offered work as an A2 after spending time shadowing, and he says of the experience:

> Shadowing the A2s and A1s really helped a lot. Then, they started putting me on as an A2. Then, the NBA season started, and they started letting me A2 for the Nuggets [NBA basketball], and then the Avalanche [NHL Hockey]. They just put me in the rotation. Over time, I started getting so many freelance shows that I wasn't getting any time off from work, and I finally left my IT job to free-lance full-time.

Andrew landed his first opportunity to work as an A1 (mixer) because of another A1 offering to train him:

> There were two high school basketball games in two days. An A1 could do the first game but not the second one. He said, "What if I come there for the first day and you guys bring Andrew up? I'll show him how to set up the truck and then he can do it the next day." That was super generous of him, and I haven't really found many people willing to do that kind of stuff. They don't want to deal with people shadowing them, which I understand. Part of it is just the time constraint. We have so little time to set up. There's no time to show anybody anything. I learned a lot from that.

Andrew says of his schedule on a show:

> Let's say it's a 7 [pm] basketball tip-off. Typically, we'd get there at 1 pm, so six hours early. That gives us three hours or so to set up, an hour break for lunch, then an hour or so of pre-production.
>
> What makes it complicated is the number of things you have to accomplish in a very small amount of time. When I walk onto the

truck, there's no stopping from when I start until I get a lunch break (hopefully) until we start pre-production . . . If it's what we call a "set, shoot, strike," the truck arrives that day, you set it up, do the show, and then you strike the truck and leave. The A1 position on sports TV broadcast is so stressful and a time crunch.

Since broadcast work is freelance, Andrew's workload varies from month to month:

I'm still fairly new in the industry. I'll have months where I'll work five or six shows in one month, and then other months where I don't work any shows. If you're one of the higher-up people in your market, you can work a lot more than that . . . It's getting in contact with the local crews and getting on their lists, getting them to believe that you know what you're doing, and then they'll start putting you on some shows.

Here in Denver, we have a company called Crewing Source. They've been contracted by [sports networks]. They post on [a website], and we can mark whether or not we're available for that day . . . They'll choose the crew unless the network specifically requests certain people.

Andrew continues to work other gigs, including live sound, to fill his schedule:

I like doing conferences, and I do concerts on cruise ships. We call them charter music cruises. Basically, a production company will charter an entire cruise ship and we take over the whole thing. We'll set up maybe five stages on the cruise and do concerts all week long. We've done as many as 100 concerts in one week on a cruise.

The opportunity to work on cruise ships came from another alumnus of NESCom, who Andrew met as a student:

The guy that crews production [for the cruise ship work] said, "I want to help out the students, so I'm going to fly your senior level live sound class down here and do a cruise with us." That was the Blake Shelton cruise. Six or seven of us went down for that. At the end, he did a meeting with us and said, "You guys are awesome. If you ever need work, let me know." After I graduated, I contacted him, and I think now I've done almost 30 cruises for him.

In pursuit of his broadcast work, Andrew took it upon himself to learn one important skill he was lacking. Most TV trucks use Calrec audio mixing consoles, but he had not had an opportunity to work on one. Andrew signed up for a two-day training course in Phoenix, Arizona (USA), but when he got there, the first day was full and he could not participate. Andrew tells the story:

> I was really bummed, but I thought, "Well, I'll go down there the second day and check it out anyways." I went, and there were only two other guys there: the guy from Calrec, and this other guy who I had no idea who he was. It was Fred Aldous, one of the top mixers at FOX [Sports]. He has over 20 Emmy Awards. Fred was so impressed that I went down there just to learn the console that he gave me his email address and said, "I'm going to be in Denver mixing a Broncos game in a couple of weeks. Send me an email, and I'll get you in and you can come hang out with me."

> I ran into Fred a couple of times after, and then I saw him at the Advanced Audio Symposium [conference]. At the last session of the day, they were doing a panel discussion about education and how to get people to come up in the industry. Fred pointed me out [of the audience] in the middle of the panel and said, "This is Andrew. He's an up-and-coming A1," and went on with the conversation. But then, later on, Fred decided to give me the mic and said, "Why don't you tell us about your experience as a new A1?" I was talking in front of these top A1s, and Karl Malone [Director of Sound Design, NBC Sports and Olympics] was on that panel. At the end of the discussion, Karl walked right up to me, gave me his card, and said "Call me on Monday."

Through that call, Andrew landed a job working at the 2018 Winter Olympics in Korea as an A2 for ski jumping. "It was awesome. I sat in the booth right next to the announcers, so I had a perfect view of the venue," says Andrew.

Andrew says of attending conferences and training events:

> It takes my personal money to go do those things. Like, I don't have a company paying me to go to NAB. I have to pay to go. So, it's just saying, you know what? I'm going to go out there, and I might not meet anybody, or I might meet somebody really important. I'm just going to go for it and see what happens.

Andrew says what he enjoys about his work:

> The fun part to me is just doing the mix, like, doing an NBA game and you hit the net mic at just the right time, and you get a perfect

swish sound . . . Then to actually see what I did on other screens after the show. Like, I went to [dinner], I'm sitting there eating, and on *ESPN SportsCenter Top 10* is the show I just did.

Andrew looks back at his education (and working on the TV truck at school) as a one-of-a-kind chance to learn in a way not possible in the field. Andrew shares:

Today, if we get there four hours before we start producing a show, there's no time to experiment. It's like, plug it in as fast as you can. Does it work? Okay, let's go. That time to experiment was probably the biggest thing that experience gave me.

References

Bailey, Pete. "Interview with Production Sound Mixer Pete Bailey." *Rycote*. rycote.com/microphone-windshield-shock-mount/interview-with-production-sound-mixer-pete-bailey/. Accessed 2 Aug. 2021.

Bruijns, Aline. Personal interview. 21 June 2021.

Farley, Shaun. Email interview. Conducted by April Tucker, 15 July 2020.

Ferguson, Eric. Personal interview. 2 Dec. 2020.

Finan, Kate, panelist. "SoundGirls Career Paths in Film & TV Panel at Sony Studios." *YouTube*, uploaded by SoundGirls, 26 Oct. 2018, youtu.be/Y_g3drC2yyA.

Garrett, G. John. "Sound Utility Jenny Elsinger." *CAS Quarterly*, Summer 2020, pp. 12–16.

Gaston-Bird, Leslie. *Women in Audio*. Routledge, 2019, p. 120. doi.org/10.4324/9780429455940.

Gross, Jeff. Personal interview. 30 Dec. 2021.

Kennedy, Camille. Personal interview. 21 Feb. 2021.

King, Andrew. Personal interview. 15 Dec. 2020.

Kramarik, Kelly. Personal interview. 18 June 2020.

Lee, Ai-Ling. "Mixing and Sound Design in Surround Sound with Ai-Ling Lee and Paula Fairfield." *YouTube*, uploaded by SoundGirls. 15 June 2021. youtu.be/Pdv2G2gbiOM.

Leong, Mabel. Email interview. Conducted by April Tucker, 29–30 Sept. 2021.

Mazzucco, Tiziana. Email interview. Conducted by April Tucker, 30 Sept. 2021.

Mierzwa, Patrushkha. *Behind the Sound Cart: A Veteran's Guide to Sound on the Set*, Ulano Sound Services, Inc., 2021.

Mohabati, Mehrnaz. Personal interview. 29 Jan. 2021.

Motion Picture Association, "2021 Theme Report." March 2022. www.motionpictures.org/wp-content/uploads/2022/03/MPA-2021-THEME-Report-FINAL.pdf.

Ojeda, Sal. Personal interview. 27 Sept. 2020.

Palumbo, Jeri. "Broadcast Engineer." *SoundGirls*. soundgirls.org/broadcast-engineer/. Accessed 8 Aug. 2021.

Pinzon, Luisa. Personal interview. 10 Nov. 2020.

Urban, Karol. Personal interview. 17 July 2020.

Zolnai, Eliza. Personal interview. 27 Sept. 2021.

17
Spoken Word Audio

Spoken word audio is made up of podcasts, audiobooks, sports, news, talk radio, and more. Spoken word audio is a rapidly growing industry. In 2020, the global radio market was worth over US$100 billion, with Western Europe making up 50% of the market (Business Research Company). According to Broadcast Networks Europe, 47% of Europeans listened to radio daily, with radio reaching 84% of European citizens; The average listener in Germany tuned in for about four hours each day. Buzzsprout.com, a popular podcast hosting service, had 93 million downloads and 111,000 active podcasts in February 2022.

While radio has existed for over 100 years, the modern audio industry is seeing many new opportunities and uses for spoken word – and healthy budgets. More than one interviewee referred to spoken word audio (and especially podcasting) as "The Wild West." Podcast editor Tom Kelly explains:

> There's no structure in podcasting at all. It's very frustrating, but also there's a lot of opportunity to really own that space . . . my dad always tells me that in the California Gold Rush [starting in 1848], a bunch of people went to California to try to get wealthy. A lot of people got wealthy, but most of them were the ones that were selling shovels [and other supplies] to the gold miners. My thing is like, I'm selling these people shovels, and that's such a good place to be.

Jeff Dudzick, Content Creation Manager for Audible Studios in Newark, New Jersey (USA), says of spoken word as a field:

> Spoken word is the umbrella. It's a lot more modern than "audiobooks" or your grandma's "books on tape," and most of the stuff we're talking about now is not audiobooks. Audiobooks are, "There's a printed book. We turn it into audio." There's not much wiggle room around that. There's original scripted content but the script is never published as a book. Then, there's news, which is non-scripted content. There's live content. For us, we're making our stuff into premium audio storytelling . . .

DOI: 10.4324/9781003050346-20

We always say, "nobody went to school for audiobooks." Everyone that I work with (in-house) went to school to record music or they come from that [generalist] background . . . However, they found their way here.

Spoken word audio has traditional audio roles – from live engineering and mixing to post-production editing, mixing, and sound design. Having a variety of audio skills can be beneficial, as well. Jonathan Hubel, who has worked in radio and podcasting, says:

Companies are trying to streamline things as much as possible, and often aren't looking to hire a recording engineer, an editor, and a mixer. They want one person who can do it all. So, the more you can do, the better off you'll be.

One advantage of working in spoken word audio is the skills are transferrable to other types of work in the discipline. Jeff Dudzick explains:

Plenty of engineers and editors are working on spoken word, audiobooks, [and] podcasts . . . Cast a wide net. It just takes an open mind, some creative thinking, and looking around . . . I didn't see those things until I was in it.

Some gigs will be exclusively related to audio, but opportunities exist for those who are willing to take on other aspects of the work (such as marketing, producing, technical support, or consulting). Podcast editor/consultant Britany Felix explains:

The entire time I have been working as a professional editor (a little over four years now), I have had maybe two (and that might be generous) clients who have strictly wanted me to edit audio and send the file back and they do everything else . . . About 80% of the people who come to me want audio editing in addition to show notes, uploading it to their hosting service, or some other add-on services.

Work opportunities vary from being an employee at an established company to operating your own business and seeking your own clients. Radio and podcast producer Sarah Stacey says the crew size depends on the job:

In terms of the podcasts, I'm usually coordinating with one person. With the [radio] promos that I do, it's maybe two or three people at different stages . . . I prefer teamwork when it's a live broadcast situation because you get that adrenaline rush of people coming together and trying to make this program happen . . . When it comes to pre-recorded stuff, I like working alone . . . I like being able to know what I have to do and when I have to have it done, and I just do it.

Jeanne Montalvo Lucar, who has worked as a radio producer, engineer, and podcast editor, found her background in music to be an asset when working in spoken word audio:

> I have a knack for pacing with the voice. Nobody taught me that pacing. It comes from my music background because I think of it as a beat. When you speak, you're speaking with a rhythm. That came from me being able to listen to music . . . Doing an audiobook or spoken word project might not be what you have envisioned for your life, but it's work, it's valuable, and it can be fun.

Some job listings will be advertised publicly and require a formal hiring process, such as sending a resume (or CV) and doing an interview. Other opportunities will come by word of mouth, social media, or other informal paths of seeking out work. Avery Moore Kloss suggests that work exists at many different experience and price levels:

> You can always find someone on Fiverr to edit your podcast for 50 bucks for an hour-long episode. I can't do that. Just the attention to detail I give something just doesn't warrant that. But if that's what you need, it's out there. The different tiers of service are there – it just depends on what you need and what your project calls for.

One way to find opportunities is to target other companies that produce spoken word audio. Britany Felix explains:

> There are companies where they have 5–20 editors working for them under the umbrella of their company . . . It can be a great way to get started . . . but all of that work you're doing under them is technically their work. You can't always use it for your own portfolio. You're still working for somebody else. You're dependent upon somebody else to give you work and keep the money coming in. So, say a pandemic hits or there's a huge crash in the economy and they need to scale back their expenses – you might be let go and you have no control over that.

Podcast editor Tom Kelly says subcontracted work tends to be audio-only tasks:

> If you work with another producer that handles uploading, communicating, and interfacing with the client, it allows you to just do the edit, and really get comfortable with the specific needs of audio for podcasting. It's wildly different than music and a lot of other stuff. Subcontracted work would be a good place to start because they can also give you volume. They might pay you a little less, but it's consistent. I have one person I still work for, and they give me like seven shows a week.

Jeff Dudzick describes the process of finding your own work:

> You could seek out people trying to [create content] . . . [Quality is] what eludes a lot of people. Or, you somehow find talent – [find] someone who's doing a show but the technical stuff eludes them . . .

> Surely there are places out there (like churches, schools, local governments) that want to make content. You can advertise, or you can try to connect with them. Find people who are maybe struggling (and they don't even know it), or you see they could be at the next level really easily with your help . . . Offer your services. "With just a little bit of effort and knowledge, we can make this much better, and for $200 I could do that." You do that enough times, someone will bite, maybe. You become their go-to person. There are a lot of ways to get a foot in the door. You have to be imaginative.

Faith-Based Audio as a Niche: Jonathan Hubel

Jonathan Hubel (Richardson, Texas, USA) is an audio engineer (for spoken word) and podcast editor who has also worked as a board op for live radio broadcast. Jonathan discusses faith-based spoken word as a niche.

> Christian radio stations are an excellent opportunity if you're more interested in the broadcasting side of things, whether on-air talent, programming/scheduling, or editing and production. There are also other small organizations and ministries that will produce syndicated radio shows or podcasts, but it can be difficult to locate these types of positions. Is there a demand for these types of professionals? To some extent, yes, but because [positions] are limited, you're likely to face a lot of competition for those spots and have to really stand out . . .

> There are quite a few faith-based podcasts . . . I think the majority of them are DIY, but there certainly are some that are professionally produced. It's definitely a niche that someone could specialize in, but I think it would take someone who's very driven, charismatic, marketing-savvy, and is confident enough to take on a big challenge. A lot of faith-based podcasts are being produced by nonprofits, churches, and people who have little to no budget, so it's going to take a lot of work to make it financially viable, especially as a freelancer. It's definitely a calling based on shared beliefs, not just a career and money-maker.

Journalism and Spoken Word

Spoken word audio is a unique field that blends the art of journalism and story-telling with the art of audio engineering. There is a market for producers – those who can craft and tell a story; and also for audio engineers, who capture and enhance a story with sound. Historically, these two roles generally did not overlap, but the modern audio industry sees the two intertwined at times. Audio engineer and radio producer Jeanne Montalvo Lucar describes the two roles:

> "Producer" means a lot of things in different places because a producer can also be considered a reporter. The producer is the person who is put-ting together the actual segment but not literally assembling it. They're writing scripts and they're usually choosing the music. If it's an interview with two people, they're cutting it down to the time length that they want it. They're selecting the quotes of the interviewee. If it's a narrated piece, that's scripted . . . Engineering is strictly mixing and making it sound pretty. Sometimes it can be scoring and sound design . . . Most indie DIY are all doing everything – a one-person band. They're called the podcast editor.

Learning both elements can be a challenge for those seeking education specific to spoken word audio. Avery Moore Kloss, who has a degree in journalism, says of merging the two practices, "When you come from the journalism side, you're there to be a storyteller . . . I'm a storyteller who adopted audio brain. I struggled to find [audio] education that didn't require going back to school for audio engineering."

Jeanne Montalvo Lucar came into spoken word audio from music production, and learned producing on the job (at NPR's *Latino USA*). Jeanne says of having engineering and producing skills:

> The fact that I can do a lot is usually advantageous to people. They think, "If I don't have enough money to have two producers and an editor, at least Jeanne thinks editorially so she can help edit as she's assembling and producing this piece." It's an asset [to do both] because it's very rare to find someone who can engineer and produce . . . They don't normally overlap. Personally, it's a *lot* of work to do both, so I don't tend to do both. If I'm producing, I like to hand it off to have somebody else mix it. If I am mixing, I don't necessarily want to be producing. When it can be two jobs, it's good.

Sarah Stacey earned a master's degree in radio production from Bournemouth University (UK), which she found prepared her for both areas of the field. Sarah explains:

> I wasn't even sure at that stage whether I was more interested in the tech-nical side or the actual content creation aspect. So, it's definitely allowed

me to try a lot of different things and have a varied skillset as a result. [My degree] has definitely helped when it comes to taking on different bits of freelance work because I think the more that you're able to do, the more you can get asked to do.

Sarah says of her varied work (in radio and podcasts) since going freelance:

Most of what I'm doing is editing podcasts for people, freelancing with the BBC, as well [as work] from overseas. I've been making promos and radio trails, so there's a lot of technical work involved there. Then, there's also the script-writing side of that, so a lot of writing involved. My most recent job in a radio station was a lot of journalism because a lot of what I was doing was writing copy for podcasts and doing a lot of social media stuff in addition to editing audio . . .

I think it would have been possible to work in radio if I hadn't done a master's in it, but I would do it again (given the choice) . . . [College radio] didn't teach me quite as much as actually learning from former BBC producers and [professionals] that taught me. It just gives you a chance to learn about different types of program-making as well, because college radio is pretty limited to music shows.

Audio Storytelling: Avery Moore Kloss

Avery Moore Kloss is an audio storyteller, podcast producer, and award-winning radio documentarian. Avery teaches Audio Storytelling at Wilfrid Laurier University (Brantford, Ontario, Canada).

The base of everything I do is always the storytelling and the writing. Audio happens to be my medium. I can see – where should we start, and what's the middle? What should the end be, and how do we keep people engaged as we go? How do we answer the question the listener just thought of when you said something? How do we make sure that's the next thing you say? People can definitely pay me for audio editing, but I don't do a ton of those jobs. The more jobs that I do, my clients want my storyteller brain, and the end product is a podcast episode . . . [My clients] don't know how to make a podcast or put the story together. [That's] what they're really paying me the most for . . .

Telling a really good story in audio takes more time than just turning a mic on, talking, and throwing it on [a free hosting platform] . . . I know people who would say, "You spent three months on one episode?!" I love that. I love that ability to think about it . . . I love going over it again and over again . . . I feel prouder of the work I do when I have the chance to sit with audio, edit it, re-edit it, and catch something I didn't catch the last time. Then, I listen to it with my eyes closed and tap my foot and make sure it fits the cadence I want. I just like that.

Necessary Skills

Even though it's audio, the base of everything that I do is in creative writing, like, how to tell a story with words in a way that captures people's attention . . . [Take] English classes, even if it's fiction writing. Just get writing [experience]. Write things down. Figure out how to write in a way that hooks people in from the beginning and keeps them. As much as I'm an audio storyteller like that, I'm a writer first.

[If writing is a challenge], immerse yourself in the audio storytelling that's already happening. I think there's so much to learn from stuff like *Snap Judgment*, *This American Life*, and *Radiolab* . . . Listening to the different ways people tell stories in audio is super valuable . . . When I teach a [class or] workshop, I always start with audio because I think the best way to learn audio storytelling is just to listen to what people are doing and feel it.

[My students] are learning the basics of audio editing . . . I want them to know where a mic should go, how you get the best sound, what kind of sound should you be looking for . . . I want them to know how to interview when their questions are on the air. I want them to know how to edit for all these different things. Like, what's the difference between editing for an interview and editing just sounds? What's the difference between that and editing for a reported story where it's narration plus clips? I start from a storytelling place, but it's really important to me that they understand how the story that's in their brain translates into a digital audio workstation.

Podcasting

In podcasting, the minority of podcasts are making the majority of the money. According to Buzzsprout.com, in February 2022, the top 10% of podcasts (out of about 111,000 active podcasts) only received 333 downloads or more in the first week.

Podcast editor Tom Kelly explains:

> It is tragic how low the numbers are. The top .001% holds 99% of the downloads, but there are still people making crazy money with a couple hundred or couple thousand people listening. It doesn't take much to do very well.

The podcast market exists within professional radio, amateurs/hobbyists, and even businesses seeking to use a podcast as a marketing tool. Avery Moore Kloss explains:

> What's happening to podcasting now is what happened with blogging 15 years ago . . . People started blogs because of the low barrier. They had a story to tell, and it was pretty easy. You could get a Blogger account and you could just start blogging . . . Then content became king, and every business everywhere has a blog. The same thing is happening with podcasting. Every business everywhere now is like, should we have a podcast?

Podcasting has a large independent market where it is possible to land clients and perform work without ever meeting (or possibly even speaking to) a client. Tom Kelly (owner of Clean Cut Audio) explains his client interaction:

> I've never met any of my clients in person. Most of them we've never even talked on the phone. It's all been email or Facebook Messenger . . . The people I've subcontracted for –we've met up at local events. Since the medium is so non-personal (like, it's all emails) – if I'm a really good dude but my edits suck, I don't know if being a good dude matters because that's not even displayed in my emails . . . I'm not sure (especially just on the web) how far your personality gets you.

A **podcast editor** edits and mixes a podcast. In the independent market, podcast editors can carry many roles outside of audio. "With podcasts, there's this idea that an editor can do anything. A lot of people fill [extra] roles and still just call themselves an editor, so it's forced this standard of terminology," says Tom Kelly.

An indie podcast editor may be able to niche as their career and clients grow. Tom Kelly explains:

> You have to know more starting out. Over time, as your clients get a little higher profile (hopefully) and a little more serious, maybe they'll have a production team, and you can afford to play a smaller role . . . There would probably be roles like an assistant editor when there are big teams, but not so much in the amateur space.

Podcast quality varies widely. Because the industry is a mixture of professional media creators and amateurs, it may fall on the podcast editor to educate clients about audio quality, what to buy, and technical setups. Avery Moore Kloss explains:

> Original podcast audiences didn't care too much about the quality of audio . . . Now, as people with money get into podcasting – they've got the nice mic, they've got the good setup, and they've got the audio engineer . . . the average listener is now getting used to really good audio. As the ocean gets bigger, people want the boats that are more put together. But I still think if you're doing something novel, you can still break through without the really good audio, also.

The basic skills needed to be a podcast editor can be learned online. Britany Felix explains:

> You can easily come [into this work] and you become a student, and you study. You honestly just look and see what other editors are recommending. Just about every podcast editor that I work with recommends one particular hosting service because they're the best. So, you just adopt that as your recommendation, you get in there (especially if you have your own show) and you use that one. Now you know the backend, and you can repeat the process for somebody else. Certain things will come up – there'll be technical issues that you won't know how to handle because they've never happened before, but we've all been there . . . You just figure out how to fix it for your client as you're doing it, and you're also being paid to learn how to fix things that you can then use next time.

In podcasting, a demo is important to show potential clients your skill level. Sarah Stacey shares:

> A lot of job listings that I've seen, especially when it's promo stuff or podcasts, will ask you for a demo. So, if you can have something (even if it's college radio) – that is really important . . . [Personal podcasts on a professional demo] could be taken seriously. The main thing is just to show that you're really passionate about what you're doing. That really counts for a lot because it's clear that you really enjoy what you're working on.

Starting your own podcast can be a great way to learn the ins and outs of the process, and also to have as a demo/promotional tool. Tom Kelly explains:

> I get questions all the time like "Which file host should I go with?" Two years ago, I wouldn't have even known what they were talking about. By doing my own podcast, I've learned what all this stuff is, and experiment without any real consequences from a client. Especially for someone that wants to get into it, start your own podcast because there's so much [little] stuff that you don't know that you have to know. Clients will ask you about anything and everything.

Having Your Own Podcast: Britany Felix

Britany Felix has worked on over 2,000 podcast episodes including 400 episodes of her own podcasts. Her business, Podcasting for Coaches, specializes in helping women entrepreneurs, coaches, and consultants to use podcasting as a marketing tool.

> I would absolutely recommend that your podcast be professional in nature because then it serves as one example of your work, which is great in the beginning . . . the podcast serves as your portfolio. It becomes a selling tool for your services. In terms of my podcast, I'm talking about the topics that my ideal clients need to know in order to have a successful podcast. So, in every single one of those episodes, I'm covering a different question that I get asked by my clients all the time. In a Facebook group, somebody asks, "Can you guys help me?" I swoop in and say, "I just so happen to have this podcast episode."

> It shouldn't necessarily be a hobbycast. You can do that for fun for more experience . . . But if you're going to start off with just one, it should be somehow connected to your business because it needs to work for you. Also, the show notes on your website will help with SEO [search engine optimization]. It gives you built-in SEO, basically.[1]

> Now, you'll obviously have less content and some of the information could become outdated, because as with everything in the audio industry, podcasting is evolving constantly . . . it would just have to be something that absolutely could be evergreen [won't become outdated]. If it's a limited-run series, it shouldn't be 10 random episodes, but a purpose to those 10 episodes . . . It doesn't have to be this long, drawn-out thing. A podcast can be anything you want it to be.

> If part of your skillset is that you can also do video editing, that is going to help because people don't only want audio. There are a lot of people who want to have a video component to their podcast, so they can take advantage of YouTube.

Podcasting can be a career path, or gigs taken in conjunction with other disciplines. "I always joke that podcasting pays for my music habit because that's what it is," says Jeanne Montalvo Lucar. Tom Kelly adds:

> The work itself is very easy to predict – when it's due and when you're going to get it (since people are usually on a weekly schedule). It's super easy as a side hustle because one client [might be] the same job every week. Rather than like, sound designing a five-minute film – when it's done, what's the next thing? You can land one podcast client and have that work for years.

However, editing audio for a podcast takes time. Britany Felix shares:

> Even if you're a very skilled and experienced audio editor, generally at best you can count on it taking one and a half to two times the length of the audio you're editing to get through it. If you're also doing a full-time job, and you're wanting to edit as a side hustle to make it worth your time, you're not going to have any time left in your week. So, you need to either work on one show for experience or work on enough to make it a full-time business. Anything in between, you're going to be stuck with not really making enough money to make it worth it, but also feeling like you have zero personal time left.

Audio Quality and Podcasts: Tom Kelly

Tom Kelly, owner of Clean Cut Audio, is a podcast editor based in Denver, Colorado (USA). Tom has edited over 1,500 podcast episodes.

> There are people at all levels of podcasting who don't understand quality. Some of the biggest podcasts in the world have terrible audio quality. It's like, why? Every resource is available to you. It's so wild.
>
> I'm competing against this mindset of "audio doesn't matter at all." There are people literally preaching, "Don't buy a microphone. It's not worth it. Don't invest in quality. Don't try. No one cares." I think that's changing pretty dramatically as people are dumping serious money into podcasting . . .
>
> I'm trying to push everyone I know [with an audio background working in podcasting] to be in more of an educational role. There's so much bad information and these podcast gurus who don't know audio are teaching these concepts . . . I would love to see more audio professionals jumping into forums, subreddits, and any group that they can.
>
> There are so many podcasts. The more the merrier. My whole thing is audio standards, so the more great people, the better it is for me trying to champion this cause . . . So, there's room for everyone to fill a unique role in this space.

Music mastering engineer Piper Payne does podcast work partly because of her personal interest in podcasts. Piper suggests you have to be mindful of your areas of expertise when doing it as a side hustle:

I get paid fairly well for editing and mastering podcasts, but I get paid well because I demand a specific rate . . . You also have to keep in mind a music engineer is not usually good at editing podcasts. By the same token, someone who has only worked in podcasts – you wouldn't pay them to mix your record.

Podcast editors vary in skill level and price. Avery Moore Kloss says:

The beauty of podcasting is that it's got a low barrier to entry. Do you need to be a pro to do it? No, you don't. Do I charge more per hour than someone who's just going to take out someone's "umms" and "ahhs" and send you it in the format that Apple Podcasts wants? Yeah, I do.

Tom Kelly adds:

You can be an editor tomorrow. You can have income tomorrow, which is not super common in most industries. Even live sound people [can edit] while they're in the van and on the road. As long as you have the files, you don't even need internet to work on it . . . There's room for everyone.

Professional vs. DIY: Jeanne Montalvo Lucar

Jeanne Montalvo Lucar is a Grammy-nominated audio engineer, radio producer, and podcast editor. She believes emerging professionals who are learning podcasting on their own (DIY) are not at a disadvantage over those who learn at a radio or podcast company.

If you are just DIY, putting [podcasts] up online, and you are in the *New York Times*, then obviously you're doing something right. I think it's a matter of how professional you can make your DIY sound, and your following. If the content is good, it doesn't matter if you're working for an established company or not. How many podcasts have started out like that?

If you're doing your homework to find out how to make your podcast sound the best that it can sound, then what's to say that I need to judge how you're getting there? If the end product sounds good, then that's fine. If someone DIY came to me, I would just want to hear their stuff . . . Now, there are certain things that are acquired over time, like cutting out too many breaths or having double breaths . . . Pacing is another one that not everybody gets . . .

I didn't take a podcasting class in college. I just started doing the work, and then eventually landed somewhere . . . So why wouldn't

> somebody who's DIY be able to make their way? I think the crux of
> that is getting that very first actual job . . . once you have the first one
> on your resume, then that leads to the next job. That's the hard part
> is getting that first one.

Opportunities in Podcasting

The podcast industry can be broken into different markets: hobby/special
interest, corporate/for-profit, and professional podcasts.

Professional podcasts are generally produced by production companies that
specialize in podcasts and possibly radio (including radio drama). These companies can be small or large (such as the BBC or Spotify) and may have audio-
only work opportunities. Kelly Kramarik says of her former job at The Podcast
Network, "I was just recording, editing, and mixing, and we had about 30 shows
on a regular schedule. I was just doing that constantly."

Hobby/special interest (hobbycast) are podcasts created for personal fulfillment versus for-profit. These are typically DIY (do it yourself) with self-taught
creators. Sarah Stacey says of hobbycasts:

> That's actually a really good way to start out and see where it takes you
> because you can afford to make mistakes. If that's the kind of podcast you're
> doing, you can have fun with it and just learn as you go . . . A lot of the
> people I know who just [create] podcasts as a hobby wouldn't generally be
> audio people (in terms of their careers). They're really focused on the pro-
> motion and the social media side of things.

Corporate/business podcasts are podcasts created by a business or organization
for education, entertainment, marketing purposes, or internal/business use.
Britany Felix says of for-profit podcasts:

> I target business owners, and so many other podcast editors are now targeting
> business owners. They're going after corporate clients. They're going after
> entrepreneurs because their podcast isn't just for fun. It is a marketing tool
> [for them], and people are always willing to invest in their business. If you're
> wanting to go work on a hobby podcast, that's going to dictate your rates
> because this is for fun for them. There isn't a return on their investment
> unless they get a sponsor, which is ridiculously difficult to do.

Avery Moore Kloss shares her variety of clients in corporate/business podcasting:

I have a tech company where they send me all their audio, I write a script for their co-hosts, and then I'm hands off. I do one for [Wilfrid Laurier] University where I ideate, write, produce, host, interview – I do everything. I have a corporate one where it's just interviews (and I'm the voiceless interviewer from their company) and I edit it, too. I had a grant project for a women's networking group in my local community. They wanted to do a podcast, so I coached them through hosting it themselves.

Corporate Podcasts: Fela Davis

Fela Davis is co-owner/co-founder of 23dB Productions (Fort Lee, New Jersey, USA). 23dB offers podcast production services to a variety of clients (including corporate).

> Usually [clients] are scared about the technical side of it. They already have the content – they're an expert at that. They just don't know how to put together an intro or outro. That's what we walk you through. We have a podcast setup service that we give to all of our corporate clients . . . The initial setup is what we call gathering their digital assets . . . the music, logos, if I'm going to be uploading to the platform, or if I'm just giving you the finished product and your podcast producer is going to take care of the rest . . .
>
> Corporate clients want a finished product. You can do a "we're at the kitchen table" type of podcast, but that's totally different from corporate . . . It's a structured show normally . . .
>
> With corporate clients, sometimes it's just an internal podcast, which we do, too. They're just talking to their employees [through a podcast]. They're not thinking about clients. So, a lot of the time they'll have that idea of whether it's to the public or internal.

The podcast market (including business/corporate work) largely involves working with clients who are not savvy with media production or audio. This makes consulting an ideal opportunity for podcast editors to earn additional income. Tom Kelly explains:

> There are so many people who need help – a lot of help. I had a client that was like, "My audio sounds bad. I bought the microphone you told me to. What's going on?" They had a dynamic microphone like seven feet away from them. That's my fault for not being specific, but these people are

paying a ton of money for content writers and all this stuff . . . A lot of them are just looking to pay someone to ask, "What do I do?"

For my own clients, I do tackle everything from their weekly editing, doing the uploading, to "Can you look at my website and see if anything looks a little weird?" We'll hop on a call. It might be, "Show me your room and we'll look at some sound treatment options." Basically, anything but talk into the microphone for them – there is someone that will pay you to do all the rest.

Britany Felix explains how she monetizes podcast consulting:

I'll have people constantly come to me who just want to pick my brain. "Can I ask you a question or two?" . . . So, you schedule a 30-minute appointment in your calendar that requires a payment for them to book that call. You say, "Hey, I think this is something that deserves a little bit more attention. Here's this link. Why don't you go schedule an actual consultation and we can dive into this a lot deeper." Then you're getting paid for somebody to pick your brain. That's literally what you're doing as a consultant.

On the consulting side of it – your rate skyrockets while your time invest-ment goes way down. Editing is the least return on investment of my time in my business, by far. Let's look at it this way: I have about 10 editing clients that I work with on a full-time basis right now . . . I'm working 40-ish hours with those 10 clients between the backend stuff and my business. Now, I can take two launch clients [who need consulting help], work for about two or three hours a week, and make the same amount of money as I do with those 10 editing clients. So, I absolutely want to get into more of that consulting coaching, higher-level offerings, besides just audio editing.

Paid podcast consulting is typically a service provided to clients, and not directed towards colleagues. Britany Felix explains:

In terms of colleagues, I definitely have seen people who are like, "I need help. Can somebody walk me through this or spend an hour with me?" Or, "I want to learn this new DAW. Is there anybody who can help? I will pay you." There are editors coaching other editors on certain aspects of the process, but for the most part, people tend to just offer that for free [to their colleagues].

Tom Kelly adds that marketing skills can be beneficial for a podcast editor interested in consulting:

People just want growth. That's the only metric that matters – downloads. There are editors who are doing much better than me (with far less audio skills) because they know how to help their clients get more and more downloads every week; whereas, I can only help my client sound better and better every week.

There are definitely people that don't have my audio background who have gotten jobs over me because they know the marketing. They want to help with content or structure and be more of an actual producer, whereas I just want to care about audio. The people that do very well are really into helping grow the show, as well.

One challenge of finding opportunities is differentiating, especially for those with professional training or experience. Tom Kelly explains:

There are a lot of people that downloaded Audacity yesterday and say, "I'm an editor" . . . There's definitely a lot of grit with some of the new people, and they are marketing themselves as, unfortunately, something that they're not. So, the actual audio professionals then have to compete with people who can sell themselves better.

One advantage to this market is the potential for major growth. Tom Kelly explains:

Trying to find that amateur who's willing to pay and also grow with you is a really unique opportunity. Maybe that's happened for some record producers – it was just their friend, and the thing blew up, and the producer happened to get a Grammy. There's a ton of opportunity for that.

Sarah Stacey adds:

With the podcasts I'm working on, things are expanding all the time. Like, maybe the number of episodes per month might go up. Sponsorship is a big thing as well, which means more money for everybody involved. So, it's quite exciting to see how things grow as you go along.

Getting Started as a Podcast Editor: Britany Felix

Britany Felix is a podcast editor and owner of Podcasting for Coaches. Britany discusses four recommendations for new podcast editors.

1. **Have a website early on, and make it look professional.**

2. **Use an online appointment/scheduling service.** "What I see in posts in Facebook groups [for podcast editors] where 30 or 50 people are commenting, the vast majority of [responses] are, 'I can help you with this. Email me and we'll talk.' But the person

who is looking to hire an editor is doing that because they feel overwhelmed. They want to free up some of their time and they want to do less work. So, by putting it on them to contact you, you're adding more to their plate . . . Instead, you could say, 'Here's my link. Let's have a no-pressure conversation scheduled at your convenience.' In these posts where there are 30 to 50 comments, I am one of the few people who will do that, and I get the consult almost every single time."

3. **Have a niche.** "Niching is something that I see a lot of other people miss out on. They try to be everything for everyone when they shouldn't be. I don't necessarily mean niching in terms of the services you offer, but niching to the people who you're going to serve so that you can be seen as an expert."

4. **Use service agreements (contracts).** "That's one thing that a lot of new people miss . . . If [clients] don't pay you and you try to get it later, or they try to claim ownership of your work that you've done, you might not be covered for all of these different things. So, getting [a service agreement] from the start is really important."

Rates and Payments

In podcasting, setting rates can be a challenge. "The problem is, you'll find people charging US$5 an episode, you'll see people charging US$800, and [the client] can have the same experience. Asking someone else what they charge isn't even really that helpful either," says Tom Kelly.

Podcast editors typically charge a flat rate, hourly rate, or price-per-minute of audio for audio-related tasks. For example, Tom Kelly charges based on the length of the raw/unedited audio:

> When people talk about how long it takes to edit a podcast, it might be three or four hours to edit an hour-long episode, whereas I'll do an episode in like an hour and 20 minutes . . . Just by being efficient, I can charge the same rate as everyone else and make more money.

Sarah Stacey says of her rates:

> Sometimes I charge per project. It differs, really. A couple of the podcasts I work on have an hourly rate that I've agreed on with the producer. Then,

sometimes I'll set a day rate if it's a full day's work . . . If you're starting out, it's hard to know what's too much [to charge], but I think the main thing people do is they don't charge enough. They kind of hold back. I think people can get taken advantage of if you do that. So, don't undercharge.

Those with formal audio education can (and should) charge more for their work in podcasting, unlike many other disciplines of the industry. Kelly Kramarik (who has a master's degree in recording arts) explains:

Don't work for less than you want to work for . . . It's hard because there's a line that exists between amateur/self-taught people that don't use industry-standard tools, and then audio engineers (in the podcasting realm) who are using the industry-standard tools and are highly trained. They all call themselves the same thing. That's where I make sure to say "I'm a professional audio engineer" anytime that I say anything to anybody, and try and explain the difference.

Britany Felix worked in the corporate world before pivoting into podcasting in 2015, where she learned audio skills on her own. Britany suggests for those with professional audio experience:

So many times, I've had to get on someone's case because they have an actual audio engineering degree [but are not charging enough]. They have experience, they've been working in the music industry, or radio, or similar industry for 10, 15, 20 years. They come into podcast editing and think they have to charge [low] for an episode. If someone is building their business and they know [audio engineering], there is no way in hell I should ever make more money than them. People forget what that experience is worth. They feel like because it's not in podcasting, it doesn't translate to a higher rate.

If I'm hiring and I'm comparing two editors and one has been doing podcast editing for six months (because let's face it, most of the editors now are less than a year into their business), and the other has 15 years of audio editing experience or radio experience, I'm going to go with that [experienced] person, especially if they're charging the same rate. Do not undervalue your skillset, your experience, and your knowledge. It is absolutely worth something in the podcasting space.

Tom Kelly (who has a bachelor's degree in audio production) adds:

People coming from any other audio background might actually have a hard time grasping their worth in the podcast industry. I know it was really hard for me and a lot of other people I've spoken to – especially for anyone coming from music. First of all, as a musician and as someone that performed in bands, that culture is, "Just do your art." I never got paid to perform. We've paid to perform shows. Then, we're going into this space where that's not that culture at all.

In podcasting, people are asking for a lot of money day one. It was hard for me to understand my worth. In music, I couldn't get people to pay me US$10 an hour to empty their trash cans, so I severely undervalued myself for way too long . . . I see a lot of people I know that are very good at audio, but when they respond to posts they don't even mention, "I have an audio background. I've been doing voiceover work for 30 years."

Setting Rates in Podcasting: Tom Kelly

Tom Kelly is a podcast editor and owner of Clean Cut Audio (USA).

Whatever you think [your rate is], you're probably worth way more. I really wanted to do sound design for film and television, and saw an entry-level job that required 15 years of experience and paid US$10/hr. How do you do this? That really gets your mindset in this low place. Don't even think of podcasting as related to the audio world at all because it's not. It's a content world. You can charge whatever you want for content. Anything you learned from pricing in the audio industry, I wouldn't count it as relevant at all . . .

What I should have done (and what I would tell people especially if you have an audio background): Get your editing to a point where it's a 2:1 ratio, so two hours to edit a one-hour episode. Then give yourself (even starting out) US$70 an hour or something. That'd be US$140 for an hour-long episode that takes you two hours of work. That's fair – that's a very middle of the road rate. What audio engineer (who comes from music) is making US$70 an hour? The client doesn't need to know your hourly rate. They probably tried to edit the show themselves, spent 12 hours doing a one-hour episode, and it still sounded bad.

My dad always says, "Whatever your gut tells you, double it and then add 15%." I did that one day, and the client said, "Cool, let's do it." I was like, Oh my God! Can I double it again? Then I did . . . It's a killer market. It really is. The better you are at your job, the faster you can make that money . . . You can even wrap in like, "We have to do a mandatory hour coaching call every month to talk about your show," and you can get a little more money that way. There's so much opportunity.

Britany Felix says of receiving payments from clients:

> For the first month, a client pays after every episode is completed. So, they're paying on basically a weekly basis. I explain that's just to establish trust between us, so that they know I will do the work and I know that they'll pay. Then after that first month, then they pay at the end of the month for all the work that was completed prior.

Britany charges up front for consulting services – such as reviewing prior shows, show notes, websites, or helping with other technical needs. She explains:

> For my auditing clients, my consulting clients, and my launch clients, it is all up front. The payment system is built right into my scheduler, and it's all automated. They literally cannot finish scheduling that call until they've submitted their payment.

Storytelling and Audible: Jeff Dudzick

Jeff Dudzick is Content Creation Manager at Audible Studios (Newark, New Jersey, USA). At Audible, Jeff began as an Audio Mastering Engineer (in 2010), growing to supervising and managing a large team of editors on audiobook titles. Jeff later shifted into live original audio (spoken word and music content) with Audible Theater and Words + Music productions.

> Audible is not an audiobook company. We're a premium audio storytelling company. It's not books on tape like your grandma had. That's still a stigma that's associated with it all.
>
> Audible distributes books from many publishers as well as content we produce on our own . . . Some book publishers can do things in-house, but chances are they farm it out. We collaborate with third-party studios on content that we produce. There are some individuals, but mostly they are studios just because of capacity.
>
> We have to do things at scale. Way back when I was supervising, I had like 70 editors that we engaged (all individuals, and they were contractors). A book would come in and I would see who's next on my list that should be open, and I sent it to them. As things scaled and scaled, we needed more one-stop shopping. Third-party producers would have people that they trained to meet our specs and aesthetic. They handled that stuff so we could produce more content and achieve our goals.

Useful Skills

If we see someone with foreign language skills, or expert English skills, or has experience or education related to language – even someone who maybe studied theater, English, or literature – that's valuable. Given our work is so email heavy, you've got to have good written communication skills . . . It's hard to teach that. You hope people come in with it, or at least absorb it once they start being exposed to it.

Business-related software matters. I don't have time to train, so you've got to be good at learning, or teaching yourself, or just fig-uring things out . . . There's always been a need to know something about Excel [for reports and QC sheets]. You don't have to be great with formulas in certain positions, but there is lots of reporting on a daily or weekly basis, so experience with a database (on some level) is good.

Working Under Tight Deadlines

Some things have to be done fast but some things can take a lot of time . . . just because of the level of editing we do. We're using room tone for pacing (not gates), cleaning up noises in between words or sentences . . . Mistakes are easy [to fix] . . . It's the scale that makes it tough. Books are long. We're not talking about someone who narrated an entire movie for 90 minutes – it's a book.

Some of the most important things have to be done the fastest for various reasons. If a celebrity is only available at a certain time (and it's right before [the project is] going to go on sale), we have to work around the clock.

Content in Multiple Languages

Going into each country is its own world – especially from the Audible Studios perspective . . . We can't just translate our stuff and put it in their store. They prefer different things or want things a different way . . . the types of content that they like, how it sounds, how it's presented – all that stuff has to be learned in each country. There's no one-size-fits-all by any means . . . We would almost always partner closely with someone in that country, a studio and talent . . . We work with them closely because it's not just a straight translation. There's so much more to it.

Shifting into Spoken Word Entertainment

> Things have changed partly because of the pandemic, but also after we scaled so much. We (and every other media company) were really getting into original content. We're using our talented in-house people to start exercising their skills more. I would do the music on audio dramas (all with library music). Some people would be doing sound design. There was lots of creative editing and that was a lot of fun. That's how the music stuff started, and then we got into the words of music.

> Audible has an Off-Broadway theater down in the Village [in New York City]. We've had it for a few years and we're really investing in it to do live shows . . . It's theater that can be accessed by everyone – not just people with money who live in New York City. They do live shows. Many are just like three nights, and they're all recorded. Some are runs, like doing the show for an audience for two months. At some point during a run, if we don't record a few nights live, we'll do a studio recording. It's always really interesting – it's a different presentation than what we'd capture live on stage. It becomes a different thing. It's special and unique.

Broadcast Radio

The radio market still exists worldwide. Sarah Stacey says of working for Today FM, a commercial radio station in Ireland:

> I found during the pandemic, I was lucky in that my job was pretty secure because more people than ever turned to radio, especially for their information about COVID. That was one time when working on a current affairs show really came in handy.

Avery Moore Kloss says of working in news for radio (versus long-form audio documentary):

> There's definitely a rush to daily news. Every day is different. You never know going into work what's going to happen that day. There's adrenaline that you do get from that . . . You make something, move on. Make something, move on. You don't have to obsess over the small details for months, which is what I do [in long-form]. It's two different brain spaces.

Sarah Stacey adds:

> It's something new every day. The show will follow the same format, but anything could happen in that time. Things can change so quickly. I think that's part of the attraction of live radio, and you get that immediate response from listeners as well.

Board Operator: Jonathan Hubel

Jonathan Hubel (Richardson, Texas, USA) is an audio engineer (for spoken word) and podcast editor who has also worked as a board op for live radio broadcast.

> Being a board op is pretty basic, and a great way to get a start in radio. If you're comfortable with computers and audio mixing boards, you already have the basics down. It's essentially running the sound board for live broadcasts as opposed to live events. You sit in the control room, turning on and off microphones at the right time for announcers, guests, etc.; playing pre-recorded elements off the automation system in the computer; paying attention to the clock and program schedule.

> Pros: It's a great, fairly entry-level opportunity to get your feet wet in radio. You get to connect with others in the field and grow your professional network. Cons: It can get pretty monotonous and boring depending on the type of program you're working on, as it's generally a talk-radio type environment. You probably won't be able to find a full-time position as board op – it's generally a relatively low-pay, entry-level, and part-time position. It often requires working late hours or even weekends, which isn't for everyone.

> Ask yourself, "Do I work well under pressure?" I personally struggle with live mixing, because there's more at stake if I mess something up, and that causes stress which then causes more mistakes. However, if I'm sitting in an editing booth, I can really focus on what I'm doing, rather than having the pressure of thousands of fans/listeners weighing on me. On the flip side, some people thrive under pressure, and love to be surrounded by the glamor of the live world. These people would often really struggle with the isolation of a small studio.

Live broadcast radio has a speed and intensity that may appeal to those who like live sound. Jeanne Montalvo Lucar, who was an audio engineer for live radio, says:

The thing that I thought was interesting is that live sound really translated to live radio because it's the same kind of adrenaline. You have to be on your toes. I was doing a live show at Bloomberg and if the ship went off the air, I had to think quickly to be able to get things back up and running. It kind of filled the void – I can do live radio and still have that adrenaline, but I don't have to lift as much gear.

However, working on live radio may require working unusual hours. Jeanne Montalvo Lucar says of the schedule:

Radio is on 24 hours a day . . . It's non-traditional hours, so you might have overnight shifts. You might have shifts from like 2 pm to 10 pm, or 4 am to 12 pm . . . I would never steer anybody away from it, but I think it just takes a certain kind of person to be able to commit to those kinds of hours.

Live Broadcast and Attention: Jeanne Montalvo Lucar

Jeanne Montalvo Lucar has worked as an audio engineer for live radio (for NPR and Bloomberg), as a radio producer (NPR's Latino USA), and as a podcast editor.

If you're on a four-hour show live, absolutely 100% of your focus needs to be on that. There's no multitasking on what you might want to do personally – like, you can't read an article online . . . If you're going to do a live show, absolutely that won't work. My phone is upside down, and I'm not even paying attention to it . . .

Part of the reason you can't do other things (or lose focus or attention) is because there is so much multitasking that's happening on a live show. They might be in the middle of an interview segment but you're pulling the bump that's going to happen at the end. You're checking to make sure that the ad is coming up, or you're maybe pulling up something else and getting all of that ready to fire for the next ad segment. There's so much multitasking that you can't add another layer.

If you're working with a high-profile guest [live or recording], you're not doing anything else. If something happens – if the mic stops recording or the guest isn't hearing themselves or something, you need to be able to handle that quickly. If it's a high-profile guest, they're going to get pissed off that you're wasting their time.

> If you're mixing a podcast or something that you can stop and start, that's a good place if you have issues with focusing or attention. Anything that you can stop and start and that doesn't involve a third party is probably fine if you have issues with attention. You just need to watch deadlines.

Radio also has opportunities beyond live mixing and hands-on audio engineering. Ann Charles describes her job:

> A broadcast engineer looks after the transmission chain: how audio gets from the studio to the transmitter and thus to the audience. Half of their job is to help radio announcers who have broken something and are live on air.
>
> (qtd. in Gaston-Bird)

Pamela Dwyer is the Production Coordinator at CKUT 90.3 FM, a community radio station in Montreal (Canada). Pamela says of her work:

> I'm in charge of all the equipment, so I have to make sure everything is working . . . It's my role to make sure the audio that's coming out on-air is good quality. I train programmers on how to use the equipment properly and teach them about mic technique. I teach them how to mix their shows. I'm also aware of what's happening up at the antenna and I keep track of that and update it . . . Because it's community radio, a big part of the job is upskilling people. We have a bunch of portable recording devices. We teach people how to use those. We have outreach events to run. We teach people how to record events and how to do a remote recording from a show and send it to the station.
>
> (Campbell)

Radio opportunities can be employee (including full-time) or freelance. Internship opportunities exist, as well. Job listings for radio can be found online, but landing your first radio gig can be a challenge. Jeanne Montalvo Lucar explains:

> It is actually hard to get your foot in the door in radio. Once you do, then word of mouth travels. Now when people call me, they'll say, "you're really highly recommended," and it's because I have done a bunch now. It's easy to get recommended for stuff because enough people know me. That stuff just comes to me.

Relationships are important when working for companies and radio stations. "With an organization as big as the BBC . . . you could be working with the same people multiple times. To build good relationships with them is quite important," says Sarah Stacey.

Radio Promos: Sarah Stacey

Sarah Stacey (Dublin, Ireland) is a podcast editor and radio producer, including working as an Audio Creative for BBC Creative (producing promos and trailers for the BBC World Service).

With promos, people might think, "Oh, this is 30 seconds. That's nothing." But the shorter the time that you have to work with, the more difficult it is because you might be tempted to pack too much in. Your pacing has to be really good. So, I actually find it more challenging to do the short form stuff.

The official title of my role is Audio Creative. So, it's literally everything from brief to final delivery. I will get a brief for a specific program. Then, it's up to me to contact the producer to find out what kind of direction they wanted to go in. They'll send me some clips that they want to use, and I'll build a script around that. Then, we'll usually get the presenter of the program to record a voiceover. I pretty much have free rein in terms of choosing music and stuff like that, but I always have to run it by my line manager. It could be things like, "we need a different word there." I'll do a few drafts of that.

You have to think in a completely different way [working on promos] . . . There's a very specific station style, and there's a whole guide because you're dealing with an international audience where English may not be their first language. You have to be thinking about cultural sensitivity, or words or phrases that you shouldn't ever use, and that sort of thing. So, it's really interesting to think in that way . . . The last promo I did, there was a time difference involved, so I was on quite a tight deadline . . . Make sure you're very clear about the dates and the deadlines.

The Untapped Spoken Word Market: Avery Moore Kloss

Avery Moore Kloss sees a variety of opportunities in spoken word audio for those interested in new areas of the medium.

I always have some personal historian project happening in the background. Most of those end up being life stories, like someone whose kids [want to capture a parents' story] . . . We do an hour interview about their childhood and an hour interview about their adult life.

I transcribe it and then I also make them a 10-minute quick piece . . . There are definitely examples of audio businesses who have made sound portraits and sound pieces scalable, or they're at least trying to. I think that there is a lot of opportunity there. I do think people realize, in this world where everything disappears, that freezing a moment in time with a loved one is really special . . .

I think there's a power to audio that people are recognizing now because of the rise of podcasts that I think does leave space for that – Like this very intimate, very personal medium that breaks down people's barriers and their walls because they're not on video. A lot of people said to me, "why wouldn't you have video with that?" . . . There are lots of people who recoil with video but are happy to do audio-only.

I think there's lots of untapped audio to come. Honestly, the rise of voice command is going to be an interesting shift for the industry anyway . . . An audio capsule that you can create for a smart speaker I think is something that's going to come, too . . .

If you're a baker and you make cookies all day, you're not reading *The Globe and Mail* [newspaper] but you might listen to a piece that you're interested in . . . I think smart speakers are going to force it. If you can say to a smart speaker, "Play me the top five stories on *The Globe and Mail*," and it just reads it out to you . . . it's not a podcast, but I'm still paying a monthly fee to get the audio version of the [news].

Note

1   SEO is search engine optimization, which improves the chances of your site being found in a search (by others seeking information on that topic).

References

"Buzzsprout Platform Stats." *Buzzsprout*, www.buzzsprout.com/global_stats. Accessed 1 April 2022.

Campbell, Madeleine. "Pamela Dwyer, Live Sound Engineer." *Women in Sound*, no. 6, 4 Feb. 2019. www.womeninsound.com/issue-6/pamela-dwyer. Accessed 9 May 2021.

Davis, Fela. "Ask the Experts – How to Produce a Podcast." *YouTube*, uploaded by SoundGirls. youtu.be/ayWMeseQOb8. Accessed 23 Sept. 2021.

Dudzick, Jeff. Personal interview. 1 Oct. 2021.

"Europe's Radio Landscape." *Broadcast Networks Europe*, broadcast-networks.eu/radio/ . Accessed 26 June 2021.

Felix, Britany. Personal interview. 22 June 2020.

Gaston-Bird, Leslie. *Women in Audio*. Routledge, 2019, p.93. doi.org/10.4324/ 9780429455940.

Hubel, Jonathan. Email interview. Conducted by April Tucker, 22 Oct. 2021.

Kelly, Tom. Personal interview. 22 June 2020.

Kramarik, Kelly. Personal interview. 18 June 2020.

Montalvo Lucar, Jeanne. Personal interview. 12 Oct. 2021.

Moore Kloss, Avery. Personal interview. 21 Oct. 2021.

Payne, Piper. Personal interview. 29 May 2020.

"Radio Broadcasting Global Market Report 2021." *The Business Research Company*, Dec. 2020. www.thebusinessresearchcompany.com/report/radio-broadcasting-global-market-report-2020-30-covid-19-impact-and-recovery.

Stacey, Sarah. Personal interview. 19 June 2021.

18
Beyond Entertainment

Entertainment production and live entertainment will not be the right fit for everyone. Luckily, the audio industry is not exclusively for recording engineers and sound designers. What starts as an interest in audio can grow in many directions. This chapter is a brief overview and sampling of the types of opportunities available.

Product designer Ryan Tucker (who worked in music production before pivoting) says:

> Music/audio production is a service industry, and political skills are a must. You *have* to be a team player, you *have* to understand it's not about you, your workflow, your preferences – your attitude must reflect this at all times. Knowing oneself, one's skillset, one's default disposition, work/life preferences, etc., is crucial to choosing a compatible career path. If being a music engineer is something you love to do, but you know the lifestyle isn't for you, you can make the right moves early on in your career, avoiding the struggle of discovering the career you've chosen isn't right for you.

Alesia Hendley started her career in live sound but pivoted into AV and IT, where her audio background has been a major asset. She shares:

> Don't just latch on to the microphone or the soundboard. Explore what these things lead to, or what they create. There's just so much opportunity. At one point, I didn't feel like anybody considered what I was doing as sound. Then I was like, wait a second, you're pulling the strings here. You're doing sound – just in a different perspective, in a different way, and it works.

Linda Gedemer was interested in recording studio work after completing two degrees (in music engineering and electrical engineering) but found work in AV. One project Linda worked on was the emergency paging system for Los Angeles' metro rail stations. Linda explains:

DOI: 10.4324/9781003050346-21

That's how I went from music engineer, to audio engineer, and then to audio/visual control systems engineer. Along the way, I had to learn about video and control systems. Then I got hired by Disneyland Imagineering as an audio/visual engineer and went to France to work on Disneyland Paris.

(qtd. in Gaston-Bird)

Tommy Edwards has worked in audio product development for Warm Audio, Blue Microphones, Line6, and other companies. He of finding his place in the audio industry:

I didn't necessarily go into this because I wanted to produce people's records. My primary focus was my craft as a musician. Even playing with side bands, or being a backup guitar player, was just in service [of that focus] . . . In the late 1980s and early 1990s, it was this hustle of doing sessions, filling in, [studio] assisting, and working in a music store.

Working in a music store introduced me to some of the manufacturing representatives. They were really excited to talk to me because [I knew] the gear and I can speak business language . . . I started thinking, "maybe instead of doing all these little things to make money to pay rent, I can do this one thing and still pursue music." That's why I made the jump. I was doing a customer service gig at Alesis, and then quickly I moved up to more of a technical representative.

Automotive Audio

Lucy Diggle is an Audio System Engineer at Jaguar Land Rover (Warwick, UK), where she has worked for over ten years. Lucy holds a master's and bachelor's degree in acoustics from the University of Salford.

My job is to calibrate or "tune" the sound quality of in-vehicle audio systems. The inside of a car is an unusual – and far from ideal listening space, where the loudspeakers are often positioned very close to you but not pointing directly at you. To compensate for this, I am able to carefully adjust the sound output from each speaker using software in a dedicated audio amplifier, and create a great listening experience – not just for the driver's seat but also for the passengers in the front and rear seats.

Generally, a bachelor's degree is the minimum requirement to work in engineering within the automotive industry. Specifically for sound-related work, an acoustics degree would be highly desirable but not a necessity; related subjects such as music technology and audio production are also accepted, as well as broader science or engineering subjects such as physics or electronics.

How to Find Jobs in Automotive Audio

Applications for employment in the automotive audio sector are typically very traditional, with an initial submission of an application form and resume. Networking can be highly beneficial though, (that old saying "it's who you know, not what you know" definitely applies!) so membership and involvement with professional institutions can help with finding potential opportunities. For example, there is an Automotive Audio sector of the AES which includes a wide range of members from academic institutions, automotive manufacturers, and suppliers.

It is worth pointing out that the automotive industry extends a lot wider than just the vehicle manufacturers themselves. There is a huge supply base, so there may be a much wider range of employers in the automotive audio sector than many people would be aware of initially. The supply base consists of a good mixture of acoustics and audio specialist companies, as well as more general electronics and communications companies. There is a balanced mix of automotive-only suppliers and suppliers which cover a range of industries. I think it would be made clear in a job listing if a particular company was servicing the automotive industry, and also if the role was primarily in this field. Many companies advertise through specialist recruitment companies, so having a profile with multiple agencies is often the best way to find a position in this field.

Advice to Emerging Professionals

Keep an open mind about the kind of work you want to do, and to explore as many of the different opportunities that may be available to you as they can. When I graduated, I had very specific ideas about the kind of work I wanted to do (acoustic consultancy or clinical audiology – nothing like the career I have now!) but jobs in those areas were very few and far between, especially for a [recent] graduate. I found myself doing temporary jobs at the university whilst looking for a graduate position. During that time, an automotive company held a recruitment event at the university where I found out about the range of potential job opportunities they had available. Honestly, at the time I didn't think the automotive audio industry was for me as it didn't fit with the career path I had planned, but I'm not one to waste an opportunity so applied for one of the positions they had available. I'm now 10 years into my career at the very same company, having held three different positions within the audio team and doing work that I enjoy.

Sonification/Accessibility

Ashton Morris is a Sound Designer for PhET Interactive Simulations at the University of Colorado Boulder (USA). Ashton has a background as a composer and sound designer for video games.

PhET at CU Boulder creates free math and science simulations for education. They have been making them since 2002 and are used around the world. I work as a sound designer for the simulations with a focus on blind accessibility.

Landing the Job

I created a Game Audio meetup in Denver. Even though I lived two hours away from Denver, I knew it was the best, most central location to attract the most audio people. I made it because another Game Audio friend said, "If a community doesn't exist, then build it," so I was inspired to help Colorado have its own . . . At the first or second meetup, my future boss attended because she was interested in incorporating more audio into her work at PhET. She was told about the event by a coworker who saw it while browsing meetup.com.

We had a normal meetup conversation, but she had a few questions about audio, and I offered to help her . . . After that, we emailed back and forth a few times and I did my best to offer her good advice and be helpful. Eventually she asked me if I wanted to contract with the team and I was excited to work in education, so I said "yes." After a year or so she asked if I'd like to become a member of the team, and I have been working with PhET happily for 3–4 years now.

Sonification as a Field

I would say that it is still emerging as a field. Ways of doing things or ways of thinking about sonification are not codified yet. As the need for greater accessibility grows, is seen as important, and as audio-only interactions with computers continue to grow (like Alexa or Siri, or interacting with AirPods), the field will continue to expand and grow. But, it feels to me like things are still very new and unexplored in this field.

My experience is that a lot of people who have done sonification work are researchers firstly, and then secondly they are audio people. I think it's rare to have someone that comes from an audio background and not a science background in this role, but perhaps that will be less uncommon in the future.

I would consider sonification to be in the academic world and in the accessibility world. I think there are growing opportunities (in media, interactive media, and video games) for audio professionals to bring their knowledge to the world of accessibility, specifically blind accessibility. Looking toward the future, I believe that experience will be useful for hands-free, screenless interactions with AI assistants and computers in general.

Work-life balance is immensely better in this educational role vs. if I worked for a game studio. A lot of my colleagues talk about "crunch" in the game industry, and that is something that we really don't have in my field. As someone with kids, I really appreciate that!

Training and Employment Opportunities

Employment opportunities in this field (in education) are rare. Perhaps in gaming they are more common, but they would be called "accessibility" and not "sonification." It is not uncommon for the audio to be done by someone with another job title, like researcher or programmer.

[Having] the knowledge of what is possible with current audio tools made for video games helped [me] a lot. There are things we're building that have existed for years in the world of games and interactive media. So, that has been helpful. The main learning curve is improving my knowledge of actual science and math, which wasn't my strong suit when I was in school . . .

Chance favors the prepared mind. You never know what opportunities are out there and what direction life will take you in. If you keep doing your best, putting yourself out there, and act kind and polite, opportunities have a way of surprising you.

Music Technology and Audio Technology

Creating tools for the entertainment industry (that can be used for music, film, games, or beyond) is an industry in itself. This industry can be referred to as music technology, audio technology, or MI (musical instruments). Tommy Edwards explains, "It can cover everything from 'I want to learn how to make violins' to 'I want to learn how to make a mobile app that can record people.'"

Live sound engineer Michael Lawrence says of audio technology:

> Not everyone that works for an audio manufacturer is a soul-sucking, money-grubbing vampire. They're people just like you and me, and they work for companies that make the cool stuff that we get to use. They get to decide how that gear is going to work and interface with people that use that [gear]. We want to advise folks looking for their career paths in audio to not discount the manufacturer side of things, because there are some great opportunities there.

Tommy Edwards adds of professionals working in audio technology:

> Traditionally, it was made up of a lot of people who (especially on the non-engineering side of the fence) were frustrated musicians. "The band didn't work out, so I'll do this." They didn't necessarily have the means to continue their education . . . So, people are in positions of power and don't necessarily have the technical insights, and they don't even necessarily have the business insights to really be making some of the decisions that they're making. I think that's changed. You have a lot more people who, if they're motivated, have access to information and can learn things. Folks who have the technical acumen, the business acumen, and understand customers are rare. It's a very rare skill set.

Audio technology companies can be major corporations or small companies. There are stable full-time jobs with regular schedules (and possibly travel opportunities), or self-created companies. Justus Gash says of starting his own guitar pedal company, Fuzz Imp, in 2021:

> My company, while still in its infancy, is the most profitable music endeavor I've done! That being said, it took a lot of tries and hard work to find the right fit, and the connections I built in previous ventures helped pave the way. I also put an obscene amount of time in early on . . . It's really looking like I'll go full-time in less than two years, and it could be sustainable.

Langston Masingale, founder of Handsome Audio, adds:

> I didn't want to go into a situation where my ideas ceased to belong to me, and my creative control, use of those ideas, and execution ceased to belong to me . . . Somebody like me had to really create their own way into this side of the business. There was no apprenticeship, there was no incubator, there was no "how to." This is actually one of the least discussed aspects of music technology is how to make equipment . . . It's very secretive- a very closed shop.

Electronics Technician: Natalie Hernandez

Natalie Hernandez is a freelance electronics technician who worked for six years for Death By Audio in New York City (USA). Death By Audio makes handmade effects pedals.

Death by Audio is a small company. There are only nine of us, so we all do several different things . . . A regular day usually starts with coming in and checking email. I respond to a lot of technical questions people have if they have an issue with our pedals. Because we're so small, we spend a lot of time actually building pedals. We don't outsource our [circuit] boards or anything like that . . . A few days of the week, I have a design day where I sit down and work on my own projects . . . I'm full-time at Death by Audio, but I still take on a lot of freelance jobs doing synth repair . . . Most of my experience is in freelance synthesizer repair and modification. I also do maintenance work for mostly analog and some digital synthesizers.

I went to school for electrical engineering . . . I dropped out of school [with one year left] because I didn't want to work at Whirlpool or GE or something more along those lines . . . I lived in Asheville [North Carolina, USA] and I really wanted to work as an engineer at Moog. I wanted to work there so badly, but I realized the people who work there kind of stay forever. After that, I moved to New York and in my spare time, I would buy things broken on eBay, take them apart, and learn to put them back together or fix them . . .

If I were to communicate anything to someone who's interested in this sort of work, it would be to not give up . . . I still have so many unfinished projects that I just couldn't figure out and kind of just jumped ship. In my years of working in electronics, that gap has slowly started to shrink, where I have fewer projects I don't understand . . .

[To learn circuit design], start with basics. Instead of jumping in with the idea, "I want to design a fuzz pedal," or something like that, I'd say learn the different aspects of the circuit. I would also suggest trying to recreate something you know you love already because that's a great way to familiarize yourself with a circuit. There are so many schematics out there for all kinds of pedals, even synthesizers, that you can build and then alter. After you've built something, you can start to experiment. Get to know components [and what they do] – there are many of the same ones in all your favorite pedals . . . I usually do all this experimental stuff on a solderless breadboard, which you can buy online. It's extremely useful for prototyping!

There are also fixers guilds . . . It's a bunch of people who bring in their broken things. Everyone sits around a table and they fix them together. It's amazing. Usually, there's one person who's been an electrician for 30 years and then someone brings in a toaster, which is different than what they normally do, so it's fun. Reaching out to people when you don't know something is important.

(Campbell)

From Women in Sound

Audio technology has opportunities within two general paths: technical roles and business/marketing. Business/marketing-related jobs include customer support, marketing, and sales. Technical jobs include quality assurance, programming (including DSP), and research and development (R&D). Some roles incorporate both, such as product design, or management. Rachel Cruz (Program Manager, Fender) says of technical jobs:

If you are looking for more of a "give it to me, I'll do it, I'll get it done, I'll give it back to you" kind of role, looking for something that is really, really technical is often the best way to go . . . You can still be part of a team, but you have your piece of the thing that you specialize in. You'll still have to work closely with a lot of people and get alignment across lots of shareholders. But any sort of design, technical engineering, QA – those fields are really good because you never have more than one person sitting in a station doing that kind of thing.

Quality Assurance Testing: Ryan Tucker

Ryan Tucker is a product designer who started his audio technology career in quality assurance (QA). Ryan has worked on hardware and software products for companies including Universal Audio, McDSP, Warm Audio, Line6, and DTS.

Quality Assurance (QA) is something I fell into. I didn't even know it was a career when I was going to college, but my technical aptitude and comprehensive knowledge of the subject matter (music engineering, tech, and production) set me up perfectly for a career in quality assurance.

A quality assurance tester is tasked with qualifying a product (its design, feature set, usability, etc.) during the development process. If the product fails to operate as was intended, the tester must document the circumstances under which the failure occurred, and the exact steps to reproduce it. This report is referred to as a "bug report." It must be clear, concise, and exact – lacking no pertinent information.

Being a domain expert [having special knowledge] in the product's use case, and knowing a bit about how things work under the hood, makes the bug reporting process that much easier. There often isn't a lot of subjectivity in QA. It's more often black or white: the DUT (device under test) works, or it doesn't. But knowing how things *should* work before even touching the product (coupled with the tenacity, persistence, and a propensity for monotony) makes for a great QA tester. I didn't know I would be so good at it! But I was, and being open to a slight career pivot led me to a successful career in something I hadn't even considered.

Getting into Audio Technology Companies

Tommy Edwards is Vice President of Product Development at Warm Audio. He has overseen the creation of music technology and audio products for companies including Warm Audio, Blue Microphones, Logitech, and Line6.

If I was going to advise somebody who wanted to get into music technology or audio technology, the first decision on that tree is: Are you technical, or are you more on the business/marketing side? That's a big decision tree.

If you're technical, you probably want to get some background in physics, engineering, math . . . If it's more marketing, you now want to lean more towards business. Maybe behavioral economics, or even human factors, which is the psychology of like, why are buttons located in certain positions? Why is the dashboard on an aircraft monochromatic and not a bunch of random colors? There's some psychology and neuroscience behind it.

Typically, you want to make that shift early. But, it doesn't mean that somebody who's technical doesn't learn the business stuff, [or] the business person doesn't learn some of the technical stuff. It tells you what kind of program and what kind of school you want to go to.

Entry-level jobs typically are everything from [quality assurance] to customer service (you're just taking phone calls) . . . Any kind of entry-level marketing (like marketing assistant) is pretty easy to get.

Quality assurance testing is going to vary depending upon the company, and whether or not they lean toward hardware or software . . . Even at Warm, our QA team needs to have a bit of technical knowledge, and then they just follow a script and are testing things. They don't necessarily have to be electrical engineers, but it varies. A quality assurance person at a [company like] Line6 or Blue Microphones or Warm Audio is not the same as a quality assurance person at Bose or probably Sennheiser and certainly Harman. Those tend to be much more technical and might even require an engineering degree.

The Path to Product Management

[You have to] understand the physics of basic hearing. Then, you need to understand audio. Then you need to understand things like ROI [return on investment], business metrics and positioning, market research, and how you get a product in Sweetwater Sound. [Finding] all those things in a single person is wildly rare . . . If you are going to acquire those skills in a traditional college pathway, that's like two or three degrees.

Get a solid undergraduate degree . . . if you lean more sciency and more technical, a math degree or a basic physics degree are great foundations. Then, instead of getting a second degree, there are graduate certificates where you take four or five classes. You could become certified in data analytics, or there's even one for product management now. So, you got that basic foundational degree or a business degree. A business degree today is actually substantial . . . So even if you have that business degree, then you can take a technical graduate program, learn a little bit more of the engineering behind what makes an audio interface work, and so forth, without incurring the costs or the time of earning a second degree.

Entry-level – you're going to see different [job titles], but it could be an associate product manager or just a product manager. That's going to be either somebody who's making a bit of a pivot in their career to product management, or somebody who's coming right out of school. Sales seems to be another kind of entryway – doing a stint as a salesperson at one of the larger retail outlets.

Teaching Yourself: Langston Masingale

Langston Masingale (Syracuse, New York, USA) is owner of Handsome Audio and the creator of Zulu, the first passive analog tape simulator. Langston is self-taught in product development and came into the industry as a recording engineer and record producer.

Every solution to a problem hasn't come to light yet, and there are plenty of problems that never really got solved . . . the audio industry is full of needs right now . . . It could be the way that we listen to music. Who's to say that the streaming platform has been perfected, that the algorithms or the codecs that are used to encrypt and compress the audio have been perfected yet? That's still new territory. Who's to say that the way that microphones function can't be improved upon or can't be done better?

Among my peers, I'm one of the few people that I know personally that actually does PCB design, the actual electromechanical design and the electrical design. Not because I'm great – I did it out of necessity. I just learned how to take these ideas and translate them into production templates. When I didn't have a skill, I subbed it out. But my goal was, at a certain point, I have to learn how to do this myself. Otherwise, I'm going to be handicapped by the fact that I'm relying on somebody else to do this . . .

There are financial obstacles [to creating a product] . . . It costs money to hire somebody to do PCB design, to have somebody help you draw a schematic, to look at your schematics, analyze them, and help you understand if you did something right or wrong. It costs money to develop products. It costs so much money just to experiment. That was one of the reasons why I got more hands-on with product development because I didn't want to pay somebody to make something for me just to know it didn't work. I wanted to be capable enough to say, I made a PCB today . . . If it blows up, maybe the values are wrong, and if it doesn't work at all, it needs to be revised. I can do that . . . Teaching yourself how to do something isn't the answer to everything, but it usually is how you get started anyways, because you're trying to just get something to happen.

Audio DSP

Steinunn Arnardottir worked for Native Instruments (in Berlin, Germany) for over ten years, starting in DSP and working into management. Steinunn is currently Chief Technology Officer and Founder of Lottie (Berlin, Germany). Steinunn studied electrical and computer engineering in Iceland. Steinunn has a master's degree in electrical engineering (focus on audio signal processing) and a Master of Arts in Music, Science, and Technology from Stanford University (USA).

[When I was younger], I would go to a concert and would secretly dream about being the one on stage. But now in later years, I've actually also realized that I'm probably a bit too shy for it. I love being in the background a lot more. What I really love about music tech and audio DSP is that it's half engineering, half creative. That is also maybe the generalist that I am – I'm neither the hardcore bridge-building engineer, and neither am I a super-creative musician. I found my sweet spot there in the middle.

I was hands-on as a research DSP engineer for several years. Native grew as a company – I think during the time I was there it grew from 100 to 500 people or more. That was also an opportunity to grow with them and to wear different hats . . . In my last role at Native, I was leading a team of about 70 people, and they were responsible more for the delivery of the product . . . I had kind of gone out of the day-to-day listening sessions for some time . . .

What I really loved about working at a company like Native Instruments is that you were creating these tools that were used to create the music that you listen to, and that's a really rewarding job. It's not a nine-to-five job if you've got the passion – it's a bit more than that, maybe. But it's easily done with family life . . . it's more settled down than being on the road.

Job Opportunities

In the audio DSP world (in terms of audio effects – where I was), it is pretty niche. After a couple of years in the field, you kind of know everyone, or know about everyone, one way or another. But this is also really great. You meet at these conferences, and you get together every year, and it's a really great vibe in that community.

I think [job opportunities] depend on how married you are to [audio effects]. There's a lot of audio signal processing needed in hearing aids, for example . . . With the aging population, and also with hearing damage being more of a health issue, it's already a [greatly] growing field. So, there are a lot of audio signal processing jobs out there . . . it's a relatively niche knowledge, but there's always a need for bright minds.

Skills Needed

It's very math-heavy, so you should definitely enjoy that. If [you were] wanting to go into audio DSP at a music company, I would try to also do as many programming courses as you can. This is really beneficial. Basic music theory is very beneficial . . . In terms of getting a job in the industry, what many companies are looking for is someone who can hit the ground running when it comes to development. [Ear training] is not fully essential, but very important . . . Early on in the process (especially if you're doing emulations), some of it is visual – you're looking at spectrograms and frequency responses. But when it comes towards the end, it's about tuning it more . . .

You can learn everything [without a degree] especially with the internet. For a lot of math and language learning, you need discipline, like, just building up every week . . . [there is] so much great content out there and really great sources. A lot of the lectures from the Audio Developer Conference are in the public space. Julius O. Smith has really great in-depth DSP educational content, and he's always had it online for free . . .

[For conferences], the Audio Developer Conference is growing. It's got quite a diverse set of talks. Obviously, AES, ISMIR (International Society for Music Information Retrieval), and there's the Web Audio Conference (WAC) . . . I remember my first conference in the field was an AES conference in San Francisco, and it was just so inspiring, and gave me all of these life goals. To see all of this top-notch research, to see all of these people using these tools (and how they use them) was a very inspiring experience . . .

If you've finished university, it probably would be wise to sharpen your coding skills and work from that. Development skills are very sought after. Also, use the time before you have big financial obligations to do some internships or try to explore new places.

Audio Preservation

Audio preservation is the process of restoring and archiving audio. Some aspects of audio preservation fit under music production (for example, some mastering engineers specialize in restoration). Jessica Thompson is a mastering and restoration engineer, audio preservationist, and educator in the San Francisco Bay area (USA). Jessica says of her field:

> For those of us who never really wanted to be on stage, and who are super comfortable behind the scenes, archiving is for you, because you are totally behind the scenes. The best-case scenario is that when you do your best work, you're invisible. No one knows you were there, except there's this beautiful recording that previously was locked away on a DAT. So, I really appreciate that aspect of this work. I like being behind the scenes. I like the details.
>
> (Frick et al.)

Archiving and restoration work has a highly logistical end. Catherine Vericolli, chief transfer engineer and archive manager at Infrasonic Sound in Nashville (USA), explains:

> So much of what we do is just very meticulous, very tedious file management . . . There are certain people out there who really enjoy that sort of tedious, detail-oriented type of management of assets or management of audio in general. I'm one of those weird nerds who doesn't hate doing it . . . It can be a lonely experience to sort of be on your own, and it's you and an old recording . . . In the small community of archival and restoration, folks are so incredibly supportive.
>
> (Frick et al.)

Jessica Thompson adds:

> A lot of my time is data management, making sure things are labeled correctly, uploading, downloading, communicating with people, making sure I have the right files, making sure I'm sending the right file – although, that work changes day to day.
>
> (Frick et al.)

Archiving Audio: Anna Frick

Mastering Engineer Anna Frick manages all archival digitization projects at Airshow Mastering (Boulder, Colorado, USA).

> I think the demand is always going to be there as new technologies arise, as new outlets arise. It's one thing to archive projects, but when you're archiving them and then bringing them up to a modern standard . . . that's really, really fun for me. It's taking this old project that hasn't seen the light of day in forever, bringing it to life

from that original format, and then bringing it into current listening standards . . .

File management is definitely tedious . . . I get really geeky about databases. I customize our database for all our projects. That's kind of fun on a totally different brain level . . . I'm fortunate my mom was a librarian and so I learned those skills of just being methodical and cataloging and wanting to know where everything goes, and where everything exists. That has been hugely helpful to me in this world. I don't think a formal education necessarily prepares you for that, unless you're doing a library sciences degree. I don't think actual preservation of audio is a focus in library science programs. So, it's almost like a marriage of the two – that is what you need for this kind of work . . .

The most difficult for me is managing expectations for clients, especially when media is deteriorated . . . and just saying, "There's only so much we can do, this is what I'm running up against, this is what you have," and then educating clients. Like, "Now it's transferred, don't throw out the old media. This is how you should be storing it." Managing those expectations and educating clients are definitely some of the more difficult aspects of it.

(Frick et al.)

From SoundGirls

Education

There are a variety of income opportunities in education, in both the formal academic world (teaching, research) or informal (virtual, coaching, one-on-one lessons). James Clemens-Seely, who works in music production and education, says of teaching:

As a career path, educator is easily one of the small pieces that you fit into being a freelancer . . . One of the best ways that I started to make a living in audio was by teaching other people audio. The path that I took was being a part-time freelance practitioner/part-time educator. I found it to be personally fulfilling and lucrative. It seemed like, for a time, there was more money to be made charging students to learn how to make records in your studio than running your studio. The more that happens, the more instructors there are needed.

You don't need to have a wall of diplomas like a dentist [to be an educator]. There are probably many avenues for convincing people that your wisdom is worth their time/money . . . There's a huge appetite for learning how to do this stuff for oneself. If you're specialized enough to teach someone how to do it for themselves, even if it seems like you're training someone to not need to pay you, it's still a way to get paid for that skill. The bottom line is a lot of people don't want to make records for a living – they just want to record themselves. So, I'm not going to get paid to make their records all the time because they just want the skills to mix it for themselves, to track it for themselves. But I can get paid to teach them how to do that.

Nathan Lively of Sound Design Live says starting an education-based business was the best career decision he has made. He shares:

To get more of the work I really love, I started an education business that allowed me to create the perfect job for myself. I really like being an entrepreneur, teacher, and coach. My mission is to pursue growth and learning above all else.

(24)

Eyal Levi is CEO and founder of URM Academy, the largest online audio school for rock and metal producers. Eyal says of his career:

I stopped wanting to make music, and I wanted to help music and audio get better by helping other people do it better . . . I always had my sights set on doing bigger things. Like, I never wanted to just be someone that was involved in production or in a band, and there's nothing wrong with that. But, we have to be true to ourselves and understand what it is that we're here to contribute to the world. I've always felt like I'm designed to lead something that affects lots of people and helps something get better. When I was younger, I didn't know how that would materialize. I just knew that it had to materialize somehow. I definitely didn't see it being this, but here it is.

Management

Management positions exist across the audio industry. Rachel Cruz says of managers in audio technology:

While a lot of us have "manager" in our titles, there are fewer and fewer actual managers, like people managers. But all those management skills – you still need them to get the job done, so it's really important to be able to work well with people.

Catherine Vericolli owned and managed Fivethirteen Studios, which specialized in analog recording, for 15 years. She says of managing:

> In this industry, I know very few people who got into this saying, "I'm going to do this thing" and then ended up doing that exact thing. Life doesn't work that way . . . When I got out of school, I was like, I want to build a studio, and I want to make records. I was in my early 20s. Making records is all well and good and fine until you realize: I have to make money. I have to pay bills. I have to make it a career. When you hit a certain age, you're like, I am really over not knowing where the money's going to be coming from next. I need something to be a little more stable.
>
> I hired people to do the stuff I didn't want to do . . . and so I became more of a studio manager. I got to pick and choose sessions that I wanted to work on, which was great . . . We all get into this because we love music, and we love being around it. I got into this to make records. Do I make records every day? No, but do I still get to be part of that process? Yeah.

Tina Morris worked in music production (as a recording engineer and tech) at the Village Studios before she took a job as studio manager. Tina explains:

> Because I had just gotten married, I felt the need to figure out how to settle down in a way . . . I knew that working 16- to 24-hour sessions wasn't really going to fly if I was going to [have a family] . . . I never, ever was like, I would really love to run a studio . . . That definitely was not a goal. I thought my life was going to be sitting in a studio and recording for the rest of my life, and I went with it. What I'm doing now, it's not bad, and I really enjoy the community that we have.

Meegan Holmes of Eighth Day Sound (USA) was building up her career as a touring sound tech and mixer when a back injury forced her off the road. Since Meegan was already working for Eighth Day, the company worked with her to find a new position. Today, as Global Sales Manager, Meegan handles a variety of tasks, including supplying equipment for tours, writing technical riders, pricing, communication with clients, and hiring personnel. Meegan says of the job:

> Never did I think that I would be a salesperson. I didn't start out in the industry going, "I'm going to do sales." Never, ever, ever in a million years did I think I would do it. It was one of those things where as long as I could still feel like I was a part of it and still be involved in some way, that was enough for me. I didn't need to tour anymore.

Some managing and supervising opportunities come about naturally for professionals who grow with a company. Jeff Dudzick, who has worked at

Audible for over 12 years, began in a hands-on audio editing role and over time moved into supervising and managing. While he later moved out of a management role, Jeff says of the experience:

> It's really beneficial to learn the "on the ground level" work. Especially in this world, it'd be weird to have someone come in as a manager [without prior experience] because there's no way they could understand or know how things are done here . . . I did like being around the creative work. I was able to creatively solve problems or give my creative input. Someone would come to me for my input, and I made a decision. I could do that because of [prior] experience . . .

> I'm lucky to be in one of the creative departments, but there's a threshold on the ladder upward where you switch from creative to strategy . . . The more responsibility you get, the higher you go, the balance will shift to be more management heavy versus creative stuff . . . Some people are down with it, but I just was not into it.

Managing at Fender: Rachel Cruz

Rachel Cruz is Program Manager of Electrics for Fender Musical Instruments Corporation in Los Angeles (USA). After graduating from Berklee College of Music, Rachel explored different areas of the industry. Rachel worked for a jingle house in Chicago and moved to Nashville to pursue music production. She went back to school for a master's in communication sciences (with a focus on psychoacoustics) before eventually landing in audio technology.

> I work at a company where we make musical instruments, guitars, basses, amplifiers, FX pedals, accessories, etc. I specialize in the guitar and bass teams. Basically, we have a team of product managers. Their job is to look at what guitar players are doing today, look at the market and see what's trending in the market. Then, they determine what new guitars or what new instruments we're going to make three to five years in the future. Then, that team does all the design work, they help manage it through production, and they eventually bring it to market.

> I basically look across the entire portfolio. So, everybody's lines, everybody's guitars, everyone's portfolio – I coordinate. I really am more like the conductor of all the lines as they go from birth into launch . . . It's a little bit about design, it's a little bit about product knowledge, and it's a little bit about understanding sales and

> marketing needs. It touches all of the departments, but it's basically conducted work for the most part . . .
>
> You get to build an ongoing universe. You keep shaping the whole thing. To me, that's so much fun, and it's definitely not something that I knew even existed back when I started. I shouldn't be surprised that it appeals to my personality so much. For me, it's the perfect mix of strategic thinking, and a little bit of group interaction, but I still have a good mix of stuff that I do independently.

AV

AV is an ideal career path for those who enjoy staying on top of technology, solving complex problems, and are interested in seeing the application of audio in many environments. AV opportunities can be stable (such as full-time jobs) or contract positions. AV has become a necessity for businesses from offices, retail businesses, and hospitals to government facilities. AV has become an important aspect of construction, as well. "We deal with facilities management, so I have a really firm grasp on construction and the process of construction, the timelines, what has to happen before what, and all of that," says Devyn Nicholson, Manager of Audiovisual Services at McGill University (Canada).

AV overlaps with many skills and interests of those pursuing audio. "So many people don't even know this exists. They randomly find it," says Alesia Hendley. Devyn Nicholson, who holds two music degrees (including a master's in sound recording), says of the overlapping skills between the audio and AV industries:

> Pro AV is a little different than pro video on its own, or pro audio on its own. It's really kind of a niche industry now. Pro AV has a connotation of control automation that really enables this super slick thing to happen, for people who otherwise wouldn't have any idea how to operate any of this sort of stuff.

However, those with an audio background may require some additional knowledge to work in AV. Devyn Nicholson explains:

> When we post jobs, what's interesting is that a lot of the applicants are actually from live sound events. They don't really understand from the language on the job description that we're not looking for someone like them. A lot of people like that would do fine on the pro AV side, but they have a few things to learn, maybe a couple of trainings here and there. They could be some pretty-valued and high-powered people in this industry.

From Live Sound to IT and AV: Alesia Hendley

Alesia Hendley went to a trade school in Texas (USA) for audio, but quickly recognized music production would be a difficult career path. Alesia recalls, "I was an audio engineer with nobody to record, and I wasn't making any money."

Alesia pivoted to corporate work, with a goal of working in live sound and going on tour. She explains:

> I found a company that did all kinds of things for hotels – staging, drapes, lighting, the whole nine. My goal there was to be A1, but you have to start somewhere. So, I started doing drapes, I started doing decor, and I worked my way up – load ins, load outs, and finally became A1.

Over time, Alesia realized that path was not the right fit for her, either. "My dream job was to be front of house for Beyonce, and I realized if I don't even like load ins at these hotels, no way I could go on tour," she says.

Alesia landed a part-time job at a multi-purpose facility – a stadium, arena, and conference center for the local school district's major events (such as plays, football games, proms, and graduations). It was in this job she started thinking outside of "traditional" audio roles. Alesia explains:

> When I started exploring the other components of AV, I found all of these spaces and verticals need audio. Even though it's not just me running front of house, I can still be a part of creating this overall experience, which is what I love about audio anyway.

Alesia recognized a market need for AV people who understand IT and networking:

> All these digital consoles are all connected to a network. The network goes down and nobody on our AV team knows how to fix it. We had to call the IT team of the school district, which was a language barrier because traditional IT doesn't really like to play with our AV stuff. They don't want the AV stuff on their network. So, the IT team had a learning curve as well . . . [I realized] the industry is taking a pivot. If I don't learn networking, I'm going to be out of a job in AV sooner or later.

It was a bunch of soul searching. It did take some time. I stayed at that job for an additional two and a half years. I couldn't be like, "let's quit this because I don't know about the internet" . . . I didn't want to just go join a random IT company. I needed to get with a company that understands the vertical that we're already in, or that understands IT that needs to talk to the gear that I love.

Alesia started building a relationship with the owner of Access Networks, an IT company focused on AV, and eventually landed a job. She explains:

It's basically an IT company but everybody that works for this company is an AV person. They've been an integrator in some form or fashion . . . We design networks for AV solutions. About 85% to 90% of what we do is in the home because our clients are people who have home studios or have smart homes, and everything lives on the network. The other 10% of what we do is on the commercial side in corporate environments.

The company paid for a lot of trainings. They've invested in me, and they completely took a risk and created this position for me. We both took a risk on each other, and it's grown into something that's a win-win for both sides. You have to know what you want up front when you go into situations like this so that a company can help you, if they're willing . . .

It's been very, very exciting to constantly pivot but audio led me through all of these different roles. My work is still creative because of the things that create the experience. It needs us. It doesn't exist without us. Yes, I would love to be mixing for my favorite artists or running front of house – that is more creative and more goose-bumpy. But at the end of the day, you're pressing the button. You're not pressing the physical button, but you're pressing the *overall* button. Without you, it doesn't exist. That's the biggest button on the console. You're still creating this experience, and you're still getting knowledge.

AV in Education

Devyn Nicholson is Manager of Audiovisual Services at McGill University in Montreal (Canada). Devyn has a Master of Music in Sound Recording from McGill, where he discovered an interest in system design. After graduating, Devyn was hired to work for McGill's music department on the AV infrastructure of a new building.

I realized fairly early on in my graduate degree I was really good at troubleshooting, and I was always really good at designing systems . . . There was one class project where we had to do a system design . . . it was just basically a big puzzle to me. When I saw that I was like, I could do this. This is interesting.

For a long time, I worked in the music faculty where we kept recording studios and concert venues working 24-7. Then, the next logical step along my career path was incorporating video – going from music critical spaces (and systems) to professional-grade systems that were both video and audio. For the new [music] building project, we were mapping infrastructure runs, and that's where I started cutting my teeth on reading professional [construction] plans . . . [Manager] Peter Holmes would say, "I need a cable run to go from this room to that room. Show me the most efficient way."

I wouldn't say I'm great at math or electronics or any of that. What it is – is systems thinking. The best engineers, whether they're mixing music or whether they're recording, need to understand the studio they're using (to some degree). In my mind, it's always been about understanding the facilities that I'm using as deeply as possible so that I do not have to wonder about how to accomplish a technical thing (which will then enable the creative more aptly).

For Someone with an Audio Background Interested in AV

What are you doing when you're doing a live event? You're doing a quick and dirty system design every time and you're setting it up and you're tearing it down, and then the next gig comes along . . . It's a super fast, super small version of what we do in the fixed-installation side . . .

If someone has really good systems-thinking chops already, then I would say we can start using them right away on the system side for things like technical support. Things like, this microphone is not working, or this mic is plugged into this device on this input. Where AV starts to become challenging, especially as we move forward into the future, is that everything is a protocol on the network. This mic gets plugged into this black box that's connected to that network jack over there, and I need to troubleshoot it.

Crucial Skills

InfoComm is your friend in a huge way. You want to get CTS certified. I'm not certified in CTS, but my degree is equivalent to that in a lot of ways. A lot of companies go through InfoComm. We have training budgets for our people on our teams. Certification is expensive if you're going on your own.

Learn a bit about [basic] construction/project management.

Learn enterprise networking. You've got to know how that network works, because in 10 years even it's gonna be night and day, like 10 years ago was today.

Personal Priorities and AV

If you have a low risk tolerance, then this is definitely a good fit for you . . . In pro AV for higher education, the most attractive thing is going to be your work-life balance. Hands down. You can do what you like to do, but you might not get paid quite what the private sector gives you, if you look at it strictly from a money standpoint. If you look at the whole picture, though, how many weeks of vacation do you get? How long is your work week? How flexible is your employer [when] you need to not show up? You can't beat it. You really can't, and this is where that value is. Something like this really attracts younger people who are family-focused. That's why I'm here.

It's hard to find a permanent gig. There's a lot of term stuff. Sometimes that term stuff – if you're the right person – will turn into a permanent gig. It's not impossible. The private sector can be a little different. In pro AV, if you're working in the private sector, it's more akin to the kinds of pressures as a studio tech.

Areas in Demand

It's not required for everyone on my team to know everything. I have people who are specialists in just about every area. We have a system designer, we have a programmer, and we have people who do operations. On a team like mine, there's room for that person who is a rock star in projectors, or who is the audio specialist. When we say audio, a lot of the time in AV, we're talking about DSP units. So, we're talking about rack-mounted processors that you do graphical programming inside them. These things are what drives the audio in these classrooms, and in fact, in concert venues most of the time now because they're so cost-effective.

I think a lot of people who could go and get trained – to know how to do best practices in setting up those products – could become rock stars in the AV industry. The people (by and large) who are certified to the top degrees of these products don't have the ears on them. You don't need it as you're building [systems], but you do need it if you want to tune the system properly at the very end.

I will get a call every once in a while from someone on my team who says, "Something sounds weird in this room and I need help." A lot of pro AV companies will send an installation technician, or they'll send a programmer. Neither of those individuals have the skill to be able to say, "That speaker is out of phase with that one. What the hell is that noise? It's in this frequency range." That niche/gray area is not gonna be entry-level, necessarily. It's a little bit up the ladder . . .

One of the things that we're constantly asked to comment on is acoustics. While I have a little bit of expertise there, it's not a huge amount. It's not our role to be answering these kinds of questions – It's about the performance of the space after our systems get installed. We do not have an acoustician on staff, so this is something that you're going to need an external consultant for.

There's a real divide between consultants that I see. One is music critical, where it's like, I design concert halls. There are those A-list concert people, and that's very expensive. Then there's those who are at the other end of the spectrum that are mechanical engineers or [similar]. They don't care about what the sound quality of the room is. They want to know that the noise in that room isn't making it out into the hallway.

If you get someone who's got this bit of understanding, who's got a bit of experience with the A-listers, this could be a niche market, as well. It's very much closer to construction than it is to creative music making. But, it has a huge impact on things like how long can I work in my room [comfortably], or current trends in how to deal with sound in rooms – diffusion versus just absorption and reflection.

References

Arnardottir, Steinunn. Personal interview. 30 June 2021.

Campbell, Madeleine. "Natalie Hernandez: Electronics Technician." *Women in Sound*, no. 6, 4 Feb. 2019. www.womeninsound.com/issue-6/natalie-hernandez. Accessed 12 May 2021.

Clemens-Seely, James. Personal interview. 20 Dec. 2020.

Cruz, Rachel. Personal interview. 9 Oct. 2020.

Diggle, Lucy. Email interview. Conducted by April Tucker, 1 Oct. 2020.

Dudzick, Jeff. Personal interview. 1 Oct. 2021.

Edwards, Tommy. Personal interview. 29 Oct. 2021.

Frick, Anna, et al. "Ask the Experts – Audio Restoration & Archiving." *YouTube*, uploaded by SoundGirls, 1 May 2021, youtu.be/_Mn1Y0F8mbs.

Gash, Justus. The Modern Music Creator Group. *Facebook*, 24 Feb. 2021. www.facebook.com/groups/themodernmusiccreator. Accessed 25 Feb. 2021.

Gaston-Bird, Leslie. *Women in Audio*. Routledge, 2019, pp. 219–221. doi.org/10.4324/9780429455940.

Hendley, Alesia. Personal interview. 17 Dec. 2020.

Holmes, Meegan. Personal interview. *What Makes You Stand Out*, SoundGirls, 20 May 2020, youtu.be/wYamiOK8Y6A. Accessed 21 May 2020.

Lawrence, Michael. "Advice in Hindsight." *Signal to Noise Podcast*, ep. 56, ProSoundWeb, 8 July. 2020, www.prosoundweb.com/podcasts/signal-to-noise/.

Levi, Eyal, guest. "Eyal Levi on Smashing His Demons & Fostering Gratitude." *The Modern Music Creator Podcast*, ep. 58, 3 Aug. 2020, pod.co/the-modern-music-creator.

Lively, Nathan. "Nathan Lively." *Get On Tour: A Sound Engineer's Guide*, edited by Nathan Lively, self-published, 2018. p. 24.

Masingale, Langston. Personal interview. 14 Sept. 2021.

Morris, Ashton. Email interview. 15 Sept. 2021.

Morris, Tina. Personal interview. *Once You Have the Gig – What Makes You Stand Out*, SoundGirls, 20 July 2020, youtu.be/QJhdm86kIlM. Accessed 21 July 2020.

Nicholson, Devyn. Personal interview. 23 Nov. 2020.

Tucker, Ryan. Personal interview. 5 Oct. 2020 and 27 Dec. 2021.

Vericolli, Catherine. Personal interview. *Once You Have the Gig – What Makes You Stand Out*, SoundGirls, 20 July 2020, youtu.be/QJhdm86kIlM. Accessed 21 July 2020.

Part 4
Industry Perspectives

19
The Audio Industry Around the World

Some challenges of working in the audio industry are universal, but other barriers are unique to different locations. While some countries have a variety of established industries (such as music, film, and games), others are still forming a basic structure. Some countries have limited opportunities to build a career in audio or even to specialize in a single discipline. Gear can be a major barrier – both in cost and accessibility. Some regions are oversaturated with audio school graduates while others have few or no formal education opportunities. These types of barriers can change how an emerging professional approaches job seeking, networking, the skills they target to learn, and the career paths they target.

At the same time, some countries provide unique advantages, such as language skills, government subsidies of the arts, or lower rates that can attract global opportunities. Professionals from different countries around the world discuss the unique challenges and advantages they have faced in their careers due to their location.

Brazil

Fernanda Starling, originally from Belo Horizonte, Brazil, started her career as a journalist in Brazil before pivoting into audio work. She relocated to Los Angeles in 2010 to pursue audio education, where she has since worked as a production sound mixer and recordist/Pro Tools operator for broadcast.

> There is little access to formal education in audio [in Brazil]. Besides that, the limited access to professional high-end gear may be one of the biggest differences. Brazil's tariff regime is ridiculous! Imported manufactured products are subject to a wide range of taxes at all stages of the chain. Because of that, the final price of an audiovisual product is two to three times more expensive than it would be in the US. Therefore, independent

DOI: 10.4324/9781003050346-23

studios in Brazil are not as well equipped as the American ones. One of the first lessons I learned from my first studio mentor, André Cabelo, was that gear is not the most important thing in the business: neither for making a good mix or to build and keep your clientele. What counts most is mastering the craft, having a relationship of trust between artist and the engineer, and creating a welcoming environment.

Incentives

Federal government incentives play a big role in the Brazilian audio-visual and music production world, particularly in the independent scene. Maybe because of that and other cultural aspects, independent Brazilian artists get more of a chance to perceive music as more of an art than as a product.

There are numerous kinds of tax relief, i.e., tax benefits and incentives at all levels of government (federal, state, and local) in Brazil . . . When an artist receives a grant, they can dedicate themselves to their craft, record, and promote their album without worrying about working multiple jobs to fund their musical career. Besides helping musicians directly, these policies also benefit studio owners, audio engineers, and other professionals involved in the Brazilian music industry.

I will say I was shocked when I arrived in the US in 2010. I was used to a non-stop recording environment back in Brazil, and it seemed that [in the US], very few independent artists had the budget or opportunity to go to the studio and record full albums.

Broadcast TV and Film

When we talk about TV programmers and filmmaking, it is almost unfair to compare the production capabilities of both countries. This is because of the difference in the size of their populations, and the difference in the ability to recover production costs domestically. It is often cheaper for Brazilian media companies to buy series and films from the US than to produce their own. In Brazil, the content produced outside the TV broadcasters, including film, is reduced and depends on government incentives.

Broadcast TV is an extremely concentrated sector in Brazil, dominated by Rede Globo. They are one of the largest commercial television corporations outside of the United States, and the largest producer of telenovelas (soap operas) in the world. Generally speaking, the US is famous for producing and exporting film, while Brazil is famous for producing and exporting telenovelas. It's actually really impressive what the

Brazilian TV industry has managed to create: there are three original soaps going out every evening, and each series lasts approximately 200 episodes.

Colombia

Luisa Pinzon has worked in music and post-production in Bogota, Colombia, after completing a bachelor's and master's in music with an emphasis in sound recording. She was an audio practicum at the Banff Centre for Arts and Creativity (Alberta, Canada) from 2018–2019.

In Colombia, everything is centralized either in Bogota or Medellín, but mostly in Bogota . . . There are a lot of gigs in live sound and a very small video game industry. There's a lot of TV work, but there are very few films . . . So, it's actually hard to get a job in art – I know anywhere in the world, but here it's pretty hard . . .

There are two big TV channels here . . . Those are the places that permanently have people working. Then, there are other small production companies, like the one I work for . . . Audio is always at the end of the budget, so they don't hire me that much. The rest of the job is mostly done by these two TV channels, and then there's basically nothing else.

There's a lot of work from Mexico here because of the language. Mexico has a lot of film because it's very close to Los Angeles. Sending jobs to the States is way more expensive than sending jobs to Colombia. In Colombia, we have a good standing in the industry that we're good at what we do, and we're cheap . . .

There are two major [music] studios now. There used to be one major studio where everything used to be recorded, and that's where most of these engineers would want to work one day. There are tons and tons of mini studios . . . You could get a job for mixing, mastering, for sound design, or for post. You just need to accept anything. You need to have a lot of things at the same time in order to make a life, or at least a decent income.

Everyone considers leaving. There's a saying here we have in Colombia (that is very sad) is that most of the brilliant minds – their one and only dream is to leave the country . . . People who want to stay in South America mention going to Argentina (Buenos Aires).

Challenges in Colombia

[In audio], not knowing English is super hard. Most of the terms of this profession have been created in English. Like, if someone tells me the word "loop" in Spanish, I wouldn't know what it is. The word "enable," like, enable the track to record – these kind of things. Most people just learn the terms as if it was a new word.

The box [for audio equipment] comes in English and maybe has one page in the manual that is in Spanish. You can, for instance, change Pro Tools to Spanish. But, it is so well known to do it in English that most of the people who are Spanish speakers prefer operating in English.

Getting gear in Colombia is hard mainly because of the exchange rate. There are a lot of brands that you just cannot find easily. Microphones are really hard to get. When we go to big cities like LA, New York, or Toronto, the first thing a lot of people do is go into gear stores. It's very common for Colombian sound engineering students [to do] that.

Differences Between Colombia and Canada

There is a huge difference between working in Canada and Colombia. There's the same thought that you need to work endlessly without being paid in order to be someone. But I feel that people do respect that the sound engineer is a job – a paid job – more in Canada than Colombia. In Colombia, there's a lot of the thought of "you're my friend so you help me," and "you're my friend and you have these microphones, so I can just borrow them."

In Canada, if you go out and you have a gig, you can actually charge the amount you should. In Colombia, you almost need to not charge in order to get the job because there's always someone that is doing it for free . . . I feel that happens a lot – mostly in music. There are a lot of people looking for credit (or people looking for this sort of work) thinking that you need to work forever without getting paid in order to one day get a Grammy.

Jamaica

Jason Reynolds was raised in Kingston, Jamaica, and learned audio through his church. Jason relocated to Toronto (Canada), in 2004, and works as a touring live sound engineer (front of house, monitor mixer, and tour/production manager) and consultant for houses of worship.

We have great recording studios in Jamaica . . . very professional. Currently, there's no real structured way for somebody to get into live sound. In Canada, a lot of people may go and work for a production company and freelance, learn, and get into the industry that way. There are fewer opportunities like that in Jamaica – they just start doing it and get discovered. A lot of people come through the studio environment, and they make the transition to live . . . There are still opportunities because reggae music is really Jamaica's biggest export. As soon as an artist gets to a certain point, most artists carry one engineer and then they move to two . . .

There are a few major production companies that are effective at what they do. Those companies are fairly big production companies that do most of the big shows on the island, and then there are a few smaller companies as well. There are those opportunities, but not as widespread.

Jamaica is the biggest country in the Caribbean, but some of the other bigger ones might have more opportunities (like Trinidad and Barbados). You might not have as much opportunity in a smaller country like Antigua or St. Kitts or something like that, although I do have friends that have come out of the scene in St. Kitts, for example. They're full-time professionals in the industry, so it's not unheard of. It's a proportion – just being a culture of three million people versus St. Kitts which has 50,000 people.

In terms of Jamaicans [relocating], in general, people tend to move where there's already a large community of people there from their home country. Jamaicans mostly end up in Miami, New York, London, or Toronto. The reggae scene out of New York is probably just as vibrant as the reggae scene authentically in Jamaica. New York has a very, very vibrant reggae scene.

A lot of [people in Canada] from Jamaica reach out to me. There's no negative to "community." Community is always a positive. So being a part of a community – and music being such a vibrant part of our culture, that's never a negative thing. But my route [into the industry] was through the church. It's partly influenced by my heritage, too, because people from Jamaica recommended me for certain gigs.

The Netherlands

Aline Bruijns, based in Harderwijk, Netherlands, is known for her sound work on feature films, shorts, museum audio tours, and project mapping. Aline is on the board of the VCA (Dutch film audio society), and co-chapter head of the SoundGirls Netherlands chapter.

What might be an advantage for Dutch people is for us, it's really easy to talk English most of the time. Sometimes our Dutch is really almost changing to English. We don't dub American films or English films or French films, so we hear a lot of different languages, and we get a lot of different languages in school. So that maybe makes it easy for us to change because we hear those languages really early on. In the US or England . . . everyone is speaking in English, which is fine, but it can be troublesome if you have to do something in a different language . . .

I'm right in the middle of the Netherlands. For me, it's not a disadvantage that I'm in Harderwijk . . . If I have to work with voice actors, for instance, I just try and see where they are located. If they are all located in Amsterdam (or in the area), then I just hire a studio there and I record them there. It is a lot easier for me to go there than for them all to have to come here . . .

You can find work anywhere [in the Netherlands] . . . But you have to make sure that you're seen – that you're noticed – that you find some way to get a foot in the door. It took me quite a while and I had to work really, really, long hours and did a lot of things – starting out for free just to get the projects going, to get your name flowing, to get experience, try stuff out. It's definitely possible, especially with the internet.

A lot of freelancers can work from home. They don't need all the rooms anymore as long as there's a place where they can have viewings with a director which is better located or has a nicer feel or look to it than someone's home studio. It's easier for people to work from home, and it's more accepted as well . . . The bigger post-production facilities in the Netherlands get the larger budget productions.

Singapore

Mabel Leong is a sound engineer in Singapore. She holds a diploma in Music and Audio Technology from Singapore Polytechnic and a bachelor's degree in music production and engineering from Berklee College of Music (USA). Mabel returned to Singapore in 2014 where she has worked on documentaries and television shows; for BDA Creative Singapore on commercials, advertising, and online media; and for Night Owl Cinematics, where she does production and post-production sound for online content.

One will definitely find the Singapore audio industry being way smaller than other countries . . . That's not to say there isn't an active industry here, but when your local poll [about] artists being the top non-essential job makes waves worldwide, you know where the industry stands in Singapore's society.[1]

With that said, the audio industry is tight, and most folks will know each other. Word-of-mouth holds a lot more weight than your audio education, and I like that down-to-earth attitude we have in our community. For example, I was introduced in my early post-production career to a veteran location soundman, Hussin Ismail, who saved me years later when I was the soundwoman for a foreign production. The team's overseas-rented adapter cables broke off with four hours to the next location. Without a budget for renting local cables, and the antsy producers demanding a resolution fast, I contacted Hussin in a desperate bid. He [reached out to] his network of "soundie" freelancers and found enough to cover me till my production ended. He espoused that we should help support one another within our small pool of skilled freelancers . . .

My work is 100% local these days. My previous company pitched regionally; thus, I worked on post-production projects from the Philippines, Australia, Indonesia, and other ASEAN (Southeast Asian Nations) adverts . . . I know some post-houses do work with our neighboring countries on the regular, but it all boils down to the budget and the networks made by respective parties.

Language-wise, Singapore has multiple Asian languages on hand, so we truly can operate multilingually. My first project ever (Days of Disaster – on Netflix) was a collection of documentaries that were subsequently re-narrated into our other languages for their respective broadcast channels. Language familiarity is useful, but not as much a deciding factor as one may believe.

Companies these days would prefer a jack-of-all-trades who can execute as many skills as possible within one employee, especially when it comes to media. It's telling when an audio engineer intern position here is asked to cover post-prod, location sound, sound design, studio operations *and* audio mixing. Sound designers at my previous company, Mediacorp, are required to have a strong music background and be able to play some musical instrument as well.

At this point in Singapore's media scene, I would implore everyone to explore as many avenues related or complementary to audio skills (after you've made audio your niche skill! Jack of all trades, master of one). For example, video production and livestream operation are popular skills listed in media job requirements these days in Singapore.

Challenges

Local jobs for live audio, media production, or music production are very few and far between. I recall scanning both USA and Singapore job listings for audio and found Singapore's quantity of positions paled in comparison. I landed my first job eight months after graduating and that was far too long if I did not have familial support . . . I personally do believe most audio experts and professionals have resorted to a mix of [jobs] to stay afloat: either holding a non-related day job and freelancing audio on the side, or wearing both hats in the music and audio field . . .

Gear is limited via distributorship here or [due to] international shipping costs. However, those have not stopped local studios, musicians, and creatives from grabbing what they need. Singapore's main language for business, politics, and education is English so we have no language barrier . . .

The advent of audio education avenues is beginning to ripple in the industry at this moment . . . It is a new experience for the veterans [having] increased numbers of young skilled workers coming in, but I think this is a good problem to have.

Hungary and New Zealand

Eliza Zolnai is a production sound assistant who studied sound design and production at film school before working professionally in the field in Budapest, Hungary. In 2020, she temporarily relocated to New Zealand, where she worked during the COVID-19 pandemic. She has worked on major Hollywood films including Terminator 6 and Dune (2021).

Budapest is super busy [for production]. It's pretty much the second biggest industry after London in Europe. We shoot a lot of American studio movies there, so it's always super busy if you want to work. It's not just the American productions – Hungary has quite a history in filmmaking . . .

[In Budapest], I was pretty much working already when I was [a student] pretty much from day one. There was such a shortage. I started as a trainee, obviously. I did lots of short movies and university movies, and then I was a trainee and bigger stuff. [After graduating], I started to work right away. In New Zealand, the people I met didn't come from schools. They mostly came from training each other . . . there's quite a few soundies who came from England originally . . .

When LA studio movies come to any country, they just turn that place into how they work, and you adjust, or you don't go to work there . . . There's always a bit of gap between the two crews (cultural and financial). If an American production comes to Hungary to shoot, there's always the Americans and the Hungarians. There's always going to be a bit of gap between those people. When do you start speaking in English or who do you talk to in Hungarian? There's quite a big cultural difference between big American LA actors and actresses and Hungarian actors and actresses, who you pretty much grew up together with because you went to the same university, and you've been watching them in theaters for years by then.

Budapest is so much busier than New Zealand. I think compared to others my age, I'm lucky to say that I arrived [in NZ] with a lot busier CV than a lot of people here (and just the fact that I have a diploma for what I'm doing). It was really easy for me, and I feel pretty lucky about it. It was really easy.

I think Kiwis [New Zealanders] are pretty chill people so things go a lot more casually here than it would in Budapest. In Budapest, there's a big competition, and you would be asked to work for days for free before you get into a team. They would never hire, like, "You look like a nice person, and I like your CV."

People in New Zealand are a lot more [relaxed] about their life. If I want to be with my kids for the summer, then I'll just say "no" to work, and that's fine. That's not really fine in Hungary. In Hungary, you have to say "yes" to the job otherwise people just start forgetting you. I think maybe there's better work-life balance in New Zealand. The approach to it is healthier.

[The attraction for production] is mostly the locations, and obviously that there was no COVID here. At the same time, New Zealand is an extremely expensive country to live in. Food, accommodation – everything is just crazily expensive here. Even compared to the UK or LA or New York, it's just an expensive place, which obviously means that it's expensive for productions to come in . . . It's a tricky thing for New Zealand, for sure, as it's so isolated from the rest of the world.

The Philippines

Agnes "Aji" Manalo is an audio engineer/project studio owner and audio educator in Quezon City, Metro Manila. Aji works primarily in music recording for composers, film scores, music directors, and musicians.

A huge challenge working as an audio professional here in the Philippines is the access to pro audio gear. Only the major broadcast networks and audio post-production facilities are able to acquire the professional standard audio equipment, as the cost and prices of such can only be afforded by the profitable business enterprises. Only those employed in such facilities can have access to this gear. They usually undergo training as users, paid for and sanctioned by their employer, and the training is usually given by the local or regional distributor of the brand . . .

Learning audio in the English language is not a problem in the Philippines since the official medium of instruction at all levels is English and Filipino (Tagalog) . . . As of the moment, as far as I know, there are five tertiary level institutions in Metro Manila who give out audio-related training and degrees . . .

In the major cities in the whole country, freelance audio work is mostly supported by an underground economy (undeclared businesses), creatives or producers who supply content and services to the advertising industry, the independent film industry, and to private companies which require a minimal budget cost or expenditure.

There are also a number of freelance live sound engineers who work for a lot less money to practice their craft, as most of them do not have their own gear and have not landed on a regular audio engineering post- or studio-related jobs. There are also a number of audio engineers who set up their own single proprietorship companies (or studios) to establish themselves as a business entity, to be able to work and circulate in the industry.

Canada

Mariana Hutten is a producer, mixing and mastering engineer in Toronto. Mariana worked in video games (in Montreal and Toronto) while building up her career in music production. James Clemens-Seely is Senior Recording Engineer at the Banff Centre for Arts and Creativity (Banff, Alberta). James is a graduate of McGill University (Montreal, Quebec) where he also started his audio career. Camille Kennedy is a production mixer based in Toronto and attended school at Fanshawe College in London, Ontario.

In Canada, the entertainment industry is centered around Vancouver, Toronto, and Montreal, with established industries in music, film/TV, and video games in all three cities. There are industries for content in both English and French, as French is the official language of the province of Quebec (where Montreal is located). Mariana Hutten says of working in Quebec:

> You have to be bilingual or French-speaking, especially because the studios do have a lot of French music . . . but video games are one of the few places in Montreal where the language doesn't matter a ton. There are French companies that hire [English-speaking] people – there are some bilingual operations.

Camille Kennedy says all three major cities have film and television production work and unions, which helps protect the rights of workers and conditions on set. Camille suggests it would be difficult to start a career in production sound without living in one of the major cities:

> You definitely have to be in the city. In Toronto, the studios are so separated and far apart. There's the East End studio district, and then there's all the studios way west. To get to some of these stages, it's over an hour drive, and that's just in the city.

Montreal has a lower cost of living than Toronto and Vancouver, which is an advantage for those in the creative arts. Mariana Hutten says of living in Montreal vs. Toronto:

If you're starting out [in Montreal], you probably don't need to do many hours of the job to be able to cover your expenses, whereas [in Toronto] you have to work really hard because you will need full-time work to be able to pay your bills as you work on your audio work. I have friends in Montreal who do a lot of work for six months and then don't work for six months. Or, they will work a day job for six months, and then for the other six months, they have enough saved up that they can just focus on that one thing. There are more music opportunities in Toronto, but there are artists who live in Montreal who come to Toronto once in a while to meet with an industry person or be at an industry event.

James Clemens-Seely says of working in music outside of Vancouver, Toronto, and Montreal:

It's not impossible, but most of the people that are succeeding at that are really into a niche and are pretty isolated, or were established [prior]. The people I know that are in Calgary mixing cool records in their basement worked in studios in Montreal or Vancouver and built a network [through that]. They can now sustain from somewhere else. But, building a network in the "sink or swim" phase of their careers – I can't think of anyone who hasn't been in those three cities.

Outside of those three, it would be pretty hard to find an internship. It wouldn't be hard to find a studio that has made some records you might have heard of, and people who are skilled professionals, but they might be a little close to starvation. It does seem like in Winnipeg or Calgary, the studio scene doesn't have much room to grow – just because the number of people who want to get into it every year is in the hundreds or thousands. The number of people who retire from [the studio scene] is single digits. In Vancouver, Toronto, or Montreal, there's more turnover.

The way to succeed in smaller marketplaces . . . is to attach to art that resonates. That can happen anywhere – someone from the middle of nowhere's-ville makes a record that people love. But if you're purely knocking on doors, trying to get a foot in and learn the skills, you're going to meet a lot of people who don't have time for anyone like that outside of the cities. There's more opportunity in a dense area. You can suffer 20 rejections in Vancouver and not have gone through every opportunity; whereas, you get 20 rejections in Calgary, it's pretty incredible that there were 20 [opportunities].

Canadian Content Quotas

Regulations have existed for 50 years that help Canadian content creators ("CanCon") reach Canadian audiences and international markets. Television and radio broadcasters have a quota of Canadian content it must air, plus television broadcasters are required to invest in the production of Canadian content. James Clemens-Seely explains:

> It forces people to consume [Canadian content], basically . . . The number of films or records that are made in Quebec is way more per capita than get made [in similar-sized communities] because it has its own start to finish industry. A lot of it is coming from government funding.

These quotas mean there is funding (grants and tax credits) to support creators (including sound recording, broadcast, film, and new media). While this can mean an advantage for project budgets, Mariana Hutten says the drawback is her work sometimes revolves around grants:

> What happens to me a lot is an artist is applying for a grant, and they're almost sure that they got it. Then, they don't get it, or they don't get the amount that they were looking for. The majority of people just flat out give up. Never gets made. I've lost a lot of people. It takes time to go for coffee and talk about a project and start a pre-production process. Even if it's a couple of hours, or like four hours to talk about a project, it's still time from your schedule.

> The impression that I get of artists in the US is because they don't have a granting system, they're not expecting anything given to them. They have more of a mentality, "I'm going to save [money] to afford it." They'll take a job cleaning the building by day and at night produce stuff. You hear those stories from people in the US. In Canada, it's more grant-type of stories, like, "I tried to get that grant five times and got rejected five times." Those people have day jobs, but the story always revolves around getting a grant.

Nigeria

Phebean AdeDamola Oluwagbemi is an audio engineer and production manager based in Lagos, Nigeria. She is the co-founder of Audio Girl Africa, an advocacy group helping women build careers in the audio industry in Africa. Phebean started her professional audio career interning for Azuza Productions, one of the top live sound companies in Nigeria.

> Africa is a very creative continent and it's a large continent. Over half of the African continent's population is less than 25 years old. We have a very young and vibrant population. The challenge for most developing countries is there's a sector that is super rich, and a larger percentage [of the population] in abject poverty. In Nigeria, there's a middle class – a substantial number of people making a decent income and actually doing great at

their jobs. That's why art, entertainment, and music can thrive even with a large number of people who are poor. There are still [those] who have buying power or have disposable income.

When we talk about key places in Africa, the entertainment and art tribes, we're talking about Nigeria majorly because Nigeria is the largest consumer of music products. Ghana also has very strong entertainment but it's not as strong or as big as Nigeria and South Africa. Kenya is also big on digital products (mostly digital music production). Zimbabwe has one of the largest music festivals in Africa. Lagos is the entertainment hub of Nigeria, followed by Port Harcourt then Abuja. If you really want to succeed in the industry, you have to move to Lagos, or you definitely need to have some ties or some connection with Lagos.

We really don't have much access to audio engineering education. Most of us learned through internships (working with somebody who's already working in the field). We don't have any paid internships in Nigeria in the industry, except for maybe established companies, which are very rare. If you want to become a music producer, you have to find a mentor in the field to learn from, or you just teach yourself. You go to YouTube, get your laptop, and begin to create something for yourself . . . We have record labels and record production companies (who take care of artist management) because that's the biggest revenue income stream in the Nigerian industry. The fastest way to make money is if your artists can quickly make it.

You definitely have to find something else that you can do to be able to support yourself. Let's say you're a recording engineer in a studio. The basic salary you get is about 150,000 Naira (US$365/€325) a month. If you're living in Lagos, you can maybe get a shared apartment for 150,000 a year (US$30/month, €27/month). If you are talking about locations where it's easy for you to connect with your people, you're talking about close to 300–600,000 Naira per year.

Even the music industry has other streams of income. Due to the rise of COVID [in 2020], a lot of people are building home studios to record at home. A lot of audio engineers are making money consulting and building home studios. As much as people think that Nigerians don't have money, the people who have money are able to create studios for themselves.

We have less than five big production companies. One company is Azusa Productions. When you mention that [company] in Nigeria, you know that these are the people who want to do major concerts, who do sound equipment, they are very good with live sound, and also have a studio that takes care of recording. Most of the time, a company can decide to set up a studio and also run concert production, run recording, like everything that has to do with music production. It's not like you have a company that just does recording alone, or you have one that just does concerts alone.

Film Industry

The movie industry in Nigeria is actually very big. It's growing and making a substantial amount of money right now. We do a lot of collaborations with Ghana because we're neighboring countries. South Africa has a film industry also, and Kenya. Those are the major countries when it comes to entertainment and music.

In Nigeria, over 2,000 movies are produced in a year. A few of them make it to Netflix or make it to the cinema each year. All those others are just shared on YouTube, or local markets, or black markets. It's just a small number of people producing movies [professionally], then there are a lot who are producing movies grassroots.

Because of exposure to international movies (like Hollywood movies) in recent years, we see that something is missing in our film industry, and that's led to a rise of sound designers. A lot of them had to go out of the country or study on their own to get the knowledge. High-end movies outsource sound design . . . so we still have a long way to go when it comes to post-production for films. We only have a few people who are good at sound designing, and how many movies can they do? Let's say you have 50 or 100 people compared to the thousands of movies made here in a year.

Other Challenges

Sadly, for any business in Nigeria right now, you have to consider power supply. It's a big issue. While some parts of the country have been able to stabilize the power supply . . . you still have to try and get a power generator into your budget. That takes a lot of money from a lot of startup [business].

It's very expensive to get gear. Everything is being imported. It makes it hard for somebody who's just starting out with basic income to be able to get gear.

A lot of people [have] no choice and don't have the means to move out of this country. But if young people in Nigeria have the means to leave the country to go study or go work somewhere, they would gladly take the opportunity. It's not because they can't make it here, but because our leaders have not created an environment that would enable us to build a business without frustrating [us] with so many things that are supposed to be basic amenities.

A lot of people still go to South Africa, especially when it comes to entertainment. Nigerians go to South Africa because its industry is more structured, and it's easier to network with people in the Western world. When a lot of Western artists or Western producers want to come to Africa, they think of South Africa first.

We're learning from [structured industries] every day, thanks to the internet, YouTube, the social media community, and blogs. We're able to learn and see what [others] are doing, and be able to try to replicate it, and create a structure.

China

Fei Yu is a music producer, music editor, and music supervisor in Beijing, China. Fei's company, Dream Studios, pairs Chinese filmmakers with Western composers. Fei started her career as a music editor for the China Film Group, and she is a graduate of Beijing Film Academy (Music Recording and Film Music) and McGill University in Canada (Master of Music in Sound Recording).

In China, if you want to work in the movie industry, you for sure still need to be in Beijing. Most of the production and post-production companies are in Beijing . . . Shanghai is more like the international city in China. When foreign movies want to do dubs into Chinese, they choose to work in Shanghai. For games, it's more about Hangzhou, Guangzhou, and Shenzhen. The gaming industry is actually booming in China.

In China, the relationships between people are a little bit more important because of cultural things – that when we're eating, we share food together – when we're doing things, we're doing everything together, rather than separately doing things . . .

In 2006 (the year I went to university), there were not that many schools providing this degree. Nowadays, there are dozens of schools having this degree. They are all called sound recording (or film composition), and they teach you really specific skills, like learning music recording, or film composing, or maybe boom operator, or how to be a mixer. The game industry boomed a lot during the last 10 years, so that's why they started to have an [academic] major to study game sound.

When I graduated in 2010 (from Beijing Film Academy), there weren't many audio students. We could find a job. When I went to work for the China Film Group, that was the first year they started hiring people. But nowadays, it's changed a lot. My university has almost 100 graduates a year. That's a lot. Many of them become freelancers. Many of them already don't do this job anymore . . . some students graduate from really good schools and they want a high salary rather than working longer days. That's why it's a little bit tricky for graduates. The cost of living is really high, and that's why a lot of them go back to their hometown. After several years, they try and then they give up. But, if you're really good at what you're doing, your career still can be really good . . . Most of the studios still want to hire really hard-working sound editors.

[Foreigners] who know the language could come to China and could start their own business as whatever – a recording engineer or as a mastering engineer, and that won't cost you too much. The market is big . . . The industry still needs these kinds of people.

Music

In China, our music industry is not very old vs. North America. We don't have worldwide well-known famous studios, so the studios are not attracting people to go there. For example, so many really big records were made at Abbey Road [in the UK], so it attracts people [to work there]. But in China, our industry is young comparatively, so people don't know where to go to work. They could go to some of the small private studios, but when they go there, they feel a little bit disappointed. You have to work for 12 hours every day and you aren't making much money. You're living in Beijing where it's so expensive for everything. Sometimes you have to get the coffee for everyone rather than doing the real recording job. So, that's why they may be working there three or four months, and then they are gone . . .

All of our big recording studios are also owned by the government. For example, China Records is a government record company. They used to import so many records, like the Beatles or [others from] outside China. They recorded traditional Chinese music at that time. The CCTV music recording studios are also owned by the government.

Pop music started booming in Hong Kong and Taiwan first and then back to mainland China (20 years ago). Because the industry started booming, some private recording studios started [opening], as well.

If you want to run your own business, you can provide really good service, and if your location in Beijing is good, then maybe you're going to have a really good business for yourself. Some of the younger generations have started their own private studios in Beijing. So far, they are doing really well. Their decorations look cool and their service is really good, so they get clients working with them.

Challenges for Westerners Working on Chinese Entertainment

For Western composers and directors, sometimes they don't quite understand censorship. They don't know why we have to change the picture at the very end. They all want a lock date because they have to deal with the delivery, and also because they feel like it might destroy the artistry of their music.

Some composers think that we just put some Chinese instruments on top of an orchestra and that's called Chinese music. I need to educate them more about the Chinese culture, or describe more things to them, and then maybe they will start to understand. We need more communication, and we need to educate more . . . and to lead composers to gain some education about the Chinese background.

For example, during the last five to eight years, Chinese comedies are really getting more and more popular. People [here] are working under pressure. They just want to go to the movie theater to see a comedy. But, because we have our own language and our big, deep culture, most Western people may not understand why people are laughing at something. That's why I seldom do comedy with Western composers because I do notice they can't understand why people are laughing [at something]. Or, sometimes suddenly, we're having traditional Chinese music, and they don't understand why. Most of our movies are romantic movies or big action movies (that's what I concentrate on).

Italy

Tiziana Mazzucco is a freelance dubbing engineer for post-production sound in Milan, Italy.

In Italy, audio jobs are underestimated and therefore not paid properly. It's frustrating to know that a freelancer averages 20 euros gross per hour while a babysitter earns 15 euros (with all due respect to babysitters).

In Italy, there are three untouchable weeks of vacation: the two central weeks of August and from 24 December to 1 January, so we usually try to finish the most urgent work before those days.

As far as I know, in Italy, there are no universities related to audio. There are only private (expensive) academies or courses, but no degrees. It's a pity if you consider that in Italy, the public universities have [very small] annual fees compared to other countries.

The audiovisual industry is paradoxically shrinking to a few clients, despite the increasing amount of content produced. In order to be a provider of streaming giants, like Netflix or Amazon, it is necessary to respect an endless list of standards that a studio can achieve with large investments. Only a few can afford it, and this is limiting the market. This means less studios, less employees, and more difficulty in finding work. In addition, the majority of studios have their employees, and freelancers are only used in case of emergency.

To those who want to embark on this career I banally say: don't stop trying because it is a difficult world but not impossible.

Italian native Giosuè Greco is a film composer, music producer, and musician based in Los Angeles. He studied at the Conservatory of Music in Vibo Valentia (Italy) before moving to study music production and engineering at Berklee College of Music in Boston (USA).

As far as music goes, when you think about Italian music – we're 60 million people, and only 50% listen to Italian pop music. It's such a small market that when it comes to music, a lot of them don't even work from Italy. In fact, the very big Italian pop stars all live in Los Angeles.

Italy has a very vibrant festival scene. There are a lot of music festivals. Many of the recording engineers that have a studio close the studio down in the summer. They become live sound engineers and they just do work with sound companies. Friends who do that tell me that more than 60% of their revenue comes from those three or four months.

Festivals are still retaining the "jazz festival" name, but most of the time the highlight of the night is never an actual jazz artist (for example, the Umbria Jazz Festival in Perugia). These jazz festivals are very inclusive with very popular artists.

As far as post-production goes, it is not a huge market, but there is an increasing demand for post-production audio people. For a really long time (like 40 or 50 years), we only had six channels – eight or nine if you consider the public access ones . . . We just recently gained hundreds of channels. TV is a very tough route because pretty much all the positions are constantly filled, and if there is another position, they already know who to call. It's a very small market.

[For entertainment], I would live in Rome, Milan, or Turin . . . Something is happening in Florence in Tuscany as well. But moving to Rome would be more strategic because Rome, besides being the capital, is really big and it's also very close to Florence.

Ireland vs. Spain: Javier Zúmer

Javier Zúmer grew up in Seville, Spain, where he earned a degree in sound for media from Néstor Almendros. After working freelance (in production sound, post-production sound, and music production), he relocated to County Galway, Ireland, where he worked as a sound designer and re-recording mixer at Telegael. Javier pivoted full-time from post-production into video games in 2020, when he relocated to Leamington Spa (UK) to work for Pixel Toys.

When I finished [audio school], it was 2010 and the financial crisis was very hard on Spain, especially the south. I got started during the worst economic collapse since 1929. The unemployment was crazy. So, I knew there was no hope there . . . but, I had the luxury of living at home. I was willing to live at home with my mother until I was 25, and from 20 to 25, I was just grinding and learning and trying to do something.

There is a small industry in Seville. It's a big city by Spanish standards. I started to contact every studio I knew, and just go there (without saying anything or without asking for anything) as an assistant to learn . . . I quickly saw that there wasn't much for me there. It's very hard to find a chance, because in Seville, you have studios where one person is the owner, maybe two . . . and in the province of Seville, there are three schools graduating 25 people in audio every year.

I knew I couldn't find a job, so I had to create my own job – create the opportunity. I tried to do post-production, production sound, and I was also doing mixing and mastering for music. That's very common in Spain to do everything because you can only be a specialist when there is enough industry to be a specialist in something.

My girlfriend went to Ireland for her Ph.D., and I moved with her. I saw the chance . . . I didn't have a job. I just moved and I had some savings. I was living in a small fishing town in the West of Ireland. Quite remote . . . I didn't even know that there was audio there, to be honest. I went there to mostly freelance, somehow. If I think about it now, it doesn't seem like a very good idea.

What I learned is that you need to throw a lot of seeds in the ground and see if something grows. I had some clients online that I knew from the first four years of struggle and making contacts like crazy. I was doing small video games and I was doing post-production of low-budget movies. I was doing almost anything. After two years of that, I found a job in a post house in Ireland.

I wanted to do audio libraries, so I was contacting audio people and sending emails everywhere. One guy answered me, and said, "Come over, we have a studio here." He was the sound director of a company doing post-production for international productions, especially in animation. The company was in my town, which is crazy. It is the only company within like 400 kilometers [~250 miles]. It was a crazy coincidence. I didn't even know he was interviewing me. He gave me a one-minute snippet of a show and told me to add some effects. I did that and then he offered me the job.

In my second month, someone was sick, and they asked me to mix in Irish. I don't speak Irish, and I was mixing Irish. Four months after that, I was a mixer . . . The company was built on doing things in Irish. They were dubbing Hollywood movies in Irish, and also cartoons in Irish. People will do post-production in Ireland to get tax benefits, so we were doing post-production of German movies, movies from Estonia, and all kinds of places because of that.

Working in Ireland (vs. Spain)

My feeling in Ireland was: if I work hard, I get a reward. There is enough room for me to grow here because there is business coming. The company was thinking two or three years into the future in terms of work. In Spain, you don't have that. Everything is more informal – not because of the people or the culture, but because there is not enough work for people to plan ahead. [In Ireland], it felt like I can progress.

Language

Some parts of Europe have great English, like the north of Europe, but the south is quite terrible. Spanish people – they don't speak English. Their English is very bad usually, mostly because we get everything in Spanish. Every movie we get is in Spanish, which I think is terrible. You don't know how awful the movies sound in Spain. You have the mix, and then you have the Spanish voices on top and they sound so clean. When I was like 18, I started to listen to every movie in English in the original version. It was a huge asset to learn English soon . . . My advice to Spanish people (and this extends to Italian or Greek) is to learn English. You cannot afford to not know English.

Note

1   Article being referred to: "Sunday Times survey saying artist is topmost non-essential job sparks anger in community" mothership.sg/2020/06/sunday-times-survey-artist-non-essential/.

References

Bruijns, Aline. Personal interview. 21 June 2021.

Clemens-Seely, James. Personal interview. 20 Dec. 2020.

Greco, Giosuè. Interview with April Tucker and Ryan Tucker. 28 Sept. 2020.

Hibberd, Clare. Personal interview. 7 May 2020.

Hutten, Mariana. Personal interview. 19 Nov. 2020.

Kennedy, Camille. Personal interview. 21 Feb. 2021.

Leong, Mabel. Email interview. Conducted by April Tucker, 29-30 Sept. 2021.

Manalo, Agnes "Aji". Email interview. Conducted by April Tucker, 3 Oct. 2021.

Mazzucco, Tiziana. Email interview. Conducted by April Tucker, 30 Sept. 2021.

Oluwagbemi, Phebean AdeDamola. Personal interview. 10 Jan. 2021.

Pinzon, Luisa. Personal interview. 10 Nov. 2020.

Reynolds, Jason. Personal interview. 27 July 2020.

Starling, Fernanda. "Fernanda Starling – Staying Versatile." *SoundGirls*. soundgirls.org/ fernanda-starling-staying-versatile/. Accessed 10 Sept. 2021.

Yu, Fei. Personal interview. 9 Nov. 2020.

Zolnai, Eliza. Personal interview. 27 Sept. 2021.

Zúmer, Javier. Personal interview. 24 April 2021.

20

Q&A (Common Questions and Answers)

How Do I Find Work in a New City?

Avril Martinez: "Go to meetups, go talk to people. Learn about their passions and talk to them about your passions. It's scary, especially if you're an introverted person. It's really hard, but it is the way to do it . . . Put yourself out there and talk about the things that excite you on social media. Usually, people are attracted to that. People who are excited about things that they care about, they're going to be fascinated by it, and they will follow you."

Nathan Lively: "The first question that anyone is going to naturally have when you tell them that you're looking for work in this new city: What kind of work [do you want]? What specifically? . . . You can't just go around telling people, 'I'm looking for work, I'm looking for work,' because either they won't know how to help you, or they'll just create something in their minds that they think is what you want, and it won't be right . . . If you just show up with this idea that 'getting work is just based on luck,' then you'll probably just be standing there, and no one's going to call you because you don't know anyone yet . . . Hope is not a strategy."

Should I Go to School for Audio (or Further My Audio Education)?

Kevin McCoy (theatrical sound mixer): "If you can go to college, go to college. College is great for a lot of things, but the most important thing it's good for is networking. I'm a college dropout, so, I'm someone who believes that you can make it work either way."

John McLucas (music producer): "School offered me a sense of structure and introduction into it. I would not have had the self-discipline at age 19 or 20 to execute on my own. I could not have just rolled up to a new city

DOI: 10.4324/9781003050346-24

and been like, 'Now I'm going to teach myself a new craft, get really good at this craft on my own, and then start a business.' I wouldn't have been able to do that. If you're going to take on student loans, that's a serious, serious commitment. That's hard for people that are young to understand how that can totally impact your life. If you want to take on that risk, you should expect to have a regimented day job to cover your loans and your overhead, because that problem world is terrible. I've watched a lot of people fall victim to that."

Megan Frazier (VR sound designer): "I was full starting a career and I didn't know what a compressor was. So, you do not need to know it all. I would point out that it is a good idea to know what a compressor is."

Jeff Gross (music producer/Foley mixer): "Take psychology classes . . . You're going to deal with a wide variety of people who think they are right, who think they know more than you, who think that you're below them. You might be pretty even keel and someone might come into the room like a tornado. You'd better know how to switch gears quickly to be able to deal with that. If you understand human nature more, then you're going to approach each person differently."

Jonathan Hubel (audio engineer/podcast editor): "If you think you ever might be a freelancer, take some business classes. You'll end up having to deal with contracts, marketing, accounting, and other things you'll never learn in an arts and technology degree. [For podcast editing], learn more about social media, website building/publishing, analytics/SEO, etc., so you can be more knowledgeable about the back end of podcasting and promoting your company's/clients' programs."

The Value of Business Education: Tommy Edwards

Tommy Edwards is Vice President of Product Development at Warm Audio in Austin, Texas (USA). He has overseen the creation of music technology and audio products for companies including Warm Audio, Blue Microphones, Logitech, and Line6.

I encourage almost everybody that I talk to or mentor to really dig into business. It doesn't matter what you do – some fundamental skills are applicable.

[In college], I took an intro to business class . . . I [became] really cognizant of this duality: there's my art, but my art is also a product, and it has to be commercialized. I never had that internal struggle that a lot of artists have because of that one single class.

I have an oldest daughter who just got done with a physics degree, and I have a 17-year-old who's about to embark on college . . . If she wanted to do this, I'd be telling her to go to college, get a business degree, and take some [audio and music technology] courses. You can learn a ton from YouTube now.

In my analysis, there are only about 10 undergraduate degrees that really correlate to a job and making a really good living. Almost all of them are technical, like an electrical engineering degree . . . Anything in medicine, nursing – those are pretty solid. The only non-technical or science degree that really has a high ROI [return on investment] now is business. Things like music, philosophy, psychology – there are reasons to go after those degrees, but they're not financial. They're more if you're passionate about it . . . It's just important for people, young people especially, to pursue these things understanding what the *real* reality is in terms of ROI – how much that degree is going to cost you versus how much you're going to earn the first 10 years.

How Can I be Better Prepared for the Field When I'm Still in School?

Javier Zúmer (game sound designer): "Don't do the bare minimum to pass your exams. Do everything. Find the three, four, or five people that are the most motivated in your class. I'm sure you know who they are. They're always there. Stay with them and get into every other project. Do more than the stuff that is in the course. When you really learn is when there is no safety net, and you need to solve problems and think outside the box."

Shaun Farley (film sound designer): "You still have access to whatever equipment and software your school makes available to you. Use them. Borrow those recorders and mics and record everything you can. Doesn't matter if you don't have time to 'master' them for your personal library right now – you will later. Just make sure you keep good records. In lieu of birthday or holiday gifts, ask for money towards your own recorder or microphones. Connect with as many film or game development students (whichever your preferred direction) as you can, and get involved in their projects. Make sure you stay connected with them after school. They could end up providing some good leads on work for you in the future."

Evaluating In-Person Schools and Programs

Audio programs (vocational schools, certificates, or degree programs) can be designed with different career paths in mind. Educator Leslie Gaston-Bird says of finding a program:

> Do you perceive yourself to be an artist, or are you a facilitator? If you like to facilitate, do you like to facilitate in terms of arrangement? Or do you want to facilitate in terms of setting up mics and being behind the glass? If you said, "I want to write and produce," then I say you need [a program with] business acumen.

Programs will vary in the balance of technical skills and studying the artistic (or musical) side of the craft. James Clemens-Seely explains:

> Trade schools (and a huge number of undergrad programs) are basically a performance degree with technology electives rather than a recording degree. Even the recording degree ones – a lot of them are built for composers more than they're built for [emerging] engineers.

Coursework is not guaranteed to align with employable skills. Mariana Hutten shares:

> It's easy to sell the dream, right? There's all this glamour about working in a studio. Come be an engineer. Be the next producer. It's easier to sell that dream than sell a dream of being a network security engineer. Stay in front of your computer looking at a screen for eight hours a day.

One educator discussed meeting with prospective students and parents:

> How many times have I done a studio tour and the parents are like, "So what happens? What are they going to do when they're done?" I say they need to have a degree in music production, a marketing degree to market themselves, and maybe [more courses] to have some chops that music production doesn't give them. It's really difficult. They have to have a wide set of skills to be able to cover themselves.

When evaluating a school, ask how many students stay in the field after graduation and what percentage become established professionals. What jobs and fields do they work in now, where are they located, and what did they do during their emerging years? Schools will naturally share about their most successful graduates, but what is the *typical* graduate doing? For master's degrees, are graduates landing the same opportunities as someone with a bachelor's degree, or what is the advantage?

Ask about the program's alumni. Do many graduates stay in touch with professors and each other? Sal Ojeda credits the alumni at the Berklee College of Music for his earliest professional opportunities:

> The only three contacts I had moving to LA were Berklee people. Before I left Boston, the dean gave me three phone numbers. My three first gigs were because of the Berklee people that I called, and they were kind enough to give me a chance.

While formal education can be an excellent way to learn fundamentals, explore, and experiment, an education will always be an introduction to a job or field. Robert Scovill (concert sound mixer, recording engineer, and producer) explains:

> In the modern era, don't be afraid to go to school. Time is on your side. But after you graduate, don't confuse education with knowledge, and don't confuse knowledge with experience. They're mutually exclusive . . . Don't be in too big a hurry to sit in the big chair. You need to get out in the world, work, and make a bunch of mistakes in order to lose your fear of making mistakes, and then gain an identity. It's better to make those mistakes when there aren't 18,000 pairs of eyes and ears watching and listening.
>
> (73)

What If I'm Getting Started Older than My Peers?

> Sal Ojeda: "It's not a problem to start out older if you use your time well. You have to be conscious that you don't have all the time in the world anymore, so you have to be smarter with your decisions."
>
> Tina Morris: "I was 31 when I applied at The Village and started working there as a runner. So, I don't think it's ever too late, as long as [you] apply for a bunch of stuff."
>
> Catherine Vericolli: "Apply for what it is that you really want to do, especially if you're making a career change at an older age, because you're even more apt to really know what that is. I don't think there should be any embarrassment about (or judgment about) whether that position is entry-level or not."

I Am a Minority in the Industry. Will That Affect My Career?

Samantha Potter (live sound engineer): "I do my best every day to make sure that I'm cultivating a diverse community, and I'm helping push this industry forward. I have the ability to do that. We're all making those very conscious efforts until they no longer have to be conscious efforts."

Kevin McCoy (theatrical mixer): "I try to be very cognizant of diversity and inclusiveness in the industry. I'm often trying to look out for opportunities for [people] who may not have the same sort of opportunities . . . It's important for us all to be thinking of making the industry more diverse."

Avril Martinez (VR sound designer): "When I send a cover letter out, I always put my name and I put my pronouns right after (they/them). I'm sure that I'm getting turned down for certain jobs because of that specifically, but I'm okay with that because I wouldn't want to work for a company that wouldn't treat me with my right pronouns – the most basic thing. Maybe I just fell into the right community, but everybody is very welcoming and open and understanding. If anything, I'm very proud of the game audio community because we're so open, happy to share, happy to help people, to help them grow. The game audio industry is getting more and more diverse. There are groups online that you can join. There's even a group that's dedicated to people who are women, non-binary, and people of color, that is there to lift other people who are in game audio."

Finding a Path: Langston Masingale

Langston Masingale (Syracuse, New York, USA) started learning audio in the 1990s as a teenager working for his uncle's recording studio. When he later worked for other studios, Langston found his race affected the opportunities others gave him. This led Langston to build businesses himself – including two recording studios and Handsome Audio, where Langston designs and manufacturers his own products.

One of the things that I experienced as a person of color, as an audio engineer, as one of a few: I knew how to record more styles of music than most of my white peers . . . [The studios I worked for said], "You're going to be our hip hop engineer . . . [we know] you're more advanced, but we want you to do the hip hop sessions because we feel like you'd be better with those clients." Those were two-hour sessions, and [the other engineer] was going to be on 16-hour sessions on a five-day block. So, when I encountered that in the music industry, I only had so much of a tolerance for it . . . I'd already gone through an apprenticeship as an engineer when I was 16 . . .

I kept dealing with these situations where I would only get so far, or opportunities were not coming to me that were going to others. I kept looking in avenues and in directions that made more sense for me . . . I [decided], "I'll open my own business." So, I did. I did that numerous times over the years. In addition to that, I have an education, so I was able to parlay that into many other fields. I didn't always just have music full-time . . . I was a great studio manager. I was a great booking manager for my band. We had a record deal, and we were touring . . . I was a community school director in New York City for three years . . . The stuff that I learned in the entertainment industry I was bringing into education. Then when that would phase out, or new opportunities arose, I would go back to music.

Advice to Minorities

In terms of my barriers to success, it was my race. I'm a firm believer in that. Not to say that would be a disservice to people that are afraid to say that . . . You have to prepare to go into a situation knowing that your presence alone [can be] disruptive . . . You need to be in the mindset of: How do I best navigate this situation? You can either fly off the handle and get offended, argue with people, or you can say it's time to develop a solution . . .

Part of your armor is your support network. You need to find a mentor, you need to find an advocate, and you need to find a supporter even if they don't work in the same company as you . . . Because if you don't have those people to talk to, to console you, to encourage you, to invigorate you, you're toast . . . Everybody needs that person in their corner that can give them that jolt of re-encouragement . . .

I encourage people of color or women, etc. – you need to (in a 2022 music industry) come into this world with as many skills as you can . . . We don't all work at some long-term thing. We have a job and then we have another job right after that. But as people that don't get first looks a lot of time . . . You've got to constantly be re-examining your situation.

The most important thing is never compromise your value. You have value. If you started a year ago as a recording engineer, give yourself a raise. Every year after that, stop saying "Well, $35 is good enough." F— that. How many hours have you logged? You should be giving yourself a raise . . . Why aren't you giving yourself a raise? Because you work for yourself? Because you're a freelance engineer, a black person, a woman? Give yourself a raise . . .

No matter what they pay you, you are still giving them the value you're worth. So, even if you're worth $85 an hour but [getting paid] $40, you're still working and providing that $85 an hour value. So, when it's time to leave, don't be loyal to people that won't feed you what you need. So many people get [stuck on] loyalty. Like, we've been somewhere for a long time, so now we don't want to leave because we've been there . . . You can get new clients – but you can't pay your mortgage with good intentions.

Finding a Welcoming Culture: Kristen Quinn

Kristen Quinn is Audio Director at Polyarc (Seattle, Washington, USA) and has worked in game audio production since 2004. Kristen has experience with game publishers and developers (including major corporations and AAA game companies) as well as reviewing job candidates and hiring.

I've been at companies [in the game industry] where there have been very public lawsuits. I've had friends at different companies go through investigations. Having seen how companies operate on that level has been very important to me because I think it adds to how much they value their own culture as a company. It's very telling if companies value that or not. The companies that value [their culture] – I'm going to enjoy working there more.

I've had negative experiences, and I've had great experiences. I've had people who I felt valued me more because I was maybe less common as a human. I've had people who I think didn't care . . . I don't think I had an advantage or disadvantage. Early on in my career when I was trying to get into the music industry, I would say I had a huge disadvantage. But since getting into games, I haven't felt that way. I have felt that I am in a male-dominated industry, and I've noticed things that could be improved. But, I just want to help contribute to making that experience better for everybody.

Because companies are starting to value diversity more, it's helped me be able to have a voice in our community . . . I want to be a part of the solution and not the problem. So, I want to help – being a woman in a specific gender-dominated field, or understanding as a queer person – I just want to be sensitive to diversity, and ensure that I'm always doing my best to make sure that we're being equitable.

I always want to hire just the best person for the job, but I also care about making sure that we're hiring in a way that is fair and equitable. [Polyarc] really values it, and it's one of the things that I really valued about them. I'm doing my best as an individual to also understand ways in which I have an unconscious bias about things – and the things that I might not be aware of. Companies are starting to be more aware of that being an experience, and so I always try and do my best to make sure there's no one getting a fast track to the front of the door.

What If My Career Is Not Moving in the Direction I'd Like It?

Shaun Farley: "It probably won't be [moving where you'd like] when you start. It takes time, patience, and persistence to get where you want to be. If you're able to succeed in your goals, you'll end up developing new ones. That means you're going to find yourself desiring new directions to pursue throughout your career. Embrace that.

"For me, the important part is to take some time (every few months or every year) to sit down and assess where I'm at [in my career]. 'Am I happy with this situation?' No? That's normal. That's when it's time to figure out why. Maybe you're not finding yourself challenged by the work. Maybe you're not working on the kinds of projects that interest you. Is the work environment the problem? If you don't know why you're unhappy, it's impossible to address the problems."

Tina Morris: "When I first got to Los Angeles, work wasn't plentiful for me, so I started doing wiring and I started doing installs and then I found a job at The Village running. So, I was open to different opportunities. At this point, looking back, it was all experience that I could build [from]. So, sometimes your path isn't going to go exactly how you planned it, but it'll weave. And then eventually you'll settle into what you were really meant to be.

"I think the position that you're going to be right for will find you – as long as you're active about it. You can't just sit there and wait for somebody to be like, 'Hey, do you want a job in a studio?' You have to really go out and get it – and obviously go about it the right ways."

Meegan Holmes: "Nothing's permanent. Don't feel like any choice that you make is, 'This is the only thing I'm going to do for the rest of my life.'

You can totally change your mind, and that's okay. Nobody's going to get mad at you, or fault you, or look poorly at you if you start out in live sound and go, you know what? I really do want to do recorded sound. I don't want to pull feeder in the morning, unload a truck, be on my feet for 15 hours running around, and [around] loud music. I really do love this other thing. It's okay, to change your mind. You're not stuck."

April Tucker: "I wouldn't change paths just because you think, 'Hey, there's more work over there.' In actuality, it's going to take a huge amount of work to network and build your way, whether you're going into live sound, or music production, or TV. The people you network with are going to be different in every single little niche. So, if you want to switch from one path to another, you really are starting over. It has to be something you're really passionate about."

Not Giving Up: Scott Adamson

Front of house engineer Scott Adamson has mixed at some of the most iconic venues in the world (including Madison Square Garden, Coachella Main Stage, and the Sydney Opera House). Scott started his audio career in Chicago before moving to New York City. Even with a decade of experience, Scott found it harder to get started in New York than he expected.

I thought, oh, New York, there's going to be tons of opportunities there. There are so many cool venues and everything. But, it was a struggle. It was tough. Every venue I went to, they would say, "Look, we have a line of sound engineers out the door. We can't help you. We can't give you any shifts." I would just be talking to bands. Everyone had sound engineers. There weren't any tours that were coming my way. It was depressing and a struggle, and I really considered giving up my career and trying to go do something else. I'm really glad I didn't.

If you're having trouble and you're not finding the right opportunities, persevere, and stick with it. Keep going. Keep doing the shows that you have. Keep talking to people. Keep working at building your connections. Give yourself the opportunity to be in the right place at the right time, and get that step up to the next level.

From the Production Academy

Should I Endure a Bad Situation to Move Up in the Industry?

Camille Kennedy: "As a boom operator, sometimes you might cast a shadow on an actor or something, and the director of photography will lose it on you. They'll ream on you in front of the whole crew sometimes. You just have to eat that shit cookie, and just put that boom up again and hope to God you're not going to do the same thing [again]. Nobody deserves any amount of disrespect on a set, but things get heated and there's pressure. It isn't an excuse for that to happen, but you should never put yourself in a position where you're being belittled just to advance. There's no reason for that at all. I've worked with my share of mixers that were toxic. I would come home crying, like, I didn't want to do sound at all anymore. I put myself through it because I wanted to get there but it was hell. I would highly say don't do that."

Susan Rogers: "If your gut instinct is saying, 'Get me out of here. This is wrong,' your gut instinct is right. That's not normal . . . Normal isn't terrible."

Devyn Nicholson: "One thing that I've learned that is hugely important to me is that a paper trail means everything . . . There are some individuals that I refuse to speak to verbally because there have just been too many incidents with them where they've said one thing and then the other happened. So, I'll respond to them in an email if they catch me on the phone . . . Even though you're scared of losing your job, being reprimanded, or whatever it is, you get to the point where you realize it's just not even worth it to you. I refuse to be treated that way. If I get pulled back in that room, I quit."

I've Been Fired or Made a Mistake. How Will That Affect My Career?

Jack Menhorn: "If you aren't asked back, then find out why and learn from the experience. Getting fired isn't a death sentence by any means. Not getting asked back for a gig could have multitudes of reasons, but still reaching out to the client and making sure the relationship is positive would be a good step to take."

Patrushkha Mierzwa: "There have been a few times when I've needed to make [boom] moves and have hit someone. Over time, I have come to learn that sometimes, no matter how many times you tell people, they're not going to pay attention . . . Thirty years later, I still announce it repeatedly before a take . . . Reflecting on my youth, I was wrong to apologize all the time. I alerted everyone, I notified the AD Department, and I made sure that everyone around me was told my work for the shot; I am not responsible for the choices of others . . . Don't needlessly apologize."

(154)

Nick Tipp: "Failure is the best teacher. I have failed a lot in my life. The goal of failure is to look at it and be like, what did I do wrong? How did this happen? Where's the first point back where I could have fixed it? And really identifying it. Then just making a firm, spiritual commitment to never make that mistake again."

Dealing with High Profile or Famous People: Patrushkha Mierzwa

Patrushkha Mierzwa is a utility sound technician and one of the first women boom operators in Hollywood (USA). She has worked in production sound for films and television shows with major actors and directors (including Robert Rodriguez, Quentin Tarantino, and James Gray).

No matter the budget or size of the project, treat everyone with respect and professionalism; in the first days of *any* shoot, keep any excitement to yourself. This is not the time to become a fan over having the chance to work with someone you've admired. Give everyone a chance to figure out the dynamics of this new group; the most important thing you can do is to *watch* and *learn*.

How much interaction you want to have with talent is really a matter of personal comfort – both of yours. In general, at the basic working-relationship level, you should introduce yourself at the moment you'll need to work with someone . . . If you have the gift [of humor], God bless; use it wisely. For the rest of us, nothing except an introduction seems appropriate . . . I worked with a mega-star for seven months before I got a smile and "hello." The last time I worked with him I got a kiss in front of paparazzi . . .

If an actor has assistants, go through them for communications. Build a relationship of trust with assistants . . . ask about any specific needs or pet peeves of the actor. Use the assistant for your prep work and save your face time with the A-list actor for necessary interaction. *Be brief!* Don't feel a need to talk; it is not a time for chatting or, heaven forbid, you commenting on their personal appearance or critiquing their work. *Not even to give them a compliment.* Give them the courtesy of letting them concentrate on their job.

I have stories. Really uplifting stories of good deeds, compassion, generosity . . . an actor calling the entire crew after an earthquake to check on each one, an actor who gave very generous gifts at Christmas. I have been witness to a famous couple learning that their toddler has a serious health condition . . .

Be sensitive to events happening in actors' lives. I'm not a follower of [gossip], but if I'm working with an actor, I'll check out an article . . . I'll ask sound people and coworkers from the actor's last three years of work how the experience was, anything it would help me to know, anything that went badly or was a trigger for upset . . . Everyone's life is a story and deserves its dignity and privacy. That's one of the basic tenets of working on a film: everyone works closely and under great stresses, and the trust that is built is forever – and ever.

(187–192)

From "Behind the Sound Cart: A Veteran's Guide to Sound on the Set"

References

Adamson, Scott. "Success in Live Sound Isn't Always Easy." *YouTube*, uploaded by The Production Academy, 18 April 2019, youtu.be/0RCYtMSzCoI.

Clemens-Seely, James. Personal interview. 20 Dec. 2020.

Edwards, Tommy. Personal interview. 29 Oct. 2021.

Farley, Shaun. Email interview. Conducted by April Tucker, 15 July 2020.

Frazier, Megan. Personal interview. 11 and 21 Aug. 2020.

Gaston-Bird, Leslie. Personal interview. 2 Oct. 2020.

Gross, Jeff. Interview with April Tucker and Ryan Tucker. 14 Jan. 2021.

Holmes, Meegan. Personal interview. *What Makes You Stand Out*, SoundGirls, 20 May 2020, youtu.be/wYamiOK8Y6A. Accessed 21 May 2020.

Hubel, Jonathan. Email interview. Conducted by April Tucker, 22 Oct. 2021.

Hutten, Mariana. Personal interview. 19 Nov. 2020.

Kennedy, Camille. Personal interview. 21 Feb. 2021.

Lively, Nathan. "From the Sound Up Lesson 2: Commit to Your Goals." *Sound Design Live*, 21 March 2017. *SoundCloud*. soundcloud.com/sounddesignlive.

Martinez, Avril. Personal interview. 28 Aug. 2020.

Masingale, Langston. Personal interview. 14 Sept. 2021.

McCoy, Kevin. Personal interview. 22 Dec. 2020.

McLucas, John. "Musicians Institute Alumni – Building A Career AFTER Graduation." *The Modern Music Creator Podcast*, ep. 42, 29 April 2019, pod.co/the-modern-music-creator.

Menhorn, Jack. Email interview. 30 June and 26 July 2020.

Mierzwa, Patrushkha. *Behind the Sound Cart: A Veteran's Guide to Sound on the Set*, Ulano Sound Services, Inc., 2021. pp. 151, 154, 187–192.

Morris, Tina. Personal interview. *What Makes You Stand Out*, SoundGirls, 20 May 2020, youtu.be/wYamiOK8Y6A. Accessed 21 May 2020.

Nicholson, Devyn. Personal interview. 23 Nov. 2020.

Ojeda, Sal. Personal interview. 27 Sept. 2020.

Potter, Samantha, host. "Reimagining Our Industry Panel." *Church Sound Podcast*, ep. 17, ProSoundWeb, 8 Oct. 2020, www.prosoundweb.com/podcasts/church-sound-podcast/.

Quinn, Kristen. Personal interview. 30 Sept. 2020.

Rogers, Susan. Personal interview. 12 Oct. 2020.

Scovill, Robert. "Robert Scovill." *Get On Tour: A Sound Engineer's Guide*, edited by Nathan Lively, self-published, 2018. p. 73.

Tipp, Nick. Personal interview. 22 Sept. 2020.

Tucker, April, panelist. "SoundGirls Career Paths in Film & TV Panel at Sony Studios." *YouTube*, uploaded by SoundGirls, 26 Oct. 2018, youtu.be/Y_g3drC2yyA.

Vericolli, Catherine. Personal interview. *What Makes You Stand Out*, SoundGirls, 20 May 2020, youtu.be/wYamiOK8Y6A. Accessed 21 May 2020.

Zúmer, Javier. Personal interview. 24 April 2021.

21
Moving Forward

Starting a career in the audio industry can be fun, invigorating, and almost addictive to see your career growing. Claudia Engelhart says of her emerging years in New York City:

> I did everything, and I didn't even think twice. I did some crazy stuff like salsa gigs up in the Bronx that started at 2 am and went 'til 6 am. This is what you have to do in order to get the experience. I loved it.

Pete Reed transitioned into video game sound while working in on a theater show in the West End (London) and says of his hours:

> I worked on average 44 hours a week at *Matilda*, and then 15–20 hours a week working on my career move. At the time it didn't seem like that many hours, as I was doing what I enjoyed. I had to put the time in to learn new skills!

Nick Tipp took a low-paid internship in Los Angeles early in his career, and shares:

> I was working over 100 hours a week, so it was less than US$3 an hour. The day I would take off, Sunday, I would go to San Francisco and work on this hip hop production company with my friend. We were selling beats and trying to record artists. It was just non-stop. 125–130 hours a week of work. I did that for about 18 months.

Jeff Gross says of his career in Los Angeles, "Literally for 25 years I survived off of like four to five hours of sleep a night. Life is short, man. You want to make things happen. Every minute, every hour of the day is useful."

Working these types of hours are definitely *not* required to have success in the industry. However, it shows how far some are willing to go to pursue their interests.

DOI: 10.4324/9781003050346-25

Work-Life Balance: John McLucas

Music producer and content creator John McLucas shares his challenge of finding work-life balance while working in the music industry.

> The biggest struggle is to give the time to be happy outside of music. It's hard to not identify with the success of the thing that literally takes up every waking moment of your day. Even if I'm making food, I'm probably watching a music industry-related thing. So, it's very difficult to separate that.
>
> I know that getting out for 20 minutes every single day does wonders for my overall mental mood because otherwise I'm within the box all day. I'm still getting a full night's rest. I'm still moving five to six days a week, working out.
>
> When people want to wear some senseless pride badge [talking] about how they literally didn't sleep at all last night because they grinded through the night and did a thing, I'm like, cool, but you just sent the entire stems for the song wrong. Your ears are gone. You're not going to be producing at your best at the 17th hour. You're making mistakes. It's just not sustainable.
>
> I'd love the conversation to go more into effective grinding vs. grinding for the sake of it. If you're going to sit down for four hours and work on [a project], how can you optimize the time that you're using so maybe then it becomes three hours? Push back the movement of "grind entrepreneur" being cool to like, "Physical/mental/health. Love your family. Text your mom 'Hello.'" Hopefully, that becomes the new movement in response to "team no sleep." You need to take breaks. You need to refresh yourself.

From the Modern Music Creator Podcast

It can be difficult to maintain a healthy work-life balance, especially for freelancers – who may feel a constant pull to find work and stay busy when the future is unknown. Patrushkha Mierzwa, who has worked on major Hollywood films including Oscar-award winning films for production sound, explains:

> The life of a freelancer is inherently unbalanced, and it's a constant pull between work/art and life/family. For years, I used to keep life on hold while I was on a job, but then I was just as unbalanced and frenetic "catching up" in my personal life. Over the course of about a year, I realized that we were stringing one show after another because, well, you never know

when someone will go on strike, or there'll be an earthquake, or who knows what. I finally saw that trying to balance personal life with professional life could never happen . . . Everyone must find their own way of making life and work make sense for them; the sooner you find it, the quicker you'll put your energy to a higher purpose.

(271–272)

For those who are willing to work excessive hours, Kristen Quinn advises:

Monitor yourself, really check in, and pay attention to yourself. Am I getting enough sleep? There's a certain point where if you're working so much, you stop working and you literally can't stop thinking about work. It's like staring at something for too long and then you're kind of delirious, like when you drive for too long, you're still seeing the road in front of you even when you stopped driving. Working too much is like that.

Some disciplines of the industry are showing a change in attitude and work climate (toward better work/life balance), especially after having to adapt to the COVID-19 pandemic in 2020 and 2021. Camille Kennedy says of the shift that happened on film and television sets in Canada:

[Before COVID], you couldn't get sick or hurt because you were expected to be there every day. If you got asked to do a show, even if it was an eight-month-long show and you wanted a day off, they would say, "No, absolutely not." . . . It's broken up marriages.

That mentality is changing in the sound department . . . We're taking Fridays off. We're taking long weekends. If they offer a sixth day [of work], we're going to turn it down. We're going to live our life – as long someone can replace you . . . People are taking the whole mental health thing really seriously now because the industry just ruined people. It's totally changing for the better.

Some disciplines of the audio industry will offer better work-life balance than others, and culture can vary between different jobs and companies. Kevin McCoy explains:

If you need time off for family, personal or medical reasons, or whatever, you have to be an advocate for yourself. Some employers are going to be crappy to work with on that front, and some are going to be great. Sometimes employers are going to gaslight you and make you think that you're an awful human being for daring to ask for these things, and those are the crappy employers that we should be shunning from the industry.

If you're in a job interview and the employer says, "Do you have any questions for me?" One of the questions should be, "How did you relate with your employees during the pandemic? How did you deal with it? How did you keep in contact with them?" Hopefully, that's something that a lot of people will ask potential employers going forward.

Javier Zúmer adds:

> Having hobbies outside of audio is also quite important, I think, even to use the artistic muscle outside of there. For example, I play Dungeons and Dragons. That is out of creativity and inventing things. It kind of helps me with audio because we are professional creatives.

Advice for Freelancers: Sarah Stacey

Sarah Stacey is a freelance radio, audio, and podcast producer in Dublin, Ireland.

> I think a lot of people are afraid of the word freelance because they think you won't get regular work, but you absolutely can. Some times are just naturally busier than others. I've had a few pretty quiet weeks, and it's easy to think, why did I decide to do this? Was I crazy to leave a [full-time] job for this? But I think once you have a few clients that you've established a good working relationship with, it's definitely worth it.
>
> Be careful not to take on too much because then you need to know your limits, basically. You need to really have a good idea of how much you can afford to turn down. It's also the work-life balance, as well – just looking after your mental health (obviously that applies in any industry). But I think when you're freelancing, you can kind of forget, as well.
>
> It's great to have the freedom to set your own hours, but you need to make sure that you find a structure in your days that works for you – so that you need to be motivated, know exactly when you're going to do your work, and when you're going to take a step away. On the other end of the scale, it's easy not to switch off and to take work calls or emails at all hours of the day, so you need to set boundaries as well.

Working Through a Medical Emergency: Kristen Quinn

Kristen Quinn is Audio Director at Polyarc Games. Prior, she worked for Riot Games, Microsoft Game Studios, and Monolith Productions.

I was really burnt out for my first four years in the industry. We shipped three games in one year. I was working 80- to 100-hour work weeks. I was really exhausted. I was excited, but at the same time, I was young, and I didn't realize what I was doing to myself . . .

It actually took me having a pretty major back injury to realize what I'd been doing to my body, and to slow myself down. I couldn't walk for three or four months. Then, it took about two years to get back to a point where I could sit or stand for anything longer than a 15-minute interval.

It's hard as a person who's coming in because the instinct is that you're hungry, you really want to learn, you're just excited to be there. As an entry-level person, you want to impress everyone around you. There's a lot of pressure to not say "no," but I think it's very important.

What that experience taught me is that as long as I was setting expectations around what I was capable of doing (and asked for what I needed), I was actually surprised at the willingness of people to be understanding. I wish I'd known that earlier in my career. If you need to leave at a certain time because you need to go home to your family, or childcare is leaving at a particular time, or it's my wife's birthday, or whatever the situation is – you're setting expectations upfront. It requires keeping people aware of [details] like, what's the current situation on the projects? Are you on schedule? Bring up those issues to let people know and raise awareness when you're off track.

A company that values its people – it's not going to work them to death. That's not the goal. If a company doesn't value that kind of culture, then I would say they don't necessarily value their people.

What Drives You?

A common characteristic among successful professionals is a drive to improve. Avril Martinez says of their approach to finding work early on:

> It was first surviving, making sure I had food on the table and could pay my bills. Then, I just felt like I was never done. I knew that I needed to do more, that I could do more, and that I wanted to do more.

Even established professionals continue to carry this drive through their careers. Karol Urban shares:

> I've been doing this for over 20 years. I have a dream – I'm not there yet. I've gotten quite far along my line. Happy, satisfied with my results. But I also have aspects of my career where I also feel that the fruition of my dreams has not quite come home yet.

What keeps you interested in a career path may be something unforeseen. For Camille Kennedy, part of what she finds fun, creative, and challenging about working in production sound is wiring, or finding where to place microphones, wireless packs, and how to attach them to people, objects, or animals:

> There's this whole arts and crafts part of it. Where am I going to hide this? . . . Working with what you have is always an adventure. I love when wardrobe comes to me saying, "Camille, you're going to hate this next outfit." I love when they say that . . . You have to get really creative.

A career can be driven and balanced with personal priorities. Kelly Kramarik wanted to build her career in Denver, Colorado (USA), versus moving to Los Angeles, which had a larger market. "It's the mentality of being a little fish in a big pond or a big fish in a little pond. I am a big fish in a little pond here and I like that," says Kelly. She says of being one of the only women in the Denver area who does production and post-production sound, "Obviously I'd welcome other people and I'll help anybody that wants to do it. But at the moment, if someone's looking for [a woman], it's me every time. It's a marketing tool, which is cool."

Having a drive (and interest) can be crucial for career success, especially in competitive fields. "You need to have a willingness to learn and a natural curiosity about it. If you're coming into this with zero experience, and you don't really care about it, then it's going to be miserable," says Britany Felix.

Some professionals have the drive to overcome complex barriers and have gone on to build careers in the field. Jason Reynolds learned live sound through his mentor, Paul Price, at their church in Kingston, Jamaica. When the church's pastor said Christmas morning mass was going to be held outdoors (in a volatile area of Kingston), the crew had to set up overnight with armed guards watching over the group. Jason explains:

> We just sprung into action. That's how Paul trained us as the audio engineers. We're just there to do whatever needs to be done. So, whatever the need is, find a way . . . I think for us, we just kind of grew up with this. Violence, criminal activity, and gang violence are almost a part of life, so life didn't stop because there was violence happening. Life continues, and school continues, and everything around you continues . . . We were also raised to be incredibly grateful. Having work was something to be grateful for. It didn't really matter if it was 16-hour days or 18, or not getting much sleep. I was just always grateful to be working.

In Nigeria, electricity is inconsistent, gear is expensive, and formal education and internships are nearly non-existent. Despite these challenges, Phebean AdeDamola Oluwagbemi is still inspired to work in the music industry, and supporting other women through Audio Girl Africa. She shares:

> Production and audio engineering are what you see me do naturally . . . But when it comes to fundamental, long-term purpose, it's about educating people, developing people, and ensuring that they don't fall into the same challenges that a lot of our predecessors fell into.

Langston Masingale, who grew up in the USA and created Handsome Audio, also sees a bigger purpose in his work:

> Before my mother was a schoolteacher, I was four and we lived in the projects. I remember what it's like to live in poverty and to have dreams. I remember what it's like to be a little black kid that doesn't have positive role models, and yet to have dreams. So, no matter how good shit gets for me, I know that there are people that don't have it as good as I do, who are starting from the same place I started from, and they need a hero, or they need an example – whatever you want to call it. I'm very self-conscious of the fact that sometimes that's me – and I don't want to describe myself as that person, but that's who I am to some of these kids, like, "Wow, you're the first person of color that made something that I know."

> From the beginning, the path that I've really been on is having this company, and I'm finding different ways to give back to communities that are in need. In addition to having Handsome Audio, in the summer of 2020, I started the Black and Brown Association of Sound Engineers, which is also an organization that anybody can join . . . I have to work on both halves of my being and spirit. So, it's not just being a great recording engineer or record producer.

The Value of Entertainment: Mehrnaz Mohabati

Mehrnaz Mohabati is a supervising sound editor/re-recording mixer for television and film in Los Angeles (USA). She is originally from Tehran, Iran, where she worked professionally as a music engineer.

> When I was a child (in Iran), there was a war between Iran and Iraq. One of the things that was keeping us children happy was watching animations. So, one of the things I always think about when I do my work is if a child from anywhere in the world watched what I did and gets happy for even an hour, I'm so happy. But again, my job that's amazing.
>
> It's always important what we do because we are building a culture, and we are giving people food for thought, which is really important. Like, when I needed to work on some projects (I needed the money), I didn't accept them because of the joke behind those movies, or whatever it was that was offensive to me. I don't do that no matter how much I want to do that job. I think it's important for us to know that it can cause a war even, can cause many issues, or can bring light to many people's lives. That's why it's so important to me. It's not just pressing buttons and doing something. If it was like that, I would become an X-ray tech.
>
> We are entertaining people. We help people to stay sane during the COVID time, during weird times.

Pivoting

Pivoting (changing disciplines, jobs, or career paths) can be a means of survival, or a way to find a better fit for your interests or personal priorities. For example, Avril Martinez says of their pivot from theater into VR:

> My approach to sound design for theater was very immersive. I used to put speakers in random places and an overabundance of speakers in different places . . . Once I made that connection [to VR] in my head, I was like, why don't you just do this? It's the same thing, but you have more space. You're doing the thing that you like, and it's fascinating. It's a pioneering field right now.

Alesia Hendley, who started in live sound before moving into AV and IT for audio, says of pivoting:

I have people ask me all the time, "Do you miss running front of house? You do nothing with audio now." You can look at it that way, or you can look at it as I'm the person who's orchestrating the sound that people experience. They need the network that I've designed and without it, it's not going to work. So, you can change the perception of it and try to look at it in a different sense that's more positive than "I'm losing something" . . . I've just always been trying to stay ahead of the curve.

Pivoting can happen at any point in a career. Jeff Gross pivoted into Foley mixing after two decades in music production, and says of the experience:

The first year or two was a bit of a letdown because I felt like I had failed at music. I think a lot of people do. But then I hit a wall and just woke up and said, I'm using the exact same skills. Instead of recording a guitar, now I'm recording somebody walking on cement. It became just as fun. Going back to music after that made music more fun. I wasn't worried about the paycheck so much, so I could charge what I wanted. It afforded me the ability to turn down potentially problematic clients. It really changed my perspective.

Andrea Espinoza pursued touring live sound for years before pivoting into finance as her primary work. Andrea says:

When I pivoted, I associated shame for some reason with doing anything that was outside of touring. Touring was the one thing I spoke to everyone about, the one thing I loved. It's the only passion that I knew. I had other passions, of course, but I wasn't actively pursuing them because touring was my be all end all. I tied my sense of self-worth to my touring . . . but I'm not a tour. That's not who I am.

Pivoting from Classical Music to Spoken Word: Jeff Dudzick

Jeff Dudzick was building a career in classical music and building prestigious credits, including the Boston Symphony Orchestra (at the Tanglewood Music Center) and Metropolitan Opera. When Jeff's personal priorities changed, he pivoted to spoken word entertainment, taking a full-time job with Audible. Jeff's prior music experience turned into an unexpected opportunity in the future.

I had a good thing going at Sony doing the Met Opera work. I was working alongside and with these big classical people (big to me). I was in very good company and doing really cool work. It was perfect for me, and I got to go to Tanglewood in the summer . . . I was very happy with what I was doing.

It was probably getting married that started getting me thinking differently. I've always been someone who wants more stability, anyway. The contractor life never suited me . . . I knew I was lucky enough to get a full-time job at Audible (a good company and under Amazon). I was giving up on classical music . . . It was tough. I thought, if I really want to, I could probably freelance (even though I hate freelance) but I could probably get to do something if I wanted to. At the very least, I can listen to music . . .

I was initially in the Audio Production department. I was like a traffic coordinator just moving files around . . . then two years after that, I did a trial position [in Audible Studios] . . . It worked out and I stayed. Cut to a few years later – now I'm doing music and live recording projects, which didn't exist before. It was just the right place at the right time, the right person with the right background. I'm the only one with any significant live recording experience because that's what I did at Tanglewood. So, I'm perfectly suited for that. When Audible came out with the *Word + Music* series, I was right there . . . . Now it's branched out and I head up the live recording at Audible's Minetta Lane Theater, which is Off-Broadway [in New York City]. It's all connected . . .

I figured my best shot at having a career was going to be in what I've studied and worked in. So do audio, and it'll just go from there. Hopefully, you don't have to do something too grueling (or the lower-level stuff) for too long.

Eric Ferguson spent 15 years working with major artists as a recording engineer in Los Angeles and toured around the world doing live sound. When Eric's personal priorities shifted, he decided to move across the country for a teaching job:

I was all about music when I was younger, and now I'm not so much. I still love music, but I'm much more about my kid. I'm much more into electronics, building things, complicated systems, and some software . . . Ultimately, you're going to reach a point in your age where you want to own a home, or be married, or support children. The bucket list will be in the past, and you're going to want to have a living [while] doing something you enjoy.

Jeanne Montalvo Lucar has learned to balance the work she wants to pursue with being a parent:

I hate hearing about people who feel like they have to leave [the industry] to have a family, but it's understandable, also. There are so many different facets of this career that you can do, and you don't necessarily have to leave it. Where you end up may not be what you set out in your heart when you first started your first day of recording class, but it doesn't mean that you can't still dabble in audio.

Steinunn Arnardottir (Berlin, Germany) is Chief Technology Officer and founder of Lottie, a company focused on changing the way young children experience technology. Steinunn founded Lottie after spending over 15 years studying and working in audio DSP. Steinunn says of her career shift:

I realized I'm the happiest when I'm learning something new that interests me, and it doesn't need to be yet another audio effect. It can also be a broader thing. This is maybe something that changes over time. But, it's about knowing yourself and what makes you tick, and what makes you happy. That's often the hardest thing to find out . . . [Lottie] is also part of my narrative – that I'm [creating] tools for a younger audience . . . It's a break from the music industry, and I'll be back.

Meegan Holmes pivoted from live sound touring to management after an injury:

Whatever you set out to do, whatever you think your path is, be ready and okay to change it . . . It's okay to try something for a few years and be like, this is not for me – I want to do something else. That's okay. All of those things are [going] to lead you down a path . . . So many of us start out thinking it needs to be something that's decided, and it doesn't. It's very fluid.

A pivot can also be shifting employment types. Jeanne Montalvo Lucar says of shifting in and out of freelancing:

There was a point in my life when I realized that I was really tired of free-lancing. I liked my freedom, but I hated keeping up with my finances and managing my money . . . At one point it became, "Alright, I need to get a full-time job and just call it a day." That's when I found my first real full-time job with benefits (at Bloomberg). I really valued the stability, getting a paycheck every two weeks, and insurance . . . so thinking about going back to freelance was scary. What I said after [a residency at] Spotify Studios was, "Let me give it six months to freelance, and see if anything comes of all of this work that I put in." Six months turned into a year and now I'm still there . . . I'm technically part-time [at a company], so that is offering me stability, but now I use the extra time to fill in with as much freelance work as I can. That's like the ideal situation, really.

While pursuing a Ph.D. (in her 40s), Susan Rogers decided to change paths away from her dream to a more practical path due to her professor's advice:

> This advice was suggesting "you probably shouldn't go for what you want." I thought about it, and thought about it, and it was the right move. So, I did it. I went for what was practical, and it ended up much better. So, I didn't get the dream I wanted. I got a different dream. I have every reason to believe that this dream is more rewarding than what might have happened if I had gone for what I wanted [originally].

Pivoting from the West End Theater to AAA Video Games: Pete Reed

Pete Reed pivoted from top of the field in theater to top of the field in video games. It took three years from when Pete started to pursue video games to landing a job on a AAA title.

> I had missed too many family events due to antisocial hours that come with theater. That was fine when I was younger, but as I approached my 30s, my priorities changed. I had always been a gamer, and while I was considering my options, I literally had a light bulb moment while playing a FPS [first person shooter]. I heard a shotgun, and everything just changed for me. My ears had been opened, if you will. The more I investigated [games as a] career, the more I knew it is what I wanted to do.
>
> After my light bulb moment, I went on a crusade of knowledge. I was still working in the West End, and I had about 20 hours a week that I could use. I would go to job adverts and look at the requirements. I used it as a checklist of sorts, slowly learning as much as I could. Along with skills, reels, and starting to make a portfolio website, I researched other sound designers' sites and aimed to stand out above them. I also looked at networking online and attended events and talks as much as I could . . . To build my portfolio I started looking for work on small games, mobile games – anything I could.
>
> To start with, I didn't know anyone . . . I got on Twitter and started connecting with people. I saw a tweet from a well-respected designer saying he would happily give feedback on showreels to help. I reached out and before I knew it, I had two industry professionals looking at my work, and giving feedback. I really used social media as a way of promoting myself and networking. I have since found out that a

video I made reached a now colleague (way before I knew them), so it does work!

After two years, I started to hear back from applications [for AAA jobs] and getting phone/Skype interviews. After every knock-back, I would update my portfolio and go again. In my third year of applying, I was getting to the test stage of the interview process and then the face-to-face interviews. I received three AAA offers in practically the same week. I was blown away but also incredibly proud. I had come a long way in that time.

There are a few things I've learned throughout my career that are relevant in different industries:

1. You never know everything. Realizing this does two things: opens you up to learning new approaches and stops you from becoming big-headed. Don't go into a job thinking you know everything; you will set yourself up to fail.

2. Everyone in the team is important, from the bottom to the top – the receptionist to the CEO, the cleaner to the producer. Treat them all with the same respect.

3. Don't be afraid to ask.

4. If no one knows what you can do, no one will ask you to do it. Put yourself out there. Connect. Network. Show people what you are capable of.

Audio is audio. It is just air vibrating at different speeds. No matter what path you choose, there are many transferable skills . . . In the start, if you can, I wouldn't pigeonhole yourself into one job role. Think about what you really want and where you see yourself (not just your career) in 10 years. In this day and age, people pivot careers all the time. At the end of the day, you can change your mind.

Career Paths Outside of Audio: Rob Jaczko

Rob Jaczko is Chairman of the music production and engineering (MP&E) department at the Berklee College of Music (USA). Berklee has over 100 students a year who graduate from their MP&E bachelor's degree program.

There's always that subset [of students] at the top where the very first time you meet them, you know clearly they're going to be successful. Charlie Puth is a great example. He came into college with a million [social media] followers.

Then, on the flip side, there are those that go through the whole program, and many that I can cite have completely changed professions. We have some that are very high-level professional photographers. It's a technical, artistic, creative outlet. It's not making records, but they took what they learned about audio recording and transposed it to capturing images on film.

One of our alumni is doing the lighting for *Hamilton* . . . We certainly didn't train her to work in theater, and certainly not in lighting in theater. But it was all of the collaborative skills, the critical thinking, the problem solving, thinking on your feet – these are all life qualities that allow you to then be successful doing something else.

We have other [alumni] that left the business of creative business entirely and became nutritionists, yoga instructors, lawyers. They went back to school and studied a different discipline, but everything that they learned about being a collaborative musician serves them well in their next endeavor. So, that's not a failure of the system. That's just a wonderful byproduct of going into higher education. You learn a lot of life skills, and you may decide at the end of the day, you know what? I don't really dig hanging out for 15 hours a day, six days a week, largely with dudes in windowless recording studios. It's not what I thought.

When Reality Collides with Dreams

Sometimes the reality of a dream is not what you expected. Ryan Tucker, who worked in music production before pivoting into audio technology, explains:

It took me a long time to come to terms with the fact that I'm not the right personality for what I thought was my dream job – music engineer. My understanding of "music engineering" failed to factor in all other disparate skill sets that are required for being an entrepreneur. I value having worked at a world-class recording studio, but as much as I wish the outcome was different, I am now quite confident that I'm not the right person to be in such a delicate service role.

Claudia Engelhart, who works in touring live sound and tour managing, adds:

> Ninety-nine percent of the work I do has nothing to do with sound, and I kind of hate it sometimes. But, it's all for the end game, which is when you get to that show and it's all working . . . It's easy and fun, and everybody's having a good time. That's really the bottom line – you've got to have fun and you've got to love it. If you don't love it, if you hate the music, don't do it, because it's going to come out. You might be genius, the best sound engineer in the world, but if you hate the music, you can't get away with that.

Income can be a challenge that can collide with dreams. Sal Ojeda left behind a music studio job for a full-time job in post-production. He shares:

> The day before they offered me the post gig, my next task at the other job was repainting the wood outside of this studio. The owner was just looking for stuff for me to do . . . I was looking at my friends who I graduated with who worked at The Village or Capitol Records, and three, four, five years later they were still runners. I was three years older than everyone else because I graduated later. I couldn't wait that long. I had to start climbing up.

Some may find the reality of a career path is not what they expected, and instead decide to pursue their interests as a hobby. Brian M. Jackson explains:

> Some people really should admit to themselves that they are passionate hobbyists. There is nothing wrong with being a hobbyist, serious or otherwise. Music, music technology, and music production are fun, fulfilling, and rewarding . . . Also, serious hobbyists can make great clients and can help the rest of us to pay the bills while we help them enjoy life.
>
> (Jackson, ch. 2)

There may also be opportunities you are interested in but cannot take on. Devyn Nicholson explains:

> My boss said to me, "Mark my words, there will be a day where you will see something really cool that you wish you were part of, but you can't because you just don't have the time. It'll probably happen to you sooner than you think it will. Try not to get too disappointed or depressed about it."

Susan Rogers shares:

> My friend, [sculptor] Tim Bruckner, has a great quote: "Sometimes you don't get the dream you want. Sometimes you get the dream you didn't even realize you had." Your life takes on a path that is much more rewarding and makes you much happier than it would have if you had followed your original plan – which is why I strongly encourage students to know themselves because your gut instinct is going to tell you.

Being Married to Your Career: Susan Rogers

Berklee College of Music Professor Susan Rogers tells the story of a talented former student who changed their mind about moving to a major entertainment city.

[My student said], I told all my family, all my friends back home, my teachers, and all my peers, "I want to be an engineer." I'm letting everyone down if I don't go and become an engineer . . . [My advice was to] arrange your priorities to get what you want the most. Do not force yourself into this career, because it won't be kind to you if you do. It's a tough business. The only way you're going to get it is if you've got that fire in the belly. The [ones] who are successful have that fire in the belly. There's no way to compete if that fire has gone out in you.

I use the marriage analogy. Look, you've loved music for a long time, and you've studied it well enough to be able to pass the audition and get into [school]. You're here spending a lot of money on tuition so that you and music can have a future together. Now, you're about to graduate and you're about to put a ring on it . . . You're going to depend on it for your livelihood. It will depend on you doing your job to keep it going. Your relationship to it will change now that you're marrying music.

You had fun together when you were just dating – when it was just a hobby. Now you're tied to it, and your relationship will change. There will be fights. There will be times when you just want to do anything other than this. There will be times when you question yourself and ask, "Did I make the right choice? Did I choose the right one? Could there have been someone else?" You might fall in love with another. It happened to me – I started getting attracted to the sciences in my 30s [while working in the recording industry] and I finally realized, "I have to make a commitment to this," and I did in my 40s. I never stopped loving music, but I'm glad I made the choice that I did to untangle my professional commitment to it, enjoy it now just as a hobby, and get into the sciences because I found a new love. I remind [my students] that their life is their own to do with what they want. If they choose (after having done all this) to just have music as a hobby and their friend, then great. It's their life. This is no shotgun wedding here. No one's forcing you to be an engineer. No one's forcing you to marry music. You can do something else for a living.

Our department tries really hard to select the right students for this major, and we do a great job. Many of the kids that we get in our major will go on to professional careers, but not necessarily this. We've had at least one medical doctor and we've had a few entertainment lawyers. We've had folks go on to grad school or do other things [like] post-production. Sometimes they do what college students do: they realize, "This was good, and I've learned how to learn, but I just don't think this lifestyle, this career is right for me."

A Supportive Industry

A common theme of established professionals is wanting to give back to the community. Professionals understand how hard it is to get over that mountain (from emerging professional to becoming established) because they have done it themselves. Live sound engineer Michael Lawrence explains:

> One thing that continues to amaze me about this field that we have, because it's not true about a lot of fields, is how open people are with their knowledge. You can just call somebody and ask them questions and they'll talk to you all day. That's really unique to our field, in a lot of ways. People don't generally feel like they have to be protective . . . I encourage people to take as much advantage of that as possible – even though it can be scary.

Game audio director Kristen Quinn adds:

> The best success in my career is being able to hire people and have them grow and become better than me. I appreciate someone who really works on growing. If I can see that they try hard and they really care, that matters.

Students may get offers of help from professors, guest lecturers, or alumni, which are highly valuable but may be short-lived. Burt Price, who has been Chief Engineer at the Berklee College of Music for over 20 years, says of offering students help to find work:

> Most of them don't really want the help, or for whatever reason, they're not interested in staying in the business and go right out . . . I think most of the time the opportunity is lost because they just don't know. I think a lot of the students aren't thinking about it.

When educator Leslie Gaston-Bird was a student, legendary music producer Eddie Kramer was a guest lecturer and offered to help her out. Leslie shares:

> I said, "I have a job. I'm not about it." I look back on that and I'm like, I'm such an a——, I should have gone with Eddie Kramer . . . If somebody's pointing their finger at you, that's the finger of destiny. Get it. Somebody at school created that opportunity. Somebody brings in a guest lecturer – That's the whole point.

The industry can support each other through sharing opportunities, particularly as you outgrow them. Karol Urban says of leaving a full-time staff job (as a re-recording mixer) to pursue her goal of working in narrative storytelling:

> I basically walked out of the type of job that people fight for decades to get. It was a great job – It is a great job. But I left it because it wasn't my job. I'm very sensitive to: What do you want? What is your job? Like, not the job you need to pay your bills right now, but what's the job that you need that will pay your bills and be what feeds you in multiple ways? . . . If [an opportunity] is somebody else's end all be all and will be everything to them, I'm kind of wrong to keep it. It's not mine.

Catherine Vericolli says of the industry supporting each other:

> I have tough times, too. It's great when people reach out like, "I'm really busy right now and I can't really pick up this gig, and I thought maybe you might be able to do it." That's how we survive in this industry: supporting one another. If you're not going into this willing to support other people, and willing to share the dinner that's on your plate with other people when they're really, really hungry, maybe this isn't for you. I don't know that you're going to be as successful as you want to be.
>
> (Vericolli, "Once You Have")

Finding Your Path: Jeanne Montalvo Lucar

Jeanne Montalvo Lucar has worked in classical music production, radio broadcast, podcasting, Latin music production, education, and more.

> I tend to find the best decisions that I have made in my career and my life have had a combination of fear and excitement. I know what that feels like at this point, so now when I have that feeling, I know I'm maybe going in the right direction. As I've gotten older, the fear part becomes scarier, and maybe I'm not doing it as much now, but when I was starting out, that was definitely a plus.
>
> Your first gig is probably not going to be the most exciting thing that you will ever do. I think being okay with that is important. You might need to spend six months to a year doing the thing that is not as glamorous and not as exciting but could provide a really, really great foundation for you to do something else. [You may] end up getting comfortable there and then spend six to ten years and then wake up one day and [decide] that's not where you want to be. It doesn't mean that you can't change and go do something else . . . Make sure that you are constantly keeping your eyes open and in tune with what's out there and what could come next.

The Next Phase

As an emerging professional, there is a tipping point where building a career becomes simpler. Mariana Hutten says of being an established professional, "I don't have time to do things on my website right now, and there's not a need. I've been so busy with word-of-mouth work or returning clients that I haven't been paying attention to that lately."

Alesia Hendley says as her career grew, she could say "no" to projects:

> Starting out, I said "yes" to everything. You want me to come run your sound every Sunday for $150 a service? Awesome, I'll do it. You want me to come walk your property with you and you're going to give me a bag of doughnuts? Awesome. You know why? Because I need the experience, and you never know what's going to happen. But as you grow, don't give your time to just anybody. It's okay to say "no." What I learned is when I said "no," the right opportunities did come because I wasn't busy doing something that I didn't even want to do in the first place.

Some opportunities will come from people whose careers are building alongside yours. James Clemens-Seely says his peers and classmates have been a crucial part of his career as an established professional:

> At first, it seemed like I needed big people to lift me up into prosperity. Now, it's the foundation I built of peers . . . By virtue of curating a collection of people (who are your core network) itself selects the kind of person that you're going to work well with. You'll arrive at the point where you can start curating instead of just being desperate to stay afloat. If you're wise with the curation, that ball will keep rolling.

Getting to that point in a career takes time. Scott Adamson advises:

> Being at peace with how much time it takes will save you a lot of frustration. Some people do have lightning-fast careers and jump into really huge stuff right away but that is a very small percentage of people. For most of us [in live sound], we do it because we love it, because we love being around the music and the energy of it, and it's a really fun way to travel and a fun way to hang out with a lot of like-minded individuals. But you can't expect that massive arena tours or stadium tours will come your way quickly – or ever. That should be fine if you're really [trying] for the right reasons.

Mariana Hutten adds:

> I know so many people who are far more talented than me (in the music business in general), and they're now working in real estate or something.

Is it because I became better? I continued doing my thing. I never stopped. That's really the secret.

Shaun Farley started working at Skywalker Sound after ten years in post-production sound. Shaun got a foot in the door through connections he met as a volunteer writer/editor for *Designing Sound* (designingsound.org). He shares:

> Some people are lucky enough to start their careers in places like [Skywalker Sound]. Others, like me, work their butts off and somehow find their way in later . . . and not everyone has the undivided attention of the sound design community like I did for several years. Even more [people] work to the bone for their entire careers and don't go to a place like that. Some leave the industry altogether because they decide they have priorities that are more important. That's a perfectly valid and respectable choice, too. This industry is hard and not for those who want to be able to coast . . . Be ready for the long haul because even if you make it in, staying in can be just as much work.

For emerging professionals, having manageable short-term goals can help the process. Guitar tech Claire Murphy shares:

> When I started, my goal was literally just to be on tour. That was such a hurdle because I didn't know how to break into it. I just wanted to be the roadie on stage. When I got to that position . . . my goal then was to do a good enough job to be asked back. Every subsequent band, I just want to do a good enough job to be asked back. That's my goal. I want to be asked to do another tour. From there, the goals change to work on bigger tours, bigger bands . . . I've just come off a 12-week stadium tour.

Mastering engineer Piper Payne adds:

> When I first started mastering . . . my ultimate dream [was] to be booked out for a week. It's something I always think back on. Now, I'm booked out for several weeks at a time . . . It took me a solid four or five years of just working my ass off, hustling so hard day and night, and not ever, ever letting an opportunity slip by without really seeing if it was going to go somewhere. It seemed like almost overnight, I went from having zero mastering work to having more than I could handle . . . At some point, you're going to be so busy that you have to pick and choose gigs, but it's going to be many, many years before that's the case . . . You have to be so rigidly disciplined to hone your craft and to never, ever, ever stop until it's actually good.

Even established professionals may take their career job by job. Jessica Paz, the first woman to win a Tony Award for Sound Design (for live Broadway theater), says:

> My biggest goal was to win a Tony. I did that and then I didn't know what to do next. For a while after that moment, I [felt] I've been working toward this for so long, and now I've done it. Now what? . . . As much as winning an award was a goal, I find that every show is a goal unto itself. Like, getting to work on these shows and tell these stories . . . having the ability to do something that I love is really the reward . . . Every show that I do challenges me in a different way because they all call for something different, and call for me to stretch in a new way that I didn't know.
>
> (Student Sound Designer)

Catherine Vericolli also suggests that there is more to being in the industry than visibility, recognition, and awards:

> I'm fortunate enough to have really great friends that have insane credits . . . At the end of the day, the thing that's most important to them is the people that are close to them and their families. Credits are great, the experience is good, but they make great friends with those clients. It's not about "we're winning Grammys." It's about, "I had this amazing experience with these people." I think that it's really, really hard to lose sight of that when it's just technology, technology, technology . . . I have the most wonderful people in my life and it's because of this industry.
>
> ("What Makes")

A career path may not be clear until looking back at it later. Rob Jaczko shares:

> Careers are much like a boat. Boats don't move in straight lines. They go back and forth in a zigzag on your way to the island. That's what your career is, and that's how we build up all these experiences . . . We need to look around and figure out how we're going to put a patchwork of events together that in hindsight we call a career.

Early in her career, Claudia Engelhart worked at the Knitting Factory in New York City with unknown jazz artists who went on to become historic musicians. For Claudia, it was just doing her job:

> It took 35 years, almost 40 years of schlepping around and doing weird crazy gigs before I got to where I am today. It doesn't just happen. I loved it the whole way. When I think back, it's been hard, but what isn't? If you love it, you want to go to work every day. If you hate it, you're not going to want to go to work every day. You really have to want to do it because it's a sacrifice, too.

Staying with It: Justus Gash

Justus Gash (Oregon, USA) explains his path from playing guitar to starting his own guitar pedal company, Fuzz Imp, in 2021.

> I tried being a gear reviewer early on in YouTube with some success, [but] mostly just spent a lot of money. Tried being a solo artist, have a decent skillset, but tough to break through with instrumental guitar. Tried being a producer for a while based on a couple of small paid gigs I got but had a hard time drumming up interest. Went back to YouTube reviewing pedals last year and really caught some momentum, but it wasn't profitable, at least not substantially monetarily.
>
> However, I met all the people I use for various tasks, and advice that I really need today through that timespan. The point being: sometimes you don't see the best path until you try a few things and see what sticks. But if you stick to it and keep trying, there's room for everyone somewhere, and eventually, the work doesn't feel like work anymore.

Closing Thoughts

Jeff Dudzick says of finding opportunities:

> Keep your eyes open, your ears open, and look in weird places. Also, don't be too stingy about it because chances are one opportunity (if you do a good job and people like to work with you) can lead to something else.

Knowing your goals and personal priorities can help determine where to best put your time and efforts. Alesia Hendley shares:

> Be conscious about your time, be conscious about what you want to do, and be conscious about your roadmap of where you're going. Whether someone gives me $150, $1,000, or a bag of bagels – is this opportunity on my roadmap leading me ultimately a step closer to where I want to be? Is that step closer to me working in a house of worship? In sports broadcasting? To being front of house for Beyonce? You need to know what your end goal is to realize which opportunities work for you.

What draws you to a discipline or job may be something you didn't expect. Kevin McCoy advises:

> Try and be in a room with people who are of different backgrounds because it's better for the world. It's also just more fun. Being in a diverse group of people is so much better than being in a group of people who are all like you. It's just a better life.

New opportunities are emerging, and what your career path turns into may not even exist right now. Piper Payne explains:

> We forget that the music industry as we know it is only a few decades old. It's not like, politics or medicine or astronomy or even space travel . . . We've been making it up as we go along this whole time. [Nobody] said, "This is how it goes, and this is how it's going to be forever." Totally not the case.

Building a career, and finding the opportunities that fit you, can take time. Sal Ojeda shares:

> I am a really big jazz fan . . . jazz celebrates maturity. It is the struggle or the path to becoming great. That doesn't come with chops; It comes with age and experience. It's hours and hours and hours and hours of doing what you love. That's how you get there, and once you get there, it's just like, let's celebrate.

Susan Rogers says about finding your career path:

> Start with what you want. Then filter that through the constraints of what you have to work with – what you want, what's possible for you, and then start climbing that mountain. When you get halfway up the mountain, you'll have a better vantage point. So sometimes you start on one career, you get a few steps down the road, and you go, now that I can see better, this isn't the career that I thought it was. I'm going to just move over to another mountain.

Finding your path in the modern audio industry will be a learning experience no matter where you end up. Clare Hibberd explains:

> I always go with the mentality that anyone can do anything . . . some people have higher barriers to climb. If you are willing to climb those barriers, identify the barrier and then climb it. It's not an equal playing field. Yes, some people will have it handed to them, and some people will mess up the opportunity when they have it handed to them. Some people will spend

years getting there. But, if that's what you really want . . . You have to at least try to go for it. There's no bad feeling about you trying, and you get to the point where you think, "You know what? I can't do this anymore, so I'm going to diverge." That's life, and we learn from the experiences that we're going through.

References

Adamson, Scott. Personal interview. 14 May 2020.

AdeDamola Oluwagbemi, Phebean. Personal interview. 10 Jan. 2021.

Arnardottir, Steinunn. Personal interview. 30 June 2021.

Clemens-Seely, James. Personal interview. 20 Dec. 2020.

Dudzick, Jeff. Personal interview. 1 Oct. 2021.

Engelhart, Claudia. Personal interview. 8 June 2020.

Espinoza, Andrea. Personal interview. 28 May 2020.

Farley, Shaun. Email interview. Conducted by April Tucker, 15 July 2020.

Felix, Britany. Personal interview. 22 June 2020.

Ferguson, Eric. Personal interview. 2 Dec. 2020.

Gash, Justus. The Modern Music Creator Group. *Facebook*, 24 Feb. 2021. www.facebook.com/groups/themodernmusiccreator. Accessed 25 Feb. 2021.

Gaston-Bird, Leslie. Personal interview. 2 Oct. 2020.

Gross, Jeff. Interview with April Tucker and Ryan Tucker. 14 Jan. 2021.

Hendley, Alesia. Personal interview. 17 Dec. 2020.

Hibberd, Clare. Personal interview. 7 May 2020.

Holmes, Meegan. Personal interview. *What Makes You Stand Out*, SoundGirls, 20 May 2020, youtu.be/wYamiOK8Y6A. Accessed 21 May 2020.

Hutten, Mariana. Personal interview. 19 Nov. 2020.

Jackson, Brian M. *The Music Producer's Survival Guide: Chaos, Creativity, and Career in Independent and Electronic Music*. Routledge, 2018, ch. 2. https://doi.org/10.4324/9781315519777.

Jaczko, Rob. Personal interview. 15 Oct. 2020.

Kennedy, Camille. Personal interview. 21 Feb. 2021.

Kramarik, Kelly. Personal interview. 18 June 2020.

Lawrence, Michael, et al. "Advice in Hindsight." *Signal to Noise Podcast*, ep. 56, ProSoundWeb, 8 July 2020, www.prosoundweb.com/podcasts/signal-to-noise/.

Martinez, Avril. Personal interview. 28 Aug. 2020.

Masingale, Langston. Personal interview. 14 Sept. 2021.

McCoy, Kevin. Personal interview. 22 Dec. 2020.

McLucas, John. "My Mental Health Struggles in the Music Industry – Mind Made Wrong Podcast," *The Modern Music Creator Podcast*, ep. 52, 8 July 2019, https://pod.co/the-modern-music-creator.

Mierzwa, Patrushkha. *Behind the Sound Cart: A Veteran's Guide to Sound on the Set*, Ulano Sound Services, Inc., 2021. p. 271–272.

Mohabati, Mehrnaz. Personal interview. 29 Jan. 2021.

Montalvo Lucar, Jeanne. Personal interview. 12 Oct. 2021.

Murphy, Claire. "Career Paths in Live Sound." *YouTube*, uploaded by SoundGirls, 1 Oct. 2019, youtu.be/dbBxX9PddhI.Nicholson, Devyn. Personal interview. 23 Nov. 2020.

Ojeda, Sal. Personal interview. 27 Sept. 2020.

Payne, Piper. Personal interview. 29 May 2020.

Price, Burt. Interview with April Tucker and Ryan Tucker. 5 Oct. 2020.

Quinn, Kristen. Personal interview. 30 Sept. 2020.

Reed, Pete. Email interview. Conducted by April Tucker, 24 Nov. 2020.

Reynolds, Jason. Personal interview. 27 July 2020.

Rogers, Susan. Personal interview. 12 Oct. 2020.

Stacey, Sarah. Personal interview. 19 June 2021.

Student Sound Designer Connection. "Live Q&A with Jessica Paz." *Facebook*, 2 April 2020, www.facebook.com/688623958/videos/10158013515468959/. Accessed 3 May 2020.

Tipp, Nick. Personal interview. 22 Sept. 2020.

Tucker, Ryan. Personal interview. 28 Sept. 2020 and 27 Dec. 2021.

Urban, Karol. Personal interview. 17 July 2020.

Vericolli, Catherine. Personal interview. *What Makes You Stand Out*, SoundGirls, 20 May 2020, youtu.be/wYamiOK8Y6A. Accessed 21 May 2020.

Vericolli, Catherine. Personal interview. *Once You Have the Gig – What Makes You Stand Out,* SoundGirls, 20 July 2020, youtu.be/QJhdm86kIlM. Accessed 21 July 2020.

Zúmer, Javier. Personal interview. 24 April 2021.

The Student Sound Designer Connection Facebook group is a public service project supported by TSDCA – Theatrical Sound Designers and Composers Association. tsdca.org/

Appendix: Bios of Interviewees

Interviews Conducted in 2020–2021

- **Scott Adamson** (New York, New York, USA) – Touring front of house sound engineer (Passion Pit, Haim, and Khalid) and founder of the Production Academy
- **Steinunn Arnardottir** (Berlin, Germany) – CTO/Founder of Lottie, formerly DSP Developer and Senior Director of Engineering at Native Instruments
- **Aline Bruijns** (Harderwijk, The Netherlands) – Sound designer (post-production); owner of AudioRally SoundDesign
- **Chris Bushong** (Nashville, Tennessee, USA) – Front of house mixer, monitor mixer, RF coordinator, and production manager (toured with Imagine Dragons and Smokey Robinson, clients include the Nashville Symphony and Alabama Symphony)
- **James Clemens-Seely** (Banff, Alberta, Canada) – Senior Recording Engineer at the Banff Centre for Arts and Creativity
- **Rachel Cruz** (Los Angeles, California, USA) – Program Manager – Electrics at Fender Musical Instruments Corporation; formerly with Xperi (DTS), Guitar Center, and Line6
- **Jeff Dudzick** (Effort, Pennsylvania, USA) – Content Creation Manager, Audible Studios at Audible, Inc. Formerly with Tanglewood Music Festival and Sony Music Entertainment
- **Tommy Edwards** (Austin, Texas, USA) – Vice President of Product Development at Warm Audio. Formerly with Blue Microphones, Logitech, Line6, and Alesis
- **Claudia Engelhart** (Valley Cottage, New York, USA) – Front of house mixer (Bill Frisell, Wayne Shorter, and Herbie Hancock), tour/production manager (Bill Frisell)
- **Andrea Espinoza** (Las Vegas, Nevada, USA, and Mexico) – Audio engineer, tour manager, and licensed financial advisor

- **Britany Felix** (Colorado Springs, Colorado, USA) – Podcast consultant and editor, owner of Podcasting for Coaches
- **Eric Ferguson** (Bangor, Maine, USA) – Assistant Professor of Audio Engineering at the New England School of Communications at Husson University
- **Gabrielle Fisher** (Hacienda Heights, California, USA) – Audio Designer for the Walt Disney Company
- **Megan Frazier** (Seattle, Washington, USA) – VR sound designer (HTV Hive, Amazon, and Hyperspace XR), owner of On Accident! Production
- **Leslie Gaston-Bird** (Brighton, East Sussex, UK) – Freelance re-recording mixer and sound editor, educator, and author of the book *Women in Audio*
- **Hiro Goto** (Los Angeles, California, USA) – Violinist, arranger, recording engineer, and live sound engineer. Originally from Japan
- **Giosuè Greco** (Los Angeles, California, USA) – Film composer, music producer, and musician. Originally from Italy
- **Jeff Gross** (Los Angeles, California, USA) – Foley mixer (Sony, Paramount, and Formosa Group), music producer, engineer, and mixer
- **Alesia Hendley** (Tampa Bay, Florida, USA) – Sales Engineer at Access Networks, multimedia journalist, and content creator
- **Clare Hibberd** (Glasgow, Scotland) – Lecturer in Sound at the Royal Conservatoire of Scotland. Former experience in musical theater (West End) and senior sound technician at the Royal Opera House Muscat in Oman
- **Meegan Holmes** (Los Angeles, California, USA) – Global Sales Manager at Eighth Day Sound. She has toured with TOOL, Queens of The Stone Age, Linkin Park, and Natalie Cole
- **Mariana Hutten** (Toronto, Ontario, Canada) – Mixing and mastering engineer; engineer/educator at Artscape (Toronto)
- **Rob Jaczko** (Boston, Massachusetts, USA) – Chairman, Music Production and Engineering Dept. at Berklee College of Music. Former staff music engineer (credits include Bruce Springsteen and James Taylor)
- **Tom Kelly** (Denver, Colorado, USA) – Podcast editor (over 1,500 podcast episodes) and content creator; owner of Clean Cut Audio
- **Camille Kennedy** (Toronto, Ontario, Canada) – Boom operator and sound utility for sound for picture
- **Karrie Keyes** (Ventura, California, USA) – Monitor mixer (Pearl Jam), Co-founder/Executive Director of SoundGirls
- **Andrew King** (Denver, Colorado, USA) – Broadcast audio engineer (A1 and A2) for sports
- **Kelly Kramarik** (Denver, Colorado, USA) – Audio engineer, podcast editor, film and TV sound specialist; audio editor for Starz (promotions)

- **Avril Martinez** (Seattle, Washington, USA) – Sound designer and implementer for video games and AR/VR (Riot Games, Tinker Design+ Storytelling, and Microsoft)
- **Langston Masingale** (Syracuse, New York, USA) – Product designer, music producer, and engineer; owner of Handsome Audio
- **Kevin McCoy** (Minneapolis, Minnesota, USA) – Head Audio, Hamilton "And Peggy" Company (based in San Francisco, California)
- **John McLucas** (Los Angeles, California, USA) – CEO of McLucas Media, pop music producer, and content creator (over 200 YouTube videos and 100 podcast episodes)
- **Mehrnaz Mohabati** (Los Angeles, California, USA) – Supervising sound editor/re-recording mixer; former music engineer in Tehran, Iran
- **Jeanne Montalvo Lucar** (Dobbs Ferry, New York, USA) – Audio engineer and radio producer; formerly with NPR, Bloomberg, Latino USA; Spotify's EQL Studio Resident at Spotify Studios and Electric Lady Studios
- **Avery Moore Kloss** (Paris, Ontario, Canada) – Audio storyteller, podcast producer, and award-winning radio documentarian
- **Tina Morris** (Los Angeles, California, USA) – Music studio manager at the Village Studios (Los Angeles)
- **Devyn Nicholson** (Montreal, Quebec, Canada) – Manager of Audiovisual Services at McGill University
- **Sal Ojeda** (Los Angeles, California, USA) – Re-recording mixer and sound editor
- **Phebean AdeDamola Oluwagbemi** (Lagos, Nigeria) – Audio engineer, production manager, and co-founder of Audio Girl Africa
- **Piper Payne** (Nashville, Tennessee, USA) – Mastering Engineer at Infrasonic Mastering; former owner/Chief Mastering Engineer of Neato Mastering in Oakland, California
- **Luisa Pinzon** (Bogota, Colombia) – Music engineer, post-production sound supervisor, and graduate student at CALARTS (Santa Clarita, California, USA)
- **Burt Price** (Boston, Massachusetts, USA) – Chief Engineer at the Berklee College of Music
- **Kristen Quinn** (Seattle, Washington, USA) – Audio Director at Polyarc Games; formerly at Riot Games, Microsoft Game Studios, and Monolith Productions
- **Jason Reynolds** (Toronto, Ontario, Canada) – Touring live sound engineer and tour/production manager and consultant for houses of worship. Jason has toured internationally with Stephen Marley, Shaggy, and Magic! Originally from Kingston, Jamaica

- **Susan Rogers** (Boston, Massachusetts, USA) – Sound engineer, producer, and educator (Berklee College of Music); former recording engineer for Prince (1983–1987), Ph.D. in music cognition and psychoacoustics
- **Sarah Stacey** (Dublin, Ireland) – Radio, audio, and podcast producer; freelance audio producer for BBC Radio. Formerly with Today FM in Dublin
- **Becca Stoll** (Norwalk, Connecticut, USA) – Freelance mixer and production audio head for musicals. Formerly with Goodspeed Musicals and American Repertory Theatre
- **Marc Thériault** (Montreal, Quebec, Canada) – Owner/Chief Mastering Engineer at Le Lab Mastering and sound/RF/electrical engineer for Celine Dion (since 1996)
- **Nick Tipp** (Los Angeles, CA, USA) – Recording engineer, mixer, and audio/video producer
- **Jack Trifiro** (New Hampshire, USA) – Front of house mixer, tour manager/production manager (Victor Wooten), and monitor engineer for Béla Fleck and the Flecktones
- **Ryan Tucker** (Los Angeles, California, USA) – Product designer and QA manager; worked on products for Warm Audio, McDSP, Universal Audio, Guitar Center, Line6, and DTS
- **Karol Urban** (Los Angeles, California, USA) – Re-recording mixer and sound editor; president of the Cinema Audio Society. Formerly with Discovery Communications
- **Catherine Vericolli** (Nashville, Tennessee, USA) – Chief transfer engineer and archival engineer at Infrasonic Transfers & Archival (Nashville); former studio owner and recording engineer at Fivethirteen Studios (Tempe, Arizona, USA). (*Interview from Tempe, Arizona*)
- **Fei Yu** (Beijing, China) – Music producer, music editor, and music supervisor; owner of Dream Studios. Former music editor for the China Film Group
- **Eliza Zolnai** (Budapest, Hungary) – Production sound assistant (credits include *Terminator* 6 and *Dune*). In 2020, she temporarily relocated to New Zealand
- **Javier Zúmer** (Royal Leamington Spa, Warwickshire, UK) – Video game sound designer. Former freelance audio engineer in Seville, Spain, and sound designer/re-recording mixer at Telegael (County Galway, Ireland)

Written Interviewees

- **Lucy Diggle** (Coventry, West Midlands, UK) – Audio Engineer at Jaguar Land Rover
- **Shaun Farley** (Albany, California, USA) – Sound Editor at Skywalker Sound; former contributing editor for DesigningSound.org
- **Jonathan Hubel** (Richardson, Texas, USA) – Audio engineer/podcast editor
- **Mabel Leong** (Singapore) – Audio Engineer at Night Owl Cinematics, former audio engineer at BDA Creative (Singapore)
- **Agnes "Aji" Manalo** (Manila, Philippines) – Assistant Professor at De La Salle-College of Saint Benilde, audio engineer
- **Tiziana Mazzucco** (Milan, Italy) – Dubbing mixer and recording engineer
- **Jack Menhorn** (Redmond, Washington, USA) – Audio Lead at Formosa Group, video game sound designer/composer. Former editor in chief of DesigningSound.org
- **Ashton Morris** (Fort Collins, Colorado, USA) – Sound Designer for PhET Interactive Simulations, University of Colorado Boulder, game sound designer/composer
- **Pete Reed** (London, UK) – Technical Sound Designer at PlayStation. Former theatrical sound engineer (London West End)

Other Significant Resources

- "Behind the Sound Cart: A Veteran's Guide to Sound on the Set" (Patrushkha Mierzwa)
- "Get On Tour – A Sound Engineer's Guide" (Sound Design Live; Nathan Lively, editor)
- "Girl on the Road: How to Break into Touring from a Female Perspective" (Claire Murphy)
- "The Bible of Getting a Job in Game Audio, 2020 Edition" (Florian Titus Ardelean, editor)
- "Women in Audio" (Leslie Gaston-Bird)
- SoundGirls Blog (soundgirls.org)
- Podcasts: The Modern Music Creator Podcast; Signal to Noise Podcast; Church Sound Podcast; Sound Design Live Podcast, SoundGirls Podcast
- YouTube Channels: The Production Academy, Akash Thakkar, SoundGirls ("Ask the Experts" series)
- GDC Vault (video): "Successful Freelancing in Game Audio" (Akash Thakkar, Speaker)

- Facebook groups: Audio Educators, Classical Music Location Recording, Hey Audio Student, Just Busters: Female Podcast Editors, Podcast Editors Club, Post Sound Mixers, The Six Figure Home Studio Community, SoundGirls Private, Student Sound Designer Connection (a public service project supported by TSDCA – Theatrical Sound Designers and Composers Association. tsdca.org)

Complete interviewee biographies and additional resources can be found at apriltucker.com/books/career-paths/

Index

Printed in the United States
by Baker & Taylor Publisher Services